26,95
ag

Eva und Walter U.
K e l l n e r
Hamannstr. 49
40882 Ratingen

D1701986

COMPLEX FUNCTION THEORY

COMPLEX FUNCTION THEORY

Anthony S. B. Holland
Associate Professor of Mathematics
University of Calgary
Calgary, Alberta, Canada

NORTH HOLLAND
New York • Oxford

Elsevier North Holland, Inc.
52 Vanderbilt Avenue, New York, New York 10017

Distributors outside the United States and Canada:

Thomond Books
(A Division of Elsevier/North-Holland Scientific Publishers, Ltd.)
P.O. 85
Limerick, Ireland

© 1980 by Elsevier North Holland, Inc.

Library of Congress Cataloging in Publication Data

Holland, Anthony S.B.
    Complex function theory.

    Bibliography: p.
    Includes index.
    1. Functions of complex variables. I. Title.
QA331.H65   515'.9   79-24796
ISBN 0-444-00342-8

Desk Editor Philip Schafer
Design Edmée Froment
Art rendered by Vantage Art, Inc.
Production Manager Joanne Jay
Compositor Computype
Printer Haddon Craftsmen

Manufactured in the United States of America

Contents

Preface
Preliminary Definitions and Symbolism

Chapter 1	Complex Numbers		1
	1.1–1.3	The Complex-Number System	1
	1.4–1.5	Conjugate, and Absolute Values and Properties	4
	1.6–1.7	Inequalities of Schwarz, Hölder, and Minkowski	5
	1.8–1.10	Geometric Interpretation, Polar Form	8
	1.11–1.15	Exponential Function, DeMoivre's Theorem, Roots of a Complex Number, Roots of Unity	10

Chapter 2	Definitions and Properties of Sets in the Complex Plane		16
	2.1–2.9	Intervals, Upper and Lower Bounds, Diameter, Neighborhood, Point of Accumulation, Closure, Boundary Point, Separation	16
	2.10	Sequences, Convergence	20
	2.11	Properties of Real Systems: Dedekind, Completeness, Bolzano–Weierstrass, Nested Intervals; Cauchy Sequences	22
	2.12	Compact Sets, Heine–Borel Theorem	23
	2.13–2.14	Series of Complex Numbers, Tests, Limits	25
	2.15–2.16	Mappings, Homeomorphism; Curves in the Plane, Domain, Contour, Jordan Curve; Simple and Multiple Connection	27

Chapter 3 Functions of Bounded Variation   31

    3.1   Definitions, Positive and Negative Variation   31
    3.2–3.3 Rectifiable Curves, Riemann–Stieltjes Integral, Properties   34

Chapter 4 Functions of a Complex Variable   40

    4.1   Single and Multiple-Valued Functions, Limits and Continuity   40
    4.2–4.3 Differentiability, Cauchy–Riemann Equations   42
    4.4–4.6 Analytic Functions, Entire Functions, Exponential Function   45
    4.7–4.8 Harmonic Functions, Trigonometric Functions   50

Chapter 5 Curvilinear Integrals   55

    5.1–5.2 Definitions, Inequalities, and Some Special Integrals   55
    5.3   Integral Around Boundary of a Domain   58
    5.4   Cauchy's Theorem, Goursat's Lemma, Landau's Lemma, Strong Cauchy Theorem   60
    5.5   Jordan Polygon, Polygonal Approximation   66
    5.6–5.8 Cauchy's Integral Theorem, Fundamental Theorem for Complex Integrals   68
    5.9–5.11 Morera's Theorem, Cauchy's Inequalities, Liouville's Theorem   80

Chapter 6 Series of Complex Functions, Taylor's Theorem, Uniform Convergence   82

    6.1   Circle and Radius of Convergence   82
    6.2   Uniform Convergence, Theorems, Weierstrass M-Test   83
    6.3–6.4 Taylor's Theorem, Zeros of Analytic Functions, Fundamental Theorem of Algebra   86
    6.5   Cauchy Product of Two Series, Mertens Theorem, Abel's Theorem and Converse   90
    6.6   Convergence of Series and Analyticity of Functions   97
    6.7   Further Properties of Uniformly Convergent Series, Ramifications of Cauchy's Inequalities   100

Chapter 7 Maximum Modulus — 107

- 7.1–7.2 Maximum-Modulus Theorem, Minimum Principle — 107
- 7.3 Schwarz's Lemma — 109
- 7.4 Poisson's Integral Formula — 110
- 7.5 Theorem of Borel and Carathéodory — 111
- 7.6 Hadamard's Theorem — 114
- 7.7 Jensen's Theorem, Poisson–Jensen Formula — 116

Chapter 8 Analytic Continuation — 121

- 8.1–8.3 Identity Theorem for Analytic Functions, Definition of Analytic Continuation, Multiple-Valued Functions — 121
- 8.4 Principle of Reflection — 126
- 8.5 Zeros of Analytic Functions, Conjugate Harmonic Functions — 129
- 8.6 Natural Boundaries, Gap Theorems — 130
- 8.7–8.9 Logarithmic Function, General Power a^z, Binomial Theorem — 131
- 8.10 Multiply Valued Functions, Branches and Branch Points, Single-Valued Branch of Log $f(z)$, Immediate Continuation — 134
- 8.11–8.12 Examples — 138

Chapter 9 Laurent Series, Singularities — 142

- 9.1 Laurent's Theorem, Extension of Cauchy's Inequalities, Picard Theorem, Weierstrass Theorem — 142
- 9.2 Convergence of Laurent Series — 151
- 9.3 Singularities, Principal Part, Poles, Essential Singularities, Residue, Point at Infinity, Values of a Function in Neighborhood of Isolated Essential Singularity, Limit Points — 152
- 9.4 Meromorphic Function, Rational Functions, Residues, Partial Fractions, Residue at Infinity — 158
- 9.5 Evaluation of Integrals Using Residues, Jordan's Lemma, Examples and Techniques, Fresnel Integrals — 166
- 9.6 Summation of Infinite Series, Expansion of Meromorphic Functions — 196
- 9.7 Properties of Meromorphic Functions, Argument Principle, Rouché's Theorem — 200

Chapter 10 Conformal Representation — 212

10.1–10.9	Mappings Induced by Elementary Functions	212
10.10	Mappings and Conformality	220
10.11–10.12	Definitions Relating to the Point at Infinity	224
10.13–10.14	Open-Mapping Theorems	225
10.15	Simple Conformal Mappings, Schlicht (or Univalent) Functions, Inverse-Function Theorem, Darboux's Theorem	226
10.16	Bilinear Transformation, Inversion, Fixed Points, Symmetric Points	229
10.17	Riemann's Mapping Theorem, Locally Uniformly Convergent Sequences	241
10.18	Further Properties; Koebe's Constant, Bieberbach Conjecture	247

Chapter 11 Infinite Products, Expansion of Functions, Mittag–Leffler Theorem, and Gamma Function — 251

11.1–11.2	Infinite Projects	251
11.3	Expansion of a Meromorphic Function	256
11.4	Expansion of an Entire Function as an Infinite Product	257
11.5	Meromorphic Functions, Mittag–Leffler Theorem	259
11.6	Gamma Function, Beta Function, Stirling's Formula	267
11.7	Analytic Continuation of $\Gamma(z)$	271
11.8	Alternative Definitions of $\Gamma(z)$	273

Chapter 12 Infinite Product Representation: Order and Type — 276

12.1	Weierstrass Factorization Theorem	276
12.2	Order of an Entire Function	278
12.3–12.7	Type, Enumerative Functions, Exponent of Convergence	280
12.8	Genus of Canonical Product	284
12.9–12.12	Hadamard's Factorization Theorem, Genus of Entire Function	285
12.13	Calculations of Order and Type	293

Author Index — 299

Subject Index — 301

Preface

The following twelve chapters dealing with the theory of functions of one complex variable have arisen after my having taught the subject on and off for about twenty years. There are several ways to approach the subject, and I believe that the proofs and style used herein are such as to make the subject very readable to a student with a good background in introductory analysis, which basically means a course of two semesters in Introductory Calculus or Real Variable. The theorems are sequential in the sense that a complete two-semester course may be obtained by starting at Chapter 1 and proceeding to Chapter 11 (and Chapter 12, if the special topic on entire functions is required). No material should be omitted, however Chapter 3 could be lightly touched upon if less emphasis was required on the concept of rectification and functions of bounded variation.

I have thought it best to start the subject right at the very beginning and briefly set out the axiomatics. Clearly, many students could omit the first three chapters on complex numbers and begin in immediately with analytic functions in Chapter 4, so that the starting point in this text will depend to some extend on the student's background.

Proofs are kept at a fairly rigorous level and I have taken care not to use the expressions "it is easily seen" or "it can be easily shown," but prove all theorems completely.

Several excellent texts in Complex-Function Theory exist, however the main point of this text is to make the subject readable, to preserve rigor, to reduce excess verbosity, and to prepare the student for a more detailed study of those topics that confront present-day researchers.

A number of explanatory examples are included throughout the text, and several more difficult exercises are gathered at the end of the chapters.

All material has been arranged into sections with coded headings of at most three digits. Thus, 5.3.2 means the second article of section (or topic) three in Chapter 5.

The proofs presented here are those that have evolved over many years as the "best" in the opinion of myself and students. Obviously, many of the results can be and have been extended to a most erudite generality, and these may be found in the literature—a nontrivial task these days, considering the number of papers and articles forthcoming on the subject.

I wish to thank Dr. Joel Schiff of the University of Auckland, New Zealand, for his useful comments and corrections in the preparation of the manuscript. I would like to acknowledge the kind permission of the Clarendon Press, Oxford, to refer to results published in *The Theory of Functions* by E. C. Titchmarsh, 1939; to Academic Press, Inc. for permitting me to quote several theorems from my text *Introduction to the Theory of Entire Functions*, Volume 56, in the series of Pure and Applied Mathematics, 1973, and also to Holt Rinehart and Winston Publishing Company, New York, for permission to use one or two quotations from their book *Elements of Complex Variables* by L. L. Pennisi, 1965. Dr. L. L. Pennisi was also most kind in permitting me to quote from his above mentioned text.

A. S. B. Holland

Preliminary Definitions and Symbolism

Throughout the text, we will use the following symbols and quantifiers that are herewith defined:

$\exists$	there exists or there is
$\forall$	for all or for every
$\in$	belonging to or is an element of
$\leftrightarrow$	if and only if
$\Rightarrow$	implies ($a \Rightarrow b$ means a implies b)
$\Leftarrow$	implied by ($a \Leftarrow b$ means b implies a or a is implied by b)
$\emptyset$	empty set
$\neq$	not equal to
$\notin$	does not belong to
$\not<$	not less than
$N(a, \rho)$	open circle with center a and radius ρ
$\vec{a}$	vector a (whose components or elements are usually defined in the body of the text)
$\mathbb{C}$	set of all complex numbers
$\mathbb{R}$	set of all real numbers
$C[a, b]$	class of all functions continuous on the closed interval $[a, b]$
$O(g)$	in $f(x) = O\{g(x)\}$ implies that $|f(x)| \leq A|g(x)|$ as x tends to some limit and A is a finite constant
$o(g)$	in $f(x) = o\{g(x)\}$ implies that $\left|\dfrac{f(x)}{g(x)}\right| \to 0$ as x tends to some limit

$\rho(a,b)$	Euclidean distance from a to b or $\|a-b\|$
$\rho(a,B)$	$\inf\{\|a-b\|, b \in B\}$
$\overline{S}$	closure of set S
S^c	complement of set S
$\cup$	union of sets
$\cap$	intersection of sets
$\triangle$	net
cis	$= \cos + i\sin$
cr Jc	closed rectifiable Jordan curve
i.e.s.	isolated essential singularity
iff	if and only if

**COMPLEX
FUNCTION
THEORY**

1
Complex Numbers

1.1

The system of complex numbers $\mathcal{C}$ is the set of all ordered pairs of real numbers $[a, b]$ endowed with two binary operations $+$ and $\cdot$. By *ordered* we mean: $[a, b] = [b, a]$ iff $a = b$; further, $[a, b] = [c, d]$ iff $a = c$ and $b = d$. The two *binary operations* $+$ and $\cdot$ may be defined as follows:

$$[a, b] + [c, d] = [a + c, b + d] \quad \text{(addition of complex numbers)}, \quad (1.1)$$

$$[a, b] \cdot [c, d] = [ac - bd, ad + bc] \quad \text{(multiplication of complex numbers)}. \quad (1.2)$$

1.1.1

Theorem *Given* $\alpha = [a, b]$, $\beta = [c, d]$, $\gamma = [e, f]$, *we have*

$$\alpha + (\beta + \gamma) = (\alpha + \beta) + \gamma \quad \text{(associative addition)},$$

$$\alpha + \beta = \beta + \alpha \quad \text{(commutative addition)},$$

$$\exists \tau = [0, 0]$$

such that $\alpha + \tau = \tau + \alpha$, $\forall \alpha$

$$\alpha \cdot \beta = \beta \cdot \alpha \quad \text{(commutative multiplication)}$$

(from now on we drop the $\cdot$ operation and write $\alpha\beta$ for $\alpha \cdot \beta$),

$$\alpha(\beta\gamma) = (\alpha\beta)\gamma \quad \text{(associative multiplication)},$$

$$\alpha(\beta + \gamma) = \alpha\beta + \alpha\gamma \quad \text{(distributive multiplication)}.$$

PROOF. To prove commutative multiplication for example, we use the definitions of 1.1:

$$\alpha\beta = [a, b][c, d]$$
$$= [ac - bd, ad + bc] \quad \text{by (1.2)}.$$

Now
$$\beta\alpha = [c, d][a, b]$$
$$= [ca - bd, cb + da] \quad \text{by (1.2)}$$
$$= [ac - bd, ad + bc]$$
$$= \alpha\beta. \qquad \square$$

1.1.2
Theorem *Given $\alpha = [a, b]$ and $\beta = [c, d]$, there exists γ such that $\alpha + \gamma = \beta$.*

PROOF. If we write $\gamma = [e, f]$, then
$$\alpha + \gamma = \beta \Longrightarrow \begin{cases} a + e = c, \\ b + f = d, \end{cases}$$
and so
$$e = c - a, \qquad f = d - b.$$
Thus
$$\gamma = [e, f] = [c - a, d - b] = \beta - \alpha. \qquad \square$$

1.2
A subset of the complex numbers may be set up in a one-to-one correspondence with real a. Thus $[a, 0] \leftrightarrow a$.

This mapping has the following properties:
$$[a, 0] + [b, 0] \leftrightarrow a + b$$
and
$$[a, 0] \cdot [b, 0] \leftrightarrow ab.$$
Thus we have an isomorphism between a subfield of the complex numbers and the reals. We make the identification $[a, 0] = a$ for all reals a.

1.3
Let us define $i = [0, 1]$. Thus,
$$i^2 = [0, 1] \cdot [0, 1] = [-1, 0] = -1.$$
We take
$$i = +\sqrt{-1}.$$
Also,
$$[a, 0] + [0, 1][b, 0] = [a, 0] + [0, b] = [a, b] = a + ib.$$

From now on we use the form $a + ib$ to mean the complex number whose real part is a and imaginary part is b. Thus, given $z = a + ib$, we write $R\{z\} = a$ and $I\{z\} = b$. It is now easy to verify that $0[a, b] = 0$ and $k[a, b] = [ka, kb]$ for all real k.

Complex Numbers

The following theorem characterizes a quotient of two complex numbers.

1.3.1
Theorem Given $\alpha = [a + ib]$ and $\beta = [c + id] \neq 0$, there exists $\delta = [g + ih]$ such that $\beta\delta = \alpha$.

PROOF. If $\beta\delta = \alpha$ then
$$cg - dh = a \quad \text{and} \quad dg + ch = b.$$

These two simultaneous equations (in g and h) may be solved to give a unique solution, provided
$$\begin{vmatrix} c & -d \\ d & c \end{vmatrix} \neq 0.$$

However this is precisely the condition $\beta \neq 0$. Thus δ is unique and $\beta\delta = \alpha$. □

1.3.2
Definition Given α, $\beta(\neq 0)$ and δ as above, we define the *quotient* $\delta = \alpha/\beta$ such that $\alpha = \beta\delta$; thus
$$\delta = \frac{ac + bd}{c^2 + d^2} + i\frac{bc - ad}{c^2 + d^2}.$$

This definition gives rise to a process called *rationalization*, whereby we can write a quotient $(p + iq)/(r + is)$ in the form $A + iB$. Thus
$$\frac{p + iq}{r + is} = \frac{pr + qs}{r^2 + s^2} + i\frac{qr - ps}{r^2 + s^2}.$$

1.3.3
Proposition If the product of two complex numbers is zero, at least one of these numbers is zero.

PROOF. If $\alpha\beta = 0$ and $\beta \neq 0$, then
$$\alpha = (\alpha\beta)\left(\frac{1}{\beta}\right) = 0\frac{1}{\beta} = 0. \quad \square$$

1.3.4
Exercises

1: Show that $z = \dfrac{1 - i\sqrt{3}}{2}$ satisfies $\dfrac{3}{z + 1} - \dfrac{1}{z} = 1$.

2: Evaluate (in the form $A + iB$): $\left(\dfrac{3i}{1 + i}\right)^6$, $\left(\dfrac{1 - i}{1 + i}\right)^3$.

3: Evaluate $\dfrac{1}{1 + i + i^2 + i^3 + i^4 + i^5}$.

1.4

Definition The *conjugate* of $\alpha = a + ib$, is given by $\bar{\alpha} = a - ib$. (We sometimes write α^* for $\bar{\alpha}$.) The conjugate $\bar{\alpha}$ has the following properties:

$$\alpha\bar{\alpha} = a^2 + b^2,$$
$$\overline{\alpha + \beta} = \bar{\alpha} + \bar{\beta},$$
$$\overline{\left(\frac{\alpha}{\beta}\right)} = \frac{\bar{\alpha}}{\bar{\beta}},$$
$$\overline{(\bar{\alpha})} = \alpha,$$
$$\overline{\alpha\beta} = (\bar{\alpha})(\bar{\beta}).$$

Verification is left to the reader.

1.5

Definition The *modulus (absolute value)* of a complex number $\alpha = a + ib$ is given by

$$|\alpha| = \sqrt{a^2 + b^2}.$$

Thus the modulus of a complex number is a nonnegative quantity. If we write $z = x + iy$, where $x = R\{z\}$ and $y = I\{z\}$, then the following theorem is easily verified.

1.5.1

Theorem With the above notation and given $z = x + iy$, we have

$$|z| = |\bar{z}|,$$
$$|z|^2 = x^2 + y^2 = z\bar{z},$$
$$z + \bar{z} = 2R\{z\},$$
$$z - \bar{z} = 2iI\{z\},$$
$$|R\{z\}| \leq |z|,$$
$$|I\{z\}| \leq |z|,$$
$$|z| \leq |R\{z\}| + |I\{z\}|.$$

All are left as exercises for the reader.

The following lemmas deal with properties of the absolute value.

1.5.2

Lemma $|z_1 z_2| = |z_1||z_2|$.

Proof is left to the reader.

Complex Numbers

1.5.3
Lemma $|z_1 + z_2| \leq |z_1| + |z_2|$ (*triangle inequality*).

PROOF

$$\begin{aligned}
|z_1 + z_2|^2 &= (z_1 + z_2)(\overline{z_1 + z_2}) = (z_1 + z_2)(\bar{z}_1 + \bar{z}_2) \\
&= z_1\bar{z}_1 + z_1\bar{z}_2 + z_2\bar{z}_1 + z_2\bar{z}_2 \\
&= |z_1|^2 + \{z_1\bar{z}_2 + (\overline{z_1\bar{z}_2})\} + |z_2|^2 \\
&= |z_1|^2 + 2R\{z_1\bar{z}_2\} + |z_2|^2 \\
&\leq |z_1|^2 + 2|z_1\bar{z}_2| + |z_2|^2 \\
&= |z_1| + 2|z_1||z_2| + |z_2|^2 \\
&= (|z_1| + |z_2|)^2,
\end{aligned}$$

and the lemma follows. Thus, by induction

$$\left|\sum_{i=1}^{n} z_i\right| \leq \sum_{i=1}^{n} |z_i|$$

for all positive integral n. Equality occurs when

$$|z_1 + z_2|^2 = (|z_1| + |z_2|)^2,$$

that is,

$$2R\{z_1\bar{z}_2\} = 2|z_1\bar{z}_2|$$

or when

$$\arg(z_1\bar{z}_2) = \arg(R\{z_1\bar{z}_2\}) = 0.$$

Thus $z_1\bar{z}_2$ is a real constant, say α, and

$$z_1 = \frac{\alpha z_2}{\bar{z}_2 z_2} = \frac{\alpha z_2}{|z_2|^2}$$

or $z_1 = kz_2$, k real constant. $\square$

1.5.4
Lemma $||z_1| - |z_2|| \leq |z_1 - z_2|$.

PROOF. Use $-R\{z_1z_2\} \geq -|z_1z_2|$ and proceed as in 1.5.3. $\square$

We now turn to other, more complicated, inequalities concerning the absolute value of a complex number.

1.6 Inequalities of Schwarz, Hölder, and Minkowski

Consider n-vectors

$$\vec{\alpha} = (a_1, a_2, \ldots, a_n) \quad \text{and} \quad \vec{\beta} = (b_1, b_2, \ldots, b_n).$$

We define

$$\text{inner-product:} \quad \vec{\alpha} \cdot \vec{\beta} = \sum_{r=1}^{n} a_r b_r$$

and

$$l^p\text{-norm:} \quad \|\vec{\alpha}\|_p = \left\{ \sum_{r=1}^{n} |a_r|^p \right\}^{1/p}, p > 0.$$

1.6.1
Theorem *If*

$$p > 1$$

and

$$\frac{1}{p} + \frac{1}{q} = 1$$

then

$$|\vec{\alpha} \cdot \vec{\beta}| \leq \sum_{k=1}^{n} |a_k b_k| \leq \|\vec{\alpha}\|_p \|\vec{\beta}\|_q \ (\textit{Hölder's inequality}).$$

PROOF. If $\alpha > 0$ and $p > 1$ are given numbers, the maximum value of $ax - p^{-1}x^p$ for all $x \geq 0$ occurs for $x = a^{1/(p-1)}$. Thus

$$ax - p^{-1}x^p \leq a \cdot a^{1/(p-1)} - \frac{(a)^{p/(p-1)}}{p} = a^{p/(p-1)}\left\{1 - \frac{1}{p}\right\} = \frac{a^q}{q},$$

hence

$$ax \leq \frac{x^p}{p} + \frac{a^q}{q}, \ 0 \leq x < \infty.$$

Replace a by $|b_k|$ and x by $|a_k|$; $k = 1, 2, \ldots, n$. Then

$$|a_k b_k| \leq \frac{|a_k|^p}{p} + \frac{|b_k|^q}{q};$$

now add for $k = 1, \ldots, n$. Thus

$$\sum_{k=1}^{n} |a_k b_k| \leq \frac{1}{p} \sum_{k=1}^{n} |a_k|^p + \frac{1}{q} \sum_{k=1}^{n} |b_k|^q = \frac{1}{p} X^p + \frac{1}{q} Y^q$$

where

$$X = \|\vec{\alpha}\|_p; \ Y = \|\vec{\beta}\|_q.$$

Multiply the elements of $\vec{\alpha}$ by $\lambda > 0$ and multiply the elements of $\vec{\beta}$ by $1/\lambda$, then $\vec{\alpha} \cdot \vec{\beta}$ is unchanged. Thus

$$\sum_{k=1}^{n} |a_k b_k| \leq \lambda^p \frac{X^p}{p} + \frac{1}{\lambda^q} \frac{Y^q}{q}$$

and we minimize the right-hand side for $\lambda > 0$. The minimum occurs when $\lambda = Y^{1/p}/X^{1/q}$; thus,

$$\sum_{k=1}^{n} |a_k b_k| \leq \frac{1}{p} \frac{Y X^p}{X^{p/q}} + \frac{1}{q} \frac{X Y^q}{Y^{q/p}} = XY = \|\vec{\alpha}\|_p \|\vec{\beta}\|_q$$

Complex Numbers

or

$$\left\{\sum_{k=1}^{n} |a_k b_k|\right\} \leq \left\{\sum_{k=1}^{n} |a_k|^p\right\}^{1/p} \left\{\sum_{k=1}^{n} |b_k|^q\right\}^{1/q}.$$

For $p = 2$ and $q = 2$ we have what is usually called Schwarz's inequality. □

1.6.2
Theorem (Minkowski) *If $p \geq 1$, then*

$$\left(\sum_{k=1}^{n} |a_k + b_k|^p\right)^{1/p} \leq \left(\sum_{k=1}^{n} |a_k|^p\right)^{1/p} + \left(\sum_{k=1}^{n} |b_k|^p\right)^{1/p}.$$

PROOF. The proposition is true for $p = 1$, since $|a_k + b_k| \leq |a_k| + |b_k|$. Consider

$$|a_k + b_k|^p \leq (|a_k| + |b_k|)|a_k + b_k|^{p-1} = |a_k||a_k + b_k|^{p-1} + |b_k||a_k + b_k|^{p-1};$$

then

$$\sum_{k=1}^{n} |a_k + b_k|^p \leq \sum_{k=1}^{n} |a_k||a_k + b_k|^{p-1} + \sum_{k=1}^{n} |b_k||a_k + b_k|^{p-1}.$$

Using the Hölder inequality (twice), we have

$$\sum_{k} |a_k + b_k|^p \leq \left\{\sum_{k} |a_k|^p\right\}^{1/p} \left\{\sum_{k} |a_k + b_k|^{pq-q}\right\}^{1/q}$$

$$+ \left\{\sum_{k} |b_k|^p\right\}^{1/p} \left\{\sum_{k} |a_k + b_k|^{pq-q}\right\}^{1/q}.$$

Since $pq - q = p$, if we divide both sides by $\{\sum_k |a_k + b_k|^{pq-q}\}^{1/q}$, then

$$\left\{\sum_{k} |a_k + b_k|^p\right\}^{1-\frac{1}{q}} = \left\{\sum_{k} |a_k + b_k|^p\right\}^{\frac{1}{p}} \leq \left\{\sum_{k} |a_k|^p\right\}^{\frac{1}{p}} + \left\{\sum_{k} |b_k|^p\right\}^{\frac{1}{p}}.$$

Conditions under which equality occurs are left for the following exercises. □

1.7
Exercises

1: If $z = x + iy$, show that

$$\frac{|x| + |y|}{\sqrt{2}} \leq |z| \leq |x| + |y|.$$

When does equality hold?

2: Prove that $\left|\dfrac{a - b}{1 - \bar{a}b}\right| = 1$ if a, b are complex and $|a| = 1$ or $|b| = 1$.

3: Prove that $\left|\dfrac{az + b}{\bar{b}z + \bar{a}}\right| = 1$ if $|z| = 1$.

4: Show that if a polynomial $P(z)$ has real coefficients, then the values of z for which it is zero are either real or occur in pairs of complex conjugates.

5: Prove that $|z_1 - z_2|^2 + |z_1 + z_2|^2 = 2\{|z_1|^2 + |z_2|^2\}$ (parallelogram law).

6: Show that equality holds in Hölder's inequality if a_k^p is proportional to b_k^q for $k = 1, 2, \ldots$.

7: Show that equality holds in Minkowski's inequality if a_k is proportional to b_k for $k = 1, 2, \ldots$.

We now investigate another representation of a complex number.

1.8 Geometric Interpretation

The complex number $[a, b]$ has a geometrical interpretation induced by the fact that the number pair $[a, b]$ may be represented by a unique point on the plane. If we choose the x-axis to be the real axis and the y-axis to be the imaginary axis, then we generate what is called the complex, or z-plane. For example, $z = -3 + 2i$ corresponds to the point $[-3, 2]$.

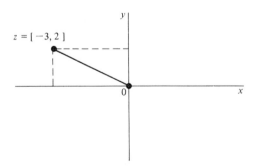

Given $z_1 = x_1 + iy_1$ and $z_2 = x_2 + iy_2$, we have the following inequalities:

$$|z_1 - z_2| = \sqrt{(x_1 - x_2)^2 + (y_1 - y_2)^2} \geq |x_1 - x_2|,$$
$$|z_1 - z_2| \geq |y_1 - y_2|,$$
$$|z_1 - z_2| \leq |x_1 - x_2| + |y_1 - y_2|.$$

If z_0 is a fixed point $(x_0 + iy_0)$ and $z = x + iy$ is arbitrary, then the locus

Complex Numbers

$|z - z_0| = k$ (real) is a circle with center z_0 and radius k, since

$$|z - z_0|^2 = k^2$$
$$\Rightarrow (z - z_0)(\bar{z} - \bar{z}_0) = k^2$$
$$\Rightarrow |z|^2 + |z_0|^2 - (z_0\bar{z} + z\bar{z}_0) = k^2$$
$$\Rightarrow x^2 + y^2 + x_0^2 + y_0^2 - (z\bar{z}_0 + \overline{z\bar{z}_0}) = k^2$$
$$\Rightarrow x^2 + y^2 + x_0^2 + y_0^2 - 2R\{z\bar{z}_0\} = k^2$$
$$\Rightarrow x^2 + y^2 + x_0^2 + y_0^2 - 2xx_0 - 2yy_0 = k^2$$
$$\Rightarrow (x - x_0)^2 + (y - y_0)^2 = k^2.$$

Similarly, the set of points given by $|z - z_0| \leq k$ is a closed disk with center z_0 and radius k.

1.9
Exercises

1: Show that z_1, z_2, z_3 are vertices of an equilateral triangle inscribed in a unit circle if $|z_1| = |z_2| = |z_3|$ and $z_1 + z_2 + z_3 = 0$.

2: If Γ is a circle through $z = \pm k$, $k > 0$, then z is exterior, on, or interior to Γ, depending on whether $z\bar{z} - 2bI\{z\} \gtreqless k^2$ (respectively), where b is real and $z = ib$ is the center of Γ.

3: Show that the points $-k, k, z_1, k^2/z_1$ are concyclic.

4: Prove that the equation of a circle in the z-plane is given by

$$az\bar{z} + b\bar{z} + \bar{b}z + c = 0,$$

where $a \neq 0$, while c is a real constant and b is a complex constant.

A point on the plane can be represented by Cartesian coordinates or polar coordinates, thus we may study the polar form of a complex number as follows.

1.10 Polar Form of a Complex Number

Given a point $P(x, y)$ on the plane, i.e., $z = x + iy$, the polar coordinates of P are generated by the equations

$$x = r \cos \theta, \quad y = r \sin \theta.$$

Thus $P(x, y) \Rightarrow P(r, \theta)$, where $r = \sqrt{x^2 + y^2}$ and $\theta = \tan^{-1}(y/x)$.

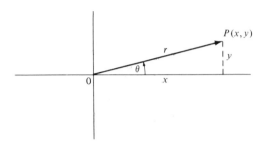

We say, r is the *length* of the radial vector $\overrightarrow{OP}$ and θ is the *argument*, or amplitude, of the vector. Thus we write $\theta = \arg z$ or amp z and measure the angle counterclockwise from the positive x-axis to the radial vector. Thus

$$z = x + iy = r(\cos\theta + i\sin\theta).$$

The principal value of $\arg z$ is in the range $-\pi < \arg z \leqslant \pi$. Given $\bar{z} = x - iy$, we have $|\bar{z}| = |z| = r$ and $\arg \bar{z} = -\theta$. Geometrically, $\bar{z}$ is the reflection of z with respect to the real axis.

Given two points z_1 and z_2, we can write $z_1 - z_2 = \rho \operatorname{cis} \varphi$, where ρ is the length of the vector $z_1 - z_2$ and φ is the angle made by this vector with the positive x-axis.

1.11 Exponential Function

Consider $e^z = e^{x+iy}$.

1.11.1

Definition $e^{x+iy} \equiv e^x \cdot e^{iy}$, where e^x is the real exponential and $e^{iy} = \cos y + i\sin y$..

This formula is known as *Euler's formula* and may be developed from the addition of two (real) absolutely convergent Taylor series for the trigonometric functions.

A vector $e^{i\theta}$ is a unit vector at an angle θ to the horizontal since

$$|e^{i\theta}| = |\cos\theta + i\sin\theta| = 1.$$

Further,

$$\begin{aligned}
e^{iy_1}e^{iy_2} &= (\cos y_1 + i\sin y_1)(\cos y_2 + i\sin y_2) \\
&= \cos(y_1 + y_2) + i\sin(y_1 + y_2) \\
&= e^{i(y_1+y_2)}.
\end{aligned}$$

Since $e^z = e^x e^{iy}$ by definition, we have

$$e^{z_1}e^{z_2} = e^{z_1+z_2}. \tag{1.3}$$

Complex Numbers

We can similarly show that:

$$\{e^{i\theta}\}^{-1} = e^{-i\theta}, \tag{1.4}$$

$$\frac{e^{i\theta_1}}{e^{i\theta_2}} = e^{i(\theta_1-\theta_2)},$$

$$e^{i(\theta+2k\pi)} = e^{i\theta}, \quad \text{where } k = 0, \pm 1, \pm 2, \ldots,$$

$$e^z \neq 0, \forall z. \text{ (Use (1.3) with } z_1 = z, z_2 = -z.)$$

We still have to show that if e^z is to be sufficiently well-behaved (in some sense to be defined), with $e^0 = 1$ and $e^{x+iy} = e^x e^{iy}$ a consistent definition, then if $e^{iy} = \varphi_1(y) + i\varphi_2(y)$, the only candidate for e^z satisfying all the required conditions is $e^z = e^x\{\cos y + i \sin y\}$. We will do this when we discuss the concept of analyticity.

Note that if

$$z_1 = r_1(\cos\theta_1 + i\sin\theta_1) = r_1 e^{i\theta_1}$$

and

$$z_2 = r_2(\cos\theta_2 + i\sin\theta_2) = r_2 e^{i\theta_2}$$

then

$$z_1 z_2 = r_1 r_2 e^{i(\theta_1+\theta_2)},$$

$$|z_1 z_2| = r_1 r_2 = |z_1||z_2|,$$

and

$$\arg(z_1 z_2) = \theta_1 + \theta_2 = \arg z_1 + \arg z_2 + 2k\pi; \quad k = 0, \pm 1, \ldots.$$

Also,

$$\arg \frac{z_1}{z_2} = \arg z_1 - \arg z_2 + 2k\pi \text{ if } z_2 \neq 0.$$

1.12 De Moivre's Theorem

If $z = e^{i\theta}$, then $z^n = (e^{i\theta})^n$. We are required to show that $(e^{i\theta})^n = e^{in\theta}$ for all n, however first we demonstrate the validity of this statement for n a positive integer.

1.12.1
Theorem (De Moivre) $(\cos\theta + i\sin\theta)^n = \cos n\theta + i\sin n\theta$, $n = 0, 1, \ldots$.

PROOF. By induction on n and by using (1.3) and (1.4) we have

$$\left(e^{i\theta}\right)^n = (\cos\theta + i\sin\theta)^n$$

$$= \cos n\theta + i\sin n\theta. \qquad \square$$

REMARK. De Moivre's theorem can be shown to be valid for all real n, in fact, for all complex n. However, the proof for any n will require the development of the logarithmic function.

1.13 Roots of a Complex Number

Given a natural number n, let us write $\omega = z^n$, where $\omega = \rho e^{i\varphi}$ and $z = re^{i\theta}$. Thus $\rho e^{i\varphi} = r^n e^{in\theta}$, and equating the real and imaginary parts, we get

$$\rho = r^n \quad \text{and} \quad \varphi = n\theta + 2k\pi, \quad k = 0, \pm 1, \ldots.$$

Consequently,

$$z = \omega^{1/n} = re^{i\theta} = \rho^{1/n} e^{i(\varphi/n + 2k\pi/n)}$$

and thus there exist n roots of the complex number $\omega \neq 0$. All nth roots have the same modulus and argument and are equally spaced along the circumference of the circle.

We can similarly show that if $z = re^{i\theta}$, then

$$z^{m/n} = r^{m/n} e^{i\{m\theta/n + 2k\pi m/n\}}, \qquad (m, n) = 1.$$

1.14 Roots of Unity

Since $1 = \cos 0 + i \sin 0 = e^{i(0 + 2k\pi)}$, then there are n roots of 1, namely, $e^{i2k\pi/n}$, $k = 0, 1, \ldots, n - 1$.

EXAMPLE 1. Find the three cube roots of 1.

SOLUTION. If $z^3 = 1 = e^{2k\pi i}$; $k = 0, \pm 1, \ldots$, then $z = e^{2k\pi i/3}$ for $k = 0, 1, 2$. For $k = 3, 4, \ldots$, we just repeat the same original three roots determined by $k = 0, 1,$ and 2. Thus

$$z = e^0, \qquad e^{2\pi i/3}, \qquad e^{4\pi i/3},$$

that is

$$1, \qquad \cos \frac{2\pi}{3} + i \sin \frac{2\pi}{3}, \qquad \cos \frac{4\pi}{3} + i \sin \frac{4\pi}{3},$$

or

$$1, \qquad -\frac{1}{2} + i \frac{\sqrt{3}}{2}, \qquad -\frac{1}{2} - i \frac{\sqrt{3}}{2}.$$

These are the three cube roots of 1. (Check: Cube each root to obtain 1.)

Note: If ω, ω^2 are the complex cube roots of 1, then $1 + \omega + \omega^2 = 0$ since $\omega^3 = 1$.

Similarly if $\omega_n = e^{2\pi i/n}$ and the nth roots of unity are $1, \omega_n, \omega_n^2, \ldots, \omega_n^{n-1}$, then $\sum_{i=0}^{n-1} \omega_n^i = 0$ since $\omega^n = 1$.

Complex Numbers

EXAMPLE 2. Show that all roots of $(z+1)^5 + z^5 = 0$ lie on the line $x = -\frac{1}{2}$.

SOLUTION

$$1 + \left(1 + \frac{1}{z}\right)^5 = 0,$$

therefore

$$1 + \frac{1}{z} = e^{i(\pi + 2k\pi)/5}, \; k = 0, \ldots, 4;$$

and thus

$$z^{-1} = -1 + e^{i(\pi + 2k\pi)/5}.$$

Designating $\theta = (\pi + 2k\pi)/5$ we can write

$$z^{-1} = \cos\theta + i\sin\theta - 1$$
$$= -2\sin^2\frac{\theta}{2} + 2i\sin\frac{\theta}{2}\cos\frac{\theta}{2}$$
$$= -2\sin\frac{\theta}{2}\left(\sin\frac{\theta}{2} - i\cos\frac{\theta}{2}\right)$$
$$= -2\sin\frac{\theta}{2}\left\{\cos\left(\frac{\pi}{2} - \frac{\theta}{2}\right) - i\sin\left(\frac{\pi}{2} - \frac{\theta}{2}\right)\right\}.$$

Thus

$$z^{-1} = -2\sin\frac{\theta}{2} e^{-i(\pi-\theta)/2}$$

and

$$z = -\frac{1}{2\sin(\theta/2)}\left\{\cos\left(\frac{\pi}{2} - \frac{\theta}{2}\right) + i\sin\left(\frac{\pi}{2} - \frac{\theta}{2}\right)\right\}.$$

Hence

$$R(z) = -\frac{1}{2\sin(\theta/2)}\sin\frac{\theta}{2} = -\frac{1}{2}.$$

Note: It is not good enough to just say

$$(1+z)^5 = -z^5,$$

hence

$$1 + z = -z, \; 2z = -1, \text{ and } z = -\frac{1}{2}.$$

This only shows that $x = -\frac{1}{2}, y = 0$ is one root.

EXAMPLE 3. Given $x_n + iy_n = (1 + i\sqrt{3})^n$, show that

$$x_{n-1}y_n - x_n y_{n-1} = 2^{2n-2}\sqrt{3}.$$

SOLUTION

$$1 + i\sqrt{3} = 2\left\{\frac{1}{2} + i\frac{\sqrt{3}}{2}\right\} = 2e^{i[(\pi/3) + 2k\pi]}.$$

Thus

$$(1 + i\sqrt{3})^n = 2^n e^{i[(n\pi/3) + 2kn\pi]}$$

and

$$x_n = 2^n \cos\frac{n\pi}{3} \;;\quad x_{n-1} = 2^{n-1} \cos\frac{(n-1)\pi}{3} \;;$$

$$y_n = 2^n \sin\frac{n\pi}{3} \;;\quad y_{n-1} = 2^{n-1} \sin\frac{(n-1)\pi}{3}.$$

Hence

$$x_{n-1}y_n - x_n y_{n-1} = 2^{2n-1}\left\{\sin\left[\frac{n\pi}{3} - (n-1)\frac{\pi}{3}\right]\right\}$$

$$= 2^{2n-1}\frac{\sqrt{3}}{2} = 2^{2n-2}\sqrt{3}.$$

EXAMPLE 4 (Enestrom's theorem). Let $n \geq 1$ and $a_0 > a_1 > \cdots > a_n > 0$. Show that all zeros of $P(z) = \sum_{i=0}^{n} a_i z^i$ are exterior to the unit circle $|z| = 1$.

SOLUTION. Consider

$$(1 - z)P(z) = a_0 - \left[(a_0 - a_1)z + \cdots + (a_{n-1} - a_n)z^n + a_n z^{n+1}\right]$$

then

$$|(1 - z)P(z)| \geq a_0 - |(a_0 - a_1)z + \cdots + a_n z^{n+1}|$$

$$\geq a_0 - (a_0 - a_1)|z| - \cdots - a_n |z|^{n+1}.$$

We have equality if $[(a_0 - a_1)z + \cdots + a_n z^{n+1}]$ is a constant multiple of a_0, that is, if $\arg\{(a_0 - a_1)z + \cdots\} = 0$. But for $z \notin [0, 1]$ we see that $\arg z$, $\arg z^2, \ldots$ differ, hence we have strict inequality. Thus

$$|(1 - z)P(z)| > 0 \text{ for } |z| \leq 1, z \notin [0, 1].$$

For $z \in [0, 1]$, $P(z)$ is clearly not zero, hence for $|z| \leq 1$, $|P(z)| > 0$ and thus all roots of $P(z)$ are in $|z| > 1$.

1.15

Exercises

1: Prove that if z_1, z_2, z_3 are vertices of an equilateral triangle, then

$$z_1^2 + z_2^2 + z_3^2 = z_1 z_2 + z_2 z_3 + z_3 z_1.$$

Complex Numbers

2: Prove that $\arg z + \arg \bar{z} = 2n\pi$; n integer.

3: Show that the two lines joining points z_1, z_2 and z_3, z_4 are perpendicular, provided that
$$\arg\left\{\frac{z_1 - z_2}{z_3 - z_4}\right\} = \pm\frac{\pi}{2},$$
that is, $(z_1 - z_2)/(z_3 - z_4)$ is purely imaginary.

4: Compute the following roots: $(-1)^{1/4}$, $(1+i)^{1/2}$, $1^{1/6}$.

5: Use De Moivre's theorem to show that:
1. $\cos n\theta = \cos^n\theta - \binom{n}{2}\cos^{n-2}\theta \sin^2\theta + \binom{n}{4}\cos^{n-4}\theta \sin^4\theta \ldots,$
2. $\sin n\theta = \binom{n}{1}\cos^{n-1}\theta \sin\theta - \binom{n}{3}\cos^{n-3}\theta \sin^3\theta \ldots.$

6: Find the real and imaginary parts and the modulus of
$$\frac{1+e^{i\theta}}{1+e^{i\varphi}}.$$

7: Show that for a positive odd integer n
$$\cos^n\theta = \frac{1}{2^{n-1}}\sum_{r=0}^{(n-1)/2}\binom{n}{r}\cos(n-2r)\theta.$$
Also, if n is an even positive integer, then
$$\sin^n\theta = \frac{(-1)^{n/2}}{2^{n-1}}\sum_{r=0}^{(n-2)/2}(-1)^r\binom{n}{r}\cos(n-2r)\theta + \frac{1}{2^n}\binom{n}{n/2}.$$

8: Show that
$$\sum_{k=1}^{n}\sin k\theta = \frac{1}{2}\cot\frac{\theta}{2} - \frac{\cos\{(n+\frac{1}{2})\theta\}}{2\sin(\theta/2)}, \qquad 0 < \theta < 2\pi.$$

2

Definitions and Properties of Sets in the Complex Plane

2.1

We briefly outline the fundamental properties of points in the complex plane and list definitions relating thereto. There is a vast amount of literature dealing with the ramifications and extensions of most of the material presented here and hosts of exercises can be found in standard works dealing with the point-set topology of the plane. The material is kept simple in order not to inundate the reader with minutiae usually found in real-analysis courses that normally precede a course on functions of a complex variable.

2.2

We assume that the reader can manipulate the basic set operations and for completeness we include the basic definitions.*

2.2.1

Definition The *union* of two sets A and B is defined as:
$$A \cup B = \{x \mid x \in A \text{ or } x \in B\}.$$

Clearly $A \cup B = B \cup A$.

2.2.2

Definition The intersection of two sets A and B is defined as:
$$A \cap B = \{x \mid x \in A \text{ and } x \in B\}.$$

Clearly $A \cap B = B \cap A$.

---

*All the elementary properties of sets may be studied for example in, Set Theory and Related Topics, by S. Lipschutz, Schaum Publishing Company, N.Y., 1964.

Definitions and Properties of Sets in the Complex Plane

We proceed with the following properties of sets of points in the plane.

2.3 Open Interval

An *open interval* on the real line is a set of points (a, b) such that it contains all t, $a < t < b$. If a, b are finite, then a, b are called *boundary points*. Similarly, a *closed interval* on the real line $[a, b]$ contains all t such that $a \leq t \leq b$.

An open interval in the complex plane $I = (a_1, a_2, b_1, b_2)$ is the set of all points $z = x + iy$ with $a_1 < x < a_2$ and $b_1 < y < b_2$, i.e., all points interior to the rectangle.

Similarly for a closed interval, we have all z in and on the boundary of the rectangle.

2.3.1 Greatest lower bound (g.l.b.)

Given a set S of points on the real line, we define a *g.l.b.* t_1 such that if

1. $t_1 \leq t$, $\forall t \in S$, then t_1 is a lower bound and
2. if $t_0 > t_1$, then t_0 is not a lower bound of S.

2.3.2 Least upper bound (l.u.b.)

Given a set S of points on the real line, we define a *l.u.b.* t_2 such that if

1. $t_2 \geq t$, $\forall t \in S$, then t_2 is an upper bound and
2. if $t_0 < t_2$, then t_0 is not an upper bound.

2.3.3

A set S of complex numbers is *bounded* if and only if we can find a k such that $|z| \leq k$ for every $z \in S$.

2.4

The *diameter* of a set S of complex numbers is the l.u.b. of the set

$$[\alpha_{ij} = |z_i - z_j| : \forall\, z_i, z_j \in S].$$

If S is bounded then the diameter is finite.

2.5 Neighborhood

We define an ϵ-neighborhood of a point z_0 to be the set of all z such that

$$|z - z_0| < \epsilon, \quad \epsilon > 0 \text{ (real)}.$$

A set S is *open* if for each $z \in S$ there exists a neighborhood of z contained in S.

Note: The set of all points on the real line is open with respect to the line but *not* open with respect to the complex plane.

2.6 Point of Accumulation

A point z_0 is a point of accumulation of a set S if every neighborhood of z_0 contains a point of S distinct from z_0. The point z_0 is not necessarily in S. For example, $z_0 = 0$ is a point of accumulation but is not in S when

$$S = \left\{1, \frac{1}{2}, \ldots, \frac{1}{n}, \ldots\right\}.$$

All points on $|z| = k$ are points of accumulation of $|z| < k$ and not in the set $S = \{z: |z| < k\}$.

2.6.1

A set S is *closed* if every point of accumulation of S is in S. For example, the set $S = \{z: |z| = M\}$, is closed. Also, every $z = x + iy$, for x and y rational, is not closed. The set $S = \{z: [|z| < 2] \cup [z = 2]\}$ is neither open nor closed. The null set is both open and closed.

2.6.2

Closure of a set S is the union of S with its set of points of accumulation.

2.7

Boundary Point of a set S is a point w such that every neighborhood of w contains at least one point $z \in S$ and one point $z' \notin S$.

2.8 Separation

We say S is *separated* into S_1 and S_2 if

1. $S_1 \neq \emptyset$,
2. $S_2 \neq \emptyset$,
3. $S = S_1 \cup S_2$,
4. $S_1 \cap S_2 = \emptyset$,
5. S_1 contains no point of accumulation of S_2,
6. S_2 contains no point of accumulation of S_1.

We say a set is *connected* if it cannot be separated.

EXAMPLE 1. We show that the union of an infinite number of closed sets can be open, whereas the union of a finite number of closed sets is always closed. Let

$$S_n = \left[\frac{1}{n}, 1 - \frac{1}{n}\right], \quad \text{then} \quad \bigcup_{n=3}^{\infty} S_n = (0, 1).$$

Note that 0 and 1 are not in the union. Also,

$$\bigcap_{n=1}^{\infty}\left(-\frac{1}{n}, \frac{1}{n}\right) = [0].$$

Definitions and Properties of Sets in the Complex Plane

EXAMPLE 2. Describe the region
$$|z + i| < |z - 1|.$$

SOLUTION
$$|z + i|^2 < |z - 1|^2$$

or
$$x^2 + (y + 1)^2 < (x - 1)^2 + y^2,$$
$$2y + 1 < -2x + 1$$
$$y < -x.$$

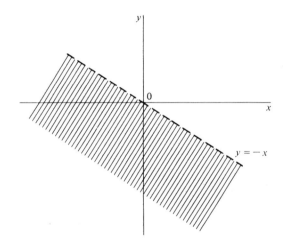

2.9

Exercises

Describe the following regions:

1: $|z - 3| + |z + 3| = 6.$

2: $|z - z_1| = |z - z_2|.$

3: $|z| = R(z) + 1.$

4: $|z| < \arg z;\ 0 \leqslant \arg z < 2\pi.$

5: $R\{1/z\} = C.$

6: $|z^2 - 1| = \lambda\, (> 0).$

2.10 Sequences

$\{z_n\}$ is a *sequence* of complex constants if to each positive integer n we assign a complex number z_n. In general, complex numbers are not ordered, i.e., we cannot say $z_1 < z_2 < \ldots$. However we could order according to some criteria, e.g., increasing argument, absolute value, or real part.

We say $\{z_n\}$ is equivalent to $\{\omega_n\}$ iff $z_n = \omega_n$ for all n.

We say $\{z'_m\}$ is a *subsequence* of $\{z_n\}$ if $z'_m = z_n$ for all m, $m \leqslant n$.

2.10.1

Definition We call $\{z_n\}$ a *Cauchy sequence* if for every $\epsilon > 0$ there exists N such that

$$|z_n - z_m| < \epsilon, \forall\, m, n, \text{ where } m > N \text{ and } n > N.$$

We can thus define a limit of a sequence.

2.10.2

Definition Given a sequence $\{z_n\}$, we say that z_0 is the *limit* of the sequence $\{z_0\}$, that is, $\lim_{n\to\infty} z_n = z_0$, if for every $\epsilon > 0$ there exists N such that $|z_n - z_0| < \epsilon$ for all $n > N$.

The next theorem now follows immediately.

2.10.3

Theorem *A necessary and sufficient condition for the sequence $\{z_n\}$ to converge to a limit z_0 (where $z_n = x_n + iy_n$ and $z_0 = x_0 + iy_0$), is that*

$$\lim_{n\to\infty} x_n = x_0$$

and

$$\lim_{n\to\infty} y_n = y_0.$$

PROOF.

Necessity: if $z_n \to z_0$, then given $\epsilon > 0$ there exists N such that $|z_n - z_0| < \epsilon$ for all $n > N$. However,

$$\left.\begin{array}{l} |x_n - x_0| \leqslant |z_n - z_0| < \epsilon \\ |y_n - y_0| \leqslant |z_n - z_0| < \epsilon \end{array}\right\} \text{ for } n > N,$$

hence necessity follows.

Sufficiency: if

$$\left.\begin{array}{l} |x_n - x_0| < \dfrac{\epsilon}{2} \\ |y_n - y_0| < \dfrac{\epsilon}{2} \end{array}\right\} \text{ for } n > N,$$

Definitions and Properties of Sets in the Complex Plane

then
$$|z_n - z_0| \leq |x_n - x_0| + |y_n - y_0| < \epsilon \text{ for } n > N,$$
and sufficiency follows. □

2.10.4

Definition We say that a sequence $\{z_n\}$ is *bounded* if there exists M such that $|z_n| \leq M$ for all n.

Note: A bounded sequence need not converge; for example, $\{(-1)\}^n$. However a convergent sequence is necessarily bounded. Also, a sequence may have a point of accumulation and not have a limit; for example, $z = 1$ is a point of accumulation of the sequence

$$\left\{ 1 + \frac{1}{2}, 2 + \frac{1}{2}, 1 + \frac{1}{3}, 2 + \frac{1}{3}, \ldots \right\}$$

and it has no limit since

$$1 + \frac{1}{n} \to 1, \quad 2 + \frac{1}{n} \to 2,$$

i.e., the limit is not unique.

2.10.5

Problem. If a sequence $\{z_n\}$ converges to z_0, then the sequence

$$\{\xi_n\} = \left\{ \frac{1}{n} \sum_{k=1}^{n} z_k \right\}$$

converges also to z_0.

SOLUTION. Choose $\epsilon > 0$ (arbitrary) and let ω be a positive number. Put $z_n = z_0 + \alpha_n$, $n \geq 1$, then there exists a positive integer $N(\epsilon)$ such that $|\alpha_n| < \omega$, $\forall n \geq N$ (assume $N \geq 2$).

For every $n > N$, we can write

$$\left| z_0 - \frac{z_1 + z_2 + \cdots + z_n}{n} \right| = \left| -\frac{1}{n}(\alpha_1 + \alpha_2 + \cdots + \alpha_n) \right|$$

$$\leq \frac{|\alpha_1| + |\alpha_2| + \cdots + |\alpha_n|}{n}$$

$$= \frac{|\alpha_1| + \cdots + |\alpha_{N-1}|}{n} + \frac{|\alpha_N| + \cdots + |\alpha_n|}{n}$$

$$< \frac{1}{n}(|\alpha_1| + \cdots + |\alpha_{N-1}|) + \frac{n+1-N}{n}\omega$$

$$< 2\omega \quad \text{whenever} \quad n \geq \max\left\{ N, \frac{|\alpha_1| + \cdots + |\alpha_{N-1}|}{\omega} \right\}$$

Choose $\omega = \epsilon/2$, then
$$\frac{z_1 + \cdots + z_n}{n} \to z_0 \quad \text{as} \quad n \to \infty.$$

2.10.6
Exercises

1: Show that if a sequence $\{z_n\}$ has a limit, then it is a Cauchy sequence.

2: Let $S_1, S_2, S_3, \ldots$ be a sequence of sets such that diameter $(S_n) \to 0$ as $n \to \infty$. Show that there cannot be more than one point common to all sets of the sequence.

3: Show that if a sequence $\{z_n\}$ has an accumulation point z_0, then there exists a subsequence $\{z'_m\}$ of $\{z_n\}$ such that $\lim_{m \to \infty} z'_m = z_0$.

We now list and prove some properties of the real-number system and show how they may or may not carry over to the complex-number system.

2.11 Fundamental Properties of the Real-Number System

1. *Dedekind cut.* If the real line is decomposed into two sets $S_1, S_2 \neq 0$ such that $t_1 < t_2$ for every $t_1 \in S_1$ and every $t_2 \in S_2$, then either l.u.b. of $S_1 \in S_1$ or g.l.b. of $S_2 \in S_2$. That is, we can find a point $t_0 \in S_1$ or S_2 such that t_0 is both the l.u.b. of S_1 and the g.l.b. of S_2.
2. *Dedekind Property.* Every bounded set of real numbers has a l.u.b. and a g.l.b.
3. *Completeness Property.* Every Cauchy sequence of real numbers has a limit.
4. *Bolzano-Weierstrass Property.* Every bounded set of real numbers containing an infinite number of points has an accumulation point.
5. *Nested-Interval Property.* Given a set of nested sequences of closed intervals $I_1, I_2, \ldots$, such that $I_{n+1} \subseteq I_n$, $n = 1, 2, \ldots$, there exists at least one point in common to all I_n.

All of the previous five properties are equivalent. Properties 1 and 2 do not exist for complex numbers since complex numbers cannot be ordered *per se*. We prove Properties 3 and 5 for complex numbers.

2.11.1

Theorem *Every Cauchy sequence (of complex numbers) has a limit.*

PROOF. If $\{z_n\}$ is a Cauchy sequence and $z_n = x_n + iy_n$, then
$$|x_n - x_m| \leq |z_n - z_m| < \epsilon$$
for all n, m such that $n > N$ and $m > N$ (similarly, for $|y_n - y_m|$). Thus $\{x_n\}$ and $\{y_n\}$ are both Cauchy sequences and hence have a limit

Definitions and Properties of Sets in the Complex Plane

x_0, y_0 (by properties of the real-number system). Thus $\{z_n\}$ has a limit $z_0 = x_0 + iy_0$. □

Now it is not difficult to prove the following theorem, which is left to the reader.

2.11.2

Theorem *Every bounded set S of complex numbers containing an infinite number of points has an accumulation point.*

2.11.3

Theorem *Let $\{I_{(n)}\}$ be a sequence of closed intervals in the plane, such that $I_{(n+1)} \subseteq I_{(n)}, n = 1, 2, \ldots$. Then $\exists$ at least one point common to all intervals.*

PROOF. Suppose $I_{(n)}$ consists of all $z = x + iy$ such that

$$a_{(n)}^1 \leq x \leq a_{(n)}^2 ; \ b_n^{(1)} \leq y \leq b_{(n)}^2.$$

Since $I_{(n+1)} \subseteq I_{(n)}$, then

$$\left. \begin{array}{l} a_{(n)}^1 \leq a_{(n+1)}^i \leq a_{(n)}^2 \\ b_{(n)}^1 \leq b_{(n+1)}^i \leq b_{(n)}^2 \end{array} \right\} \quad i = 1, 2.$$

Let $J_{(n)}$ be the set of all x such that $a_{(n)}^1 \leq x \leq a_{(n)}^2$, then $\{J_{(n)}\}$ is a nested sequence on the real line. Thus there exists a real x_0 such that $a_{(n)}^1 \leq x_0 \leq a_{(n)}^2$, for all n. Similarly, there exists a real y_0 such that $b_{(n)}^1 \leq y_0 \leq b_{(n)}^2$, for all n and the point $z_0 = x_0 + iy_0$ is thus contained in every interval $I_{(n)}$. □

2.12 Compact Sets

In the theory of functions an especially important role is played by sets that are both bounded and closed, e.g., the set of all z such that $|z| \leq K$ (constant).

2.12.1

Definition An *open covering* of a set S is a class of open sets whose union contains S.

2.12.2

Definition We say that a set S is *compact* if every open covering of the set has a finite subcover.

It is an easy exercise to show that every closed subset of a compact set is compact.

Since we are basically dealing with the topology of Euclidean spaces, we shall show that a set is compact if and only if it is bounded and closed. It turns out to be essentially no more difficult to prove this fact in a Euclidean n-space than it is in the plane. Let E^n be a Euclidean n-space. We have the following theorem.

2.12.3

Theorem I. *A set $S \subseteq E^n$ is compact, iff*
II. *S is closed and bounded.*

PROOF. We first prove that I $\Rightarrow$ II. Let $x \in S$ and B_x be any open ball with center x. Then

$$S \subseteq \bigcup_{x \in S} B_x$$

is an open cover of S. Thus we can choose $x_1, x_2, \ldots, x_n$ such that

$$S \subseteq B_{x_1} \cup \cdots \cup B_{x_n}.$$

Hence S is bounded.

To show that S is closed, we prove that the complement S^c of S is open. Let $y \in S^c$. For every $x \in S$, $x \neq y$, thus $\exists$ open balls B_x and $B_y^{(x)}$ (ball with center y dependent on the choice of x) such that $B_x \cap B_y^{(x)} = \emptyset$. Now $S \subseteq \bigcup_{x \in S} B_x$ is an open cover of S. Choose a finite subcover $S \subseteq B_{x_1} \cup B_{x_2} \cup \cdots \cup B_{x_n}$. Then $B_y^{(x_1)}, B_y^{(x_2)}, \ldots, B_y^{(x_n)}$ contain an open ball, say B_y, centered at y. We claim that $B_y \subseteq S^c$.

If $s \in S$, then $s \in B_{x_i}$ for some i, hence $s \notin B_y^{(x_i)}$ for some i and $s \notin B_y$; thus $B_y \subseteq S^c$ and S is closed.

We now wish to prove that II $\Rightarrow$ I (Heine–Borel theorem).

Let us assume that S is a bounded and closed subset of E^n. Since S is bounded, there exists a number $b > 0$, such that $S \subset B$, where B is the "cube"

$$B = \{x \mid |x_i| \leq b \quad \text{for} \quad i = 1, 2, \ldots, n\}.$$

If we consider S as a subset of B, then S is closed in B, and since every closed subset of a compact set is compact, S is compact if B is compact, Thus it is sufficient to prove that a closed "cube" B is a compact subset of E^n.

Suppose to the contrary that B is not compact. Then there exists an open cover $\mathcal{F}$ of B without a finite subcover. We shall now construct a nested chain of "cubes" in B:

$$B = B_0 \supset B_1 \supset B_2 \supset \cdots,$$

such that for each k, the "cube" B_k is not covered by a finite number of members of $\mathcal{F}$ and the side of B_i is $2b/2^i$. This is done inductively: we assume that $B_0, B_1, \ldots, B_i$ are already constructed. Subdivide B_i into 2^n "cubes" by marking the side bisecting the hyperplanes (see figure for $n = 2$). Since B_i is not covered by a finite subcollection of $\mathcal{F}$, it follows that

Definitions and Properties of Sets in the Complex Plane

at least one (say B_{i+1}) of the 2^n "cubes" also has this property. The side of B_{i+1} is $\frac{1}{2}$ that of B_i, that is, the side of B_i is $2b/2^i$.

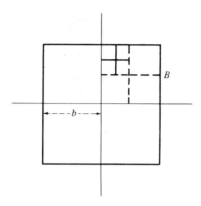

It follows that $\bigcap_{i=0}^{\infty} B_i$ is a single point (the nested-interval property in E^n), say, $p = \bigcap_{i=0}^{\infty} B_i$. Since $\mathscr{F}$ is an open cover of B, we can find a member $F \in \mathscr{F}$ such that $p \in F$, i.e., such that F is a neighborhood of p. It is clear, however, that any neighborhood of p contains B_j for a suitable j, since each open-"cube" center p has this property. Thus we have shown that for some j, B_j is covered by a single member of $\mathscr{F}$, which is a contradiction. □

2.13 Series of Complex Numbers

Write
$$S_1 = z_1,$$
$$S_2 = z_1 + z_2, \ldots,$$
$$\vdots$$
$$S_n = z_1 + z_2 + \cdots + z_n = \sum_{k=1}^{n} z_k,$$

and
$$S = \sum_{k=1}^{\infty} z_k.$$

2.13.1

Definition We say that the series $\sum_{k=1}^{\infty} z_k$ *converges* if there exists a finite S such that $\lim_{n \to \infty} S_n = S$; otherwise we say that $\sum_{k=1}^{\infty} z_k$ *diverges*.

Clearly a necessary and sufficient condition for the convergence of

$\sum_{k=1}^{\infty} z_k$ is that

$$\sum_{k=1}^{\infty} R\{z_k\} = R\{S\}$$

and

$$\sum_{k=1}^{\infty} I\{z_k\} = I\{S\}.$$

Another necessary and sufficient condition for convergence is essentially the Cauchy criterion, viz., given $\epsilon > 0$, there exists N such that $|S_n - S_m| < \epsilon$ for all $n, m > N$. The proof is left as an exercise.

Tests for convergence (e.g., the ratio, the limit of $\sqrt[n]{|a_n|}$, Raabe and Gauss tests, etc.) are much the same as for real z_n and can be found in almost all standard works on complex variables.

2.13.2

The following theorems can also be deduced in much the same way as for real variable.

1. If $\sum_1^\infty z_n$ converges, then $\lim_{n \to \infty} |z_n| \to 0$; thus if $\lim_{n \to \infty} |z_n| \not\to 0$, then $\sum_1^\infty z_n$ does not converge.
2. If $\sum_1^\infty z_n = S$ and $\sum_1^\infty \omega_n = T$ and the two series are absolutely convergent, then $\sum_1^\infty (z_1 \omega_n + z_2 \omega_{n-1} + \cdots + z_n \omega_1)$ is absolutely convergent to the number ST.

This is called the *Cauchy product of two series*. We shall study the Cauchy product more carefully in Section 6.5.*

2.14 Limits

2.14.1

Definition The l.u.b. of the set of all points of accumulation of a sequence of real numbers $\{u_n\}$ (if bounded above) is called the *limit superior* (lim sup or $\overline{\lim}_{n \to \infty}$) of the sequence. *Limit inferior*, (lim inf or $\underline{\lim}_{n \to \infty}$) is similarly defined. If $\{u_n\}$ is unbounded, we may take $\pm \infty$ as the lim sup and lim inf, respectively.

EXAMPLE 1.

$$\limsup_{n \to \infty} \quad \text{or} \quad \overline{\lim}_{n \to \infty} \left\{ 1 + \sin \frac{n\pi}{2} \right\} = 2,$$

$$\liminf_{n \to \infty} \quad \text{or} \quad \underline{\lim}_{n \to \infty} \left\{ 1 + \sin \frac{n\pi}{2} \right\} = 0.$$

---

* A particularly thorough work to consult is K. Knopp, *Theory and Application of Infinite Series*, Blackie, 1951.

EXAMPLE 2.
$$\overline{\lim_{n \to \infty}} \{[(-1)^{n+1}n] - n\} = 0,$$
$$\lim_{n \to \infty} \{[(-1)^{n+1}n] - n\} = -\infty.$$

EXAMPLE 3. Show that $\sum_{n=1}^{\infty} \frac{(1+i)^n}{3}$ is absolutely convergent.

SOLUTION. Using the ratio test, we write
$$\left|\frac{a_{n+1}}{a_n}\right| = \left|\frac{(1+i)^{n+1}}{3}\right| \left|\frac{3}{(1+i)^n}\right|$$
$$= |1 + i| = \sqrt{2} > 1;$$
thus the series converges absolutely.

2.15 Mappings

In order to study the concept of a *curve* in the complex plane we need the idea of a mapping in the following sense. Given two sets A and B, if to every point $z \in A$ there is assigned a unique point $\omega \in B$, we say that we have a *mapping f* of A *into* B and we write

$$f : A \to B.$$

Thus we can write $\omega = f(z)$ to be the point $\omega \in B$ assigned by f to $z \in A$. We say that ω is the *image* of z and z is the *inverse image* of ω.

We call A the *domain* of the function. If $X \subset A$, the set of values $f(z)$ for $z \in A$ is denoted by $f(X)$. A mapping $f : A \to B$ is said to be *one-to-one* if $f(z_1) = f(z_2)$ only for $z_1 = z_2$; it is said to be *onto* if $f(A) = B$.

A considerable number of properties and theorems concerning mappings can be deduced, and we cite several useful results, the proofs of which are standard and given in most works on the subject.

We will assume that the sets used in the rest of the book belong to the complex plane, unless specifically stated otherwise.

2.15.1

Definition A mapping f of a set A into a set B is said to be *continuous*, if for each $z_0 \in A$ and any neighborhood $N(\omega_0)$ of its image $\omega_0 \in B$ there exists a neighborhood of z_0 mapped by f into $N(\omega_0)$. We call a continuous one-to-one mapping of a set A onto a set B, such that the inverse mapping f^{-1} of B onto A is also continuous, a *homeomorphism*; A and B are said to be *homeomorphic*.

2.15.2

Theorem Let f be a continuous mapping of a set A onto a set B. If $\lim_{n \to \infty} z_n = z_0$ where the sequence of points $\{z_n\}$ and z_0 are in A, then

$$\lim_{n \to \infty} f(z_n) = f(z_0).$$

Conversely, if f maps A into B and $\lim_{n \to \infty} f(z_n) = f(z_0)$ whenever $\lim_{n \to \infty} z_n = z_0$, and $z_0, z_n \in A$, then f is continuous.

2.15.3

Theorem Let f be a continuous mapping of a set A onto a set B. If A is a connected set, then B is also a connected set.

2.15.4

Theorem If f is a continuous mapping of a compact set A onto a set B, then B is also compact.

We now have enough apparatus with which to discuss the concept of continuous curves in the plane.

2.16 Curves in the Plane

2.16.1 Continuous Curves (Arcs)

Let $I = [t_1, t_2]$ be a closed interval on the real line, i.e., $t_1 \leq t \leq t_2$; where t_1 and t_2 are finite; also let C be a continuous mapping of I onto the complex plane. Then C is a mapping representing a continuous *curve*, *path*, or *arc* in the plane. The image z of any point in I is given by $z = z(t) = g(t) + ih(t)$, where $g, h \in C[t_1, t_2]$. For example,

1. Given $z(t) = \cos 2\pi t + i \sin 2\pi t$, $0 \leq t \leq 1$; then as t moves from 0 to 1, the point moves once around the unit circle in the counterclockwise direction.
2. Given $z(t) = \cos 4\pi t + i \sin 4\pi t$, $0 \leq t \leq 1$; the point moves twice around the unit circle as t moves from 0 to 1.

We say that the path in case (2) is different from that in case (1).

2.16.2

Definition A subset A of the complex plane is *pathwise connected* if and only if, given any two points $a, b \in A$, $\exists$ a continuous curve $w: I \to A$ such that $w(t_1) = a$ and $w(t_2) = b$.

REMARK. It can be proved that if A is also open, then A is pathwise connected iff each two distinct points of A can be joined by a polygonal line. In this case we can say that A is *arcwise connected*.

2.16.3

Definition A *domain* is an open connected set.

2.16.4

Definition A *region* is a domain with all, some, or none of its boundary points.

2.16.5

Definition C is a *contour* if C is a continuous curve represented by $z = z(t)$, $\alpha \leq t \leq \beta$, $g(t)$ and $h(t)$ have piecewise continuous first derivatives and f and g are not zero for the same t.

2.16.6

Definition C is a *simple curve (Jordan curve)* or *arc* if C is a continuous curve and $[g(t_1), h(t_1)] \neq [g(t_2), h(t_2)]$ for $t_1 \neq t_2$.

2.16.7

Definition C is a simple *closed* curve if C is a continuous simple curve and $z(\alpha) = z(\beta)$ for $\alpha \leq t \leq \beta$.

2.16.8

Definition We say that $z_1 = z(t_1)$ is a *multiple point* on the curve $z = z(t)$ if z_1 can be determined for values of t other than $t = t_1$.

We now state the *Jordan-curve theorem* which is fundamental to the study of complex integration.*

2.16.9

Theorem *Let C be a simple closed curve, then C divides the plane into two disjoint domains D_1 and D_2, and C is the boundary of D_1 and D_2.*

Alternatively, the points not on C form two disjoint domains D_1 and D_2. Every point on C is a point of accumulation of D_1 as well as D_2; domains D_1 and D_2 have no boundary points other than points of C.

2.16.10

Definition In a domain bounded by a simple closed curve, the closed curve divides the plane into two disjoint domains. The domain containing the point at infinity is the *exterior domain*, and we say that an *interior point* is in the domain not containing the point at infinity. Alternatively,

---

* The proof may be found, for example, in *Elements of the Topology of Plane Sets of Points* by M.H.A. Newman, Cambridge, 1964.

the interior point is in the bounded domain. More precisely, p is an interior point of a domain D if D is a neighborhood of p.

2.16.11

Definition A domain D is said to be *simply connected* if every simple closed curve in D contains only points of D in its interior (i.e., it has no "holes").

For example, the set $1 < |z| < 2$ is not simply connected. If there exist simple closed curves in D containing points in their interiors other than the points of D, we say that D is *multiply connected*.

3
Functions of Bounded Variation

3.1

Let $f(x) = g(x) + ih(x)$ be a real or complex function of a real variable x in $[a, b]$.

Take a net Δ over $[a, b]$ as follows:
$$a = x_0 < x_1 < \cdots < x_n = b,$$
and set
$$\sigma(\Delta) = \sum_{r=1}^{n} |f(x_r) - f(x_{r-1})|.$$

3.1.1

Definition (i) If there exists K such that $\sigma(\Delta) \leq K$, $\forall \Delta$, we say that $f(x)$ is of *bounded variation* in $[a, b]$. (ii) We then call the l.u.b. $\{\sigma(\Delta)\}$ for all Δ the *total variation* of $f(x)$ in $[a, b]$ and denote it by $V(a, b)$.

It can be shown that $V(a, b) = \int_a^b |df(x)|$. Let us write $f \in B$ to mean "$f(x)$ is of bounded variation."

3.1.2 Properties of $\sigma(\Delta)$

1. $\sigma(\Delta) \leq V(a, b)$, $\forall \Delta$, and for every $\epsilon > 0$, there exists Δ such that $\sigma(\Delta) > V(a, b) - \epsilon$.
2. $\sigma(\Delta)$ does not decrease if extra points of subdivision are introduced.

EXAMPLES OF FUNCTIONS $f \notin B$
1.
$$f(x) = \begin{cases} 0 \text{ if } x \text{ rational} \\ 1 \text{ if } x \text{ irrational} \end{cases} \text{ in } [0, 1].$$

2.
$$f(x) = \begin{cases} x \sin \pi/x & \text{if } x \neq 0 \\ 0 & \text{if } x = 0 \end{cases} \text{ in } [0, 1].$$

3.1.3 Properties of Functions $f \in B$ in $[a, b]$

We sketch a few proofs of the following simple properties and leave the completion of these proofs as exercises.

1. If $f \in B$ in $[a, b]$ then f is bounded in $[a, b]$.

   [Hint: Take Δ: $a \leqslant x \leqslant b$.

2. If $f, F \in B$ in $[a, b]$, then
   i. $\lambda f + \mu F \in B$ in $[a, b]$ μ, λ constant.
   ii. $fF \in B$ in $[a, b]$.

   [Hint: For (ii) show that if $P(x) = fF$, then $\sigma_P(\Delta) \leqslant K[\sigma_f(\Delta) + \sigma_F(\Delta)]$, where
   $$K = \max\left\{ \underset{x \in [a,b]}{\text{l.u.b.}} |f|; \underset{x \in [a,b]}{\text{l.u.b.}} |F| \right\}.$$

3. If $f \in B$ in $[a, b]$ and $a \leqslant \alpha \leqslant \beta \leqslant b$, then $f \in B$ in $[\alpha, \beta]$ and $\int_\alpha^\beta |df(x)| \leqslant \int_a^b |df(x)|$.
4. If $f \in B$ in $[a, c]$ and $[c, b]$, and $a < c < b$, then $f \in B$ in $[a, b]$ and
   $$\int_a^b |df(x)| = \int_a^c |df(x)| + \int_c^b |df(x)|.$$
5. If $f \in B$ in $[a, b]$ and $V(a, x) = \int_a^x |df(x)|$ for $a \leqslant x \leqslant b$, then $V(a, x)$ is a nondecreasing function of x. Note: $V(a, x_1) = V(a, x_2)$ iff $f = \text{const}$ for $x_1 \leqslant x \leqslant x_2$.
6. If $f \in B$ and $f \in C[a, b]$, then $V(a, x) \in C[a, b]$.

PROOF OF (6). Take c such that $a \leqslant c < b$; take $h > 0$ such that $c + h \leqslant b$. Then $V(a, c + h) = V(a, c) + V(c, c + h)$. We shall prove that $V(c, c + h) \to 0$ as $h \to +0$ (proves right-hand continuity of $V(a, x)$ at c). We have
$$0 \leqslant V(c, c + h) = V(a, b) - V(a, c) - V(c + h, b)$$
$$\leqslant V(a, b) - \sigma(\Delta_1) - \sigma(\Delta_2),$$
where Δ_1 is over $[a, c]$ and Δ_2 is over $[c + h, b]$.

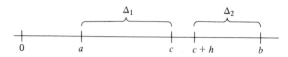

Take $\epsilon > 0$; then there exists Δ over $[a, b]$ that $\sigma(\Delta) > V(a, b) - \epsilon/2$. Choose $h < \eta$ so small that $c, c + h$ lie in the same subinterval $|x_{r-1}, x_r|$ of Δ. Let Δ^* be the result of adding $c, c + h$ to Δ. Take Δ_1 as a subnet of Δ^*

over $[a, c]$ and Δ_2 over $[c + h, b]$. Then,
$$\sigma(\Delta^*) = \sigma(\Delta_1) + |f(c + h) - f(c)| + \sigma(\Delta_2) \geq \sigma(\Delta),$$
$$> V(a, b) - \frac{\epsilon}{2}.$$
Thus,
$$V(a, b) - \sigma(\Delta_1) - \sigma(\Delta_2) < |f(c + h) - f(c)| + \frac{\epsilon}{2} < \epsilon$$
for $h < \delta(\epsilon)$ (this is possible since f is continuous) and
$$0 \leq V(c, c + h) < \epsilon \text{ for } h < \min(\delta, \eta).$$
Consequently $V(a, x)$ is continuous on the right for $x = c$ and hence for $a \leq x < b$.

Similarly $V(a, x)$ is continuous on the left at $x = c$ for $a < c \leq b$ and hence continuous on the left for $a < x \leq b$.

7. If $f(t) = g(t) + ih(t)$ for g, h real, then
$$f \in B \text{ iff } g, h \in B \text{ in } [a, b].$$
8. If f is nondecreasing in $[a, b]$, then $f \in B$ in $[a, b]$.
9. If $|f'(x)| \leq M$ (constant) in $[a, b]$, then $f \in B$ in $[a, b]$.

[Hint: Use the mean-value theorem for $\sigma(\Delta)$. The condition in 9 is not necessary (consider the diagram).

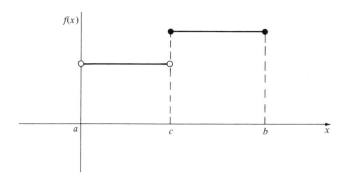

10. The *positive and negative variations* of real $f \in B$ in $[a, b]$. We write
$$p(\Delta) \equiv \sum_1 |f(x_r) - f(x_{r-1})| = \frac{1}{2} \sum_{r=1}^{n} \{|f(x_r) - f(x_{r-1})| + f(x_r) - f(x_{r-1})\}$$
$$= \frac{1}{2} \{\sigma(\Delta) + f(b) - f(a)\}$$
and
$$n(\Delta) \equiv \sum_2 |f(x_r) - f(x_{r-1})| = \frac{1}{2} \sum_{r=1}^{n} \{|f(x_r) - f(x_{r-1})| - f(x) + f(x_{r-1})\}$$
$$= \frac{1}{2} \{\sigma(\Delta) + f(a) - f(b)\}$$

Here $\sum_1$ and $\sum_2$ denote summation over the interval $[x_{r-1}, x_r]$ for which $f(x_r) - f(x_{r-1}) \geq 0$ and $f(x_r) - f(x_{r-1}) < 0$, respectively. Thus $p(\Delta) \leq \sigma(\Delta)$ and $n(\Delta) \leq \sigma(\Delta)$. □

Definition *Positive variation* $P(a, b) = \text{l.u.b.}\{p(\Delta)\}$, $\forall \Delta$. *Negative variation* $N(a, b) = \text{l.u.b.}\{n(\Delta)\}$, $\forall \Delta$. We have

$$2P(a, b) = V(a, b) + \{f(b) - f(a)\},$$
$$2N(a, b) = V(a, b) - \{f(b) - f(a)\},$$

or

$$V(a, b) = P(a, b) + N(a, b)$$

and

$$f(b) - f(a) = P(a, b) - N(a, b).$$

10a. If $f \in B$ in $[a, b]$ and $a \leq c \leq b$, then $P(a, b) = P(a, c) + P(c, b)$. Similarly for $N(a, b)$.
10b. If $f \in B$ in $[a, b]$, then $P(a, x)$ is nondecreasing for $a \leq x \leq b$. Similarly for $N(a, x)$.
10c. If $f \in B$ and continuous in $[a, b]$, then $P(a, x)$ is continuous in $[a, b]$. Similarly for $N(a, x)$.

Note: $N(a, x)$ is the positive variation over $[a, x]$ for the function $-f(x)$.

11. If f is real in $[a, b]$, then $f \in B$ in $[a, b]$ iff f can be expressed in the form $f(x) = \theta(x) - \varphi(x)$, where θ, φ are nondecreasing in $[a, b]$.

PROOF OF (11).
Necessity: If $f \in B$ in $[a, b]$, then for $a \leq x \leq b$,
$$f(x) = f(a) + P(a, x) - N(a, x)$$
$$= \theta(x) - \varphi(x),$$
where $\theta = f(a) + P(a, x)$ and $\varphi = N(a, x)$.
Sufficiency: If $f = \theta - \varphi$, where θ, φ are nondecreasing in $[a, b]$, then $\theta, \varphi \in B$ in $[a, b]$. Thus $f = \theta - \varphi \in B$ in $[a, b]$. □

Note: If $f \in C[a, b]$, so is $P(a, x)$, and hence θ and φ are continuous. Also, functions θ and φ may not be unique. If θ and φ are not unique, we could add any nondecreasing function to both. By adding a suitable constant, we ensure that θ, φ are both positive.

3.2 Rectifiable Curves

Given that the curve C is defined as $z = f(t) = g(t) + ih(t)$, $a \leq t \leq b$, construct a net Δ: $a = t_0 < t_1 < \cdots < t_n = b$.
Let $z_r = f(t_r)$, $r = 0, \ldots, n$, be points in order on C and set

$$s(\Delta) = \sum_{i=1}^{n} |z_i - z_{i-1}| = \sum_{i=1}^{n} |f(t_i) - f(t_{i-1})|.$$

Then $s(\Delta)$ is the length of the inscribed polygon with vertices z_0, $z_1, \ldots, z_n$.

3.2.1

Definition If $s(\Delta) < K$ (constant) for all Δ, we say that the curve is *rectifiable* and the length $l = \text{l.u.b.}\{s(\Delta)\}$ for all nets Δ over $[a, b]$.

Thus C is rectifiable if and only if $f \in B$ in $[a, b]$, in which case the length $s(a, b) = \int_a^b |df(t)|$.

3.2.2

Deductions

1. If $a \leq c \leq d \leq b$ and $z = f(t)$ is rectifiable over $[a, b]$, then the curve $z = f(t)$ is rectifiable over $[c, d]$.
2. If $a < c < b$, then $s(a, b) = s(a, c) + s(c, b)$.
3. $s(a, x)$ is a strictly increasing function of x.

EXAMPLE. A nonrectifiable curve is given by

$$\left. \begin{array}{l} g(t) = t \\ h(t) = \begin{cases} t \cos \pi/t & \text{if } t \neq 0 \\ 0 & \text{if } t = 0 \end{cases} \end{array} \right\} 0 \leq t \leq \pi.$$

The function $f = g(t) + ih(t)$ is continuous but not of bounded variation over $[0, \pi]$. (Prove this result.)

3.3 The Riemann–Stieltjes Integral

Let $f(x)$ and $\theta(x)$ be real-valued functions in $[a, b]$. Construct a net Δ over $[a, b]$: $a = x_0 < x_1 < \cdots < x_n = b$. Let us write

$$\delta = \delta(\Delta) = \max_r |x_r - x_{r-1}| = \text{Norm of } \Delta.$$

Choose $\xi_r \in [x_{r-1}, x_r]$ and set

$$\sigma(\Delta) = \sum_{r=1}^n f(\xi_r)\{\theta(x_r) - \theta(x_{r-1})\}.$$

3.3.1

Definition If $\sigma(\Delta)$ approaches a finite limit I as $n \to \infty$ and $\delta \to 0$ (regardless of the choice of Δ and the points ξ_r), then we call I the *Riemann–Stieltjes integral* of $f(x)$ with respect to $\theta(x)$ over $[a, b]$ and write

$$I = \int_a^b f(x) \, d\theta(x).$$

Then given $\epsilon > 0$, there exists $\eta = \eta(\epsilon) > 0$ such that $|\sigma(\Delta) - I| < \epsilon$ for $\delta < \eta(\epsilon)$ and for every choice of Δ and $\xi_r \in [x_{r-1}, x_r]$.

3.3.2

Consequences Once again we will just sketch a few proofs of the following consequences and leave the substance of the proofs as exercises.

1. If $\theta(x) = x$, then I is the Riemann integral of the function $f(x)$.
2. $\int_a^b \{\lambda f(x) + \mu g(x)\} \, d\theta(x) = \lambda \int_a^b f(x) \, d\theta(x) + \mu \int_a^b g(x) \, d\theta(x)$, if both integrals on the right exist.
3. $\int_a^b f(x) \, d\{\lambda \varphi(x) + \mu \psi(x)\} = \lambda \int_a^b f(x) \, d\varphi(x) + \mu \int_a^b f(x) \, d\psi(x)$, if both integrals on the right exist.

PROOF. Let $\theta(x) = \lambda \varphi(x) + \mu \psi(x)$. Then with the usual notation we have

$$\sum_{r=1}^{n} f(\xi_r)\{\theta(x_r) - \theta(x_{r-1})\} = \lambda \sum_{r=1}^{n} f(\xi_r)\{\varphi(x_r) - \varphi(x_{r-1})\}$$

$$+ \mu \sum_{r=1}^{n} f(\xi_r)\{\psi(x_r) - \psi(x_{r-1})\}.$$

□

The result follows when $n \to \infty$ and $\delta \to 0$.

4. $\int_a^b 1 \, d\theta(x) = \theta(b) - \theta(a)$ ($\sigma(\Delta) = \theta(b) - \theta(a)$ for all nets Δ).
5.
$$\int_a^b f(x) \, d\theta(x) = \begin{cases} 0 \text{ if } \theta(x) = \text{const in } [a, b], \\ 0 \text{ if } f(x) = 0 \text{ in } [a, b]. \end{cases}$$

6. If $a < c < b$ and $\int_a^b f(x) d\theta(x)$, $\int_a^c f(x) d\theta(x)$, $\int_c^b f(x) d\theta(x)$ all exist, then

$$\int_a^b f \, d\theta = \int_a^c f \, d\theta + \int_c^b f \, d\theta.$$

Note: The existence of $\int_a^c$ and $\int_c^b$ does not necessarily imply that of $\int_a^b$; for example, take

$$f(x) = \begin{cases} 0 & \text{if } 0 \leq x < 1, \\ 1 & \text{if } 1 \leq x \leq 2, \end{cases}$$

$$\theta(x) = \begin{cases} 0 & \text{if } 0 \leq x \leq 1, \\ 1 & \text{if } 1 < x \leq 2. \end{cases}$$

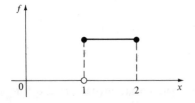
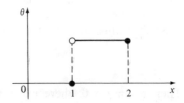

Functions of Bounded Variation

Then $\int_0^1 f \, d\theta = 0$ since θ is constant in $[0, 1]$ and

$$\int_1^2 f \, d\theta = \int_1^2 1 \, d\theta = \theta(2) - \theta(1) = 1.$$

Take any net Δ over $[0, 2]$ not containing point 1. Suppose $x_{m-1} < 1 < x_m$, then

$$\sigma(\Delta) = \sum_{r=1}^{m-1} 0\{\theta(x_r) - \theta(x_{r-1})\}$$

$$+ f(\xi_m)\{\theta(x_m) - \theta(x_{m-1})\}$$

$$+ \sum_{r=m+1}^{n} 1\{\theta(x_r) - \theta(x_{r-1})\}$$

$$= 0 + f(\xi_m) + 0 = \begin{cases} 1 & \text{if } 1 \leq \xi_m < x_m, \\ 0 & \text{if } x_{m-1} \leq \xi_m < 1. \end{cases}$$

Hence $\int_0^2$ cannot exist.

3.3.3 Conditions for Existence of the Riemann–Stieltjes Integral

1. If $f(x) \in C[a, b]$ and $\theta(x) \in C[a, b]$ and strictly increasing, then $\int_a^b f(x) \, d\theta(x)$ exists (this is sufficient but not necessary).

PROOF. Take a net Δ over $[a, b]$. This defines a net Δ' over $\alpha \leq y \leq \beta$ (see the figure). A continuous and strictly increasing function $y = \theta(x)$ defines a continuous strictly increasing inverse $x = \psi(y)$.

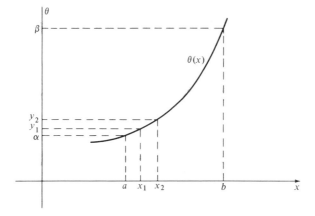

Let $y_r = \theta(x_r)$, $\eta_r = \theta(\xi_r)$. It follows from the uniform continuity of $\theta(x)$ that $\max_r |y_r - y_{r-1}| \to 0$ as $\delta = \max_r |x_r - x_{r-1}| \to 0$. (We recall that if

$f \in C[a, b]$, then it is uniformly continuous on $[a, b]$.) Thus,

$$\sigma(\Delta) = \sum_{r=1}^{n} f\{\psi(\eta_r)\}(y_r - y_{r-1}) = \sum_{r=1}^{n} F(\eta_r)(y_r - y_{r-1}).$$

F is a continuous function of a continuous function, so $F(y) \in C[\alpha, \beta]$. Thus $\sigma(\Delta) \to \int_{\alpha}^{\beta} F(y) \, dy$ as $n \to \infty$ and $\delta \to 0$ (for all possible partitions of $[a, b]$).

Consequently $\int_{a}^{b} f(x) \, d\theta(x)$ exists and equals $\int_{\alpha}^{\beta} F(y) \, dy$. □

2. If $f \in C[a, b]$ and $\theta \in C[a, b]$ and $\theta \in B$ in $[a, b]$, then $\int_{a}^{b} f(x) \, d\theta(x)$ exists and

$$\left| \int_{a}^{b} f \, d\theta \right| \leq M \int_{a}^{b} |d\theta(x)|,$$

where $M = \max_{[a, b]} |f(x)|$.

PROOF. We can write $\theta = \varphi - \psi$, where both φ, ψ are continuous and strictly increasing in $[a, b]$. Thus from part (1) and consequence (3) it follows that

$$\int_{a}^{b} f(x) \, d\theta(x) \text{ exists and equals } \int_{a}^{b} f \, d\varphi - \int_{a}^{b} f \, d\psi.$$

Also, for the given net Δ over $[a, b]$ we have

$$\left| \sum_{r=1}^{n} f(\xi_r)\{\theta(x_r) - \theta(x_{r-1})\} \right| \leq \sum_{r=1}^{n} |f(\xi_r)||\theta(x_r) - \theta(x_{r-1})|$$

$$\leq M \int_{a}^{b} |d\theta(x)|.$$

We now let $n \to \infty$ and $\delta \to 0$. □

3. If $f(x)$ and $\theta'(x) \in C[a, b]$, then

$$\int_{a}^{b} f \, d\theta = \int_{a}^{b} f(x) \theta'(x) \, dx.$$

PROOF. If $\theta'(x) \in C[a, b]$, then $\theta'(x) = O(1)$ in $[a, b]$. Thus $\theta(x) \in B$ in $[a, b]$ and $\theta(x)$ is continuous in $[a, b]$. Therefore

$$\int_{a}^{b} f \, d\theta = \lim_{\delta \to 0} \sum_{r=1}^{n} f(\xi_r)\{\theta(x_r) - \theta(x_{r-1})\}$$

exists, where $\xi_r \in [a, b]$, $n \to \infty$ and $\delta = \max_r |x_r - x_{r-1}| \to 0$ for all possible partitions of $[a, b]$.

Now, $\theta(x_r) - \theta(x_{r-1}) = \theta'(\eta_r)(x_r - x_{r-1})$, where $\eta_r \in (x_{r-1}, x_r)$. Choose the set of points $\{\xi_r\}$ such that $\xi_r = \eta_r$, then

$$\sigma(\Delta) = \sum_{r=1}^{n} f(\eta_r)\theta'(\eta_r)(x_r - x_{r-1})$$

$$\to \int_a^b f(x)\theta'(x)\,dx \text{ when } n \to \infty \text{ and } \delta \to 0$$

since both $f, \theta' \in C[a, b]$. □

Note: It is not difficult to show that the second condition of 3.3.3(1) can be reduced to read: $\theta(x) \in C[a, b]$ and is nondecreasing.

3.3.4 Extension to Complex Functions

Suppose $f(x) = g(x) + ih(x)$ and $\theta(x) = \varphi(x) + i\psi(x)$. Since

$$\sum_{r=1}^{n} f(\xi_r)\{\theta(x_r) - \theta(x_{r-1})\} = \sum_{r=1}^{n} g(\xi_r)\{\varphi(x_r) - \varphi(x_{r-1})\}$$
$$- \sum_{r=1}^{n} h(\xi_r)\{\psi(x_r) - \psi(x_{r-1})\}$$
$$+ i\sum_{r=1}^{n} g(\xi_r)\{\psi(x_r) - \psi(x_{r-1})\}$$
$$+ i\sum_{r=1}^{n} h(\xi_r)\{\varphi(x_r) - \varphi(x_{r-1})\},$$

then

$$\int_a^b f\,d\theta = \int_a^b g\,d\varphi - \int_a^b h\,d\psi + i\left\{\int_a^b g\,d\psi + \int_a^b h\,d\varphi\right\}$$

subject to the existence of all four integrals on the right. Hence,

$$\int_a^b f\,d\theta \text{ exists if } \begin{cases} g, h \in C[a, b], \\ \varphi, \psi \in B[a, b] \text{ and continuous,} \end{cases}$$

that is, $f \in C[a, b]$ and $\theta \in B[a, b]$ and is continuous in $[a, b]$.

4
Functions of a Complex Variable

4.1

Let $z = x + iy$, (where x, y are real), be any point of a nonempty open set S of the complex plane. We call S the *domain of definition* of z. We say that $w = f(z)$ is a *function* of the complex variable z if to every point $z \in S$ there corresponds at least one value $w \in T$, where T is some nonempty open set of the complex plane. Further, if whenever $z \in S$, there exists one and only one value of $w \in T$, we say that $f(z)$ is *single-valued*. If there exist more than one value of $w \in T$, we say that $f(z)$ is *multiple-valued*.

Let us write $f(z) = u(x, y) + iv(x, y)$, where u, v are real functions of the real variables x and y.

4.1.1
Definition We say that $f(z)$ tends to a *limit* l, or

$$f(z) \to l = r + is \text{ as } z \to z_0 = x_0 + iy_0$$

iff for each $\epsilon > 0$, there exists $\delta(\epsilon) > 0$ such that $|l - f(z)| < \epsilon$ for $0 < |z - z_0| < \delta$.

Thus

$$f(z) \to l = r + is \leftrightarrow \left.\begin{array}{l} u(x, y) \to r \\ v(x, y) \to s \end{array}\right\} \text{ as } (x, y) \to (x_0, y_0)$$

since

$$\left.\begin{array}{l} |u - r| \\ |v - s| \end{array}\right\} \leq \sqrt{(u - r)^2 + (v - s)^2} = |f(z) - l| \leq |u - r| + |v - s|.$$

Functions of a Complex Variable

4.1.2

Definition We say that $f(z)$ is *continuous* at z_0 iff z_0 is a limit point of S, where $z_0 \in S$, and $f(z) \to f(z_0)$ as $z \to z_0$.

Thus f is continuous at z_0 if and only if for every $\epsilon > 0$, $\exists \delta > 0$ such that $|f(z) - f(z_0)| < \epsilon$ for $0 < |z - z_0| < \delta$.

Similarly, f is said to be continuous on a set S iff $f(z)$ is continuous at all points of S.

4.1.3 Properties of Continuous Functions $f(z)$

These are similar to functions of a real variable. In particular we need the following: If f is continuous on a compact set S, then

1. $|f(z)|$ is bounded over S
2. $f(z)$ is uniformly continuous over S, that is, given $\epsilon > 0$, there exists $\delta(\epsilon) > 0$ such that $|f(z_1) - f(z_2)| < \epsilon$, $\forall z_1, z_2$ satisfying $|z_1 - z_2| < \delta$; $z_1, z_2 \in S$.

4.1.4

Theorem *If C is a Jordan curve $z = \theta(t)$, $a \leq t \leq b$, then $z = \theta(t)$ defines t as a continuous single-valued function of z over C.*

PROOF. C has no multiple points, thus $z = \theta(t)$ gives a one-to-one correspondence between points z on the closed set C and points t in the closed interval $a \leq t \leq b$; that is, $z = \theta(t)$ defines t as a single-valued function of z over C.

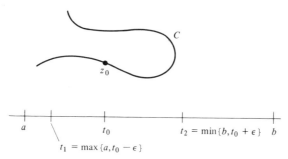

Let z_0 be a fixed point on C and t_0, the corresponding point of $[a, b]$. Let ϵ be an arbitrary positive number. It is required to prove that there exists $\eta(\epsilon) > 0$ such that $|t - t_0| < \epsilon$ for $|z - z_0| < \eta$, where $z, z_0 \in C$, that is, for $|\theta(t) - \theta(t_0)| < \eta$.

Let $t_1 = \max(a, t_0 - \epsilon)$, $t_2 = \min(b, t_0 + \epsilon)$, $z_1 = \theta(t_1)$, and $z_2 = \theta(t_2)$. Then, since t is continuous, $t \in [t_1, t_2]$ if z lies on the arc $z_1 z_2$. Take $\eta(\epsilon) > 0$ such that a neighborhood of z_0 contains no points of C not lying on the arc $z_1 z_2$. This is possible since the distance of z_0 from the closed set of points z not lying strictly between z_1 and z_2 is greater than some $\rho > 0$. Then $|t - t_0| < \epsilon$ for $|\theta(t) - \theta(t_0)| < \eta$, that is, for $|z - z_0| < \eta$.

Thus t is a continuous single-valued function of z at any point $z_0 \in C$, that is, t is a single-valued continuous function of $z \in C$. □

4.2 Differentiability

4.2.1

Definition We say that a function of two variables $u(x, y)$ is *differentiable* at (a, b) iff there exists $\rho > 0$ such that

1. i. $u(a + h, b + k)$ is defined for all h, k such that $|h| < \rho$, $|k| < \rho$;
   ii. for $|h| < \rho$, $|k| < \rho$ we have
   $$u(a + h, b + k) - u(a, b) = Ah + Bk + \epsilon(|h| + |k|),$$
   where A, B are independent of h, k and $\epsilon \to 0$ as $(h, k) \to (0, 0)$. (This implies that $A = u_x(a, b)$ and $B = u_y(a, b)$.)

Note that this is equivalent to

2. $u(a + h, b + k) - u(a, b) = \{u_x(a, b) + \epsilon_1\}h + \{u_y(a, b) + \epsilon_2\}k$, where $\epsilon_1, \epsilon_2 \to 0$ as $(h, k) \to (0, 0)$, since

$$(1) \Rightarrow (2) \text{ with } \epsilon_1 = \frac{\epsilon|h|}{h} \quad \text{and} \quad \epsilon_2 = \frac{\epsilon|k|}{k};$$

$$(2) \Rightarrow (1) \text{ with } \epsilon = \frac{\epsilon_1 h + \epsilon_2 k}{|h| + |k|}.$$

Whence we have

$$|\epsilon| \leq \frac{|\epsilon_1||h|}{|h| + |k|} + \frac{|\epsilon_2||k|}{|h| + |k|} \leq |\epsilon_1| + |\epsilon_2|$$

and $\epsilon \to 0$ as $(h, k) \to (0, 0)$.

Two results now emerge:

1. $u(x, y)$ differentiable at (a, b) implies $u(x, y)$ is continuous at (a, b) (as with a function of one variable, this latter condition is not sufficient);
2. if u_x, u_y exist through some neighborhood of (a, b) and are continuous at (a, b), then $u(x, y)$ is differentiable at (a, b) (this condition is not necessary).

Pierpont's statement of sufficient conditions are as follows: $u(x, y)$ is differentiable at (a, b) if

1. $u(x, y)$ is differentiable in some neighborhood N of (a, b);
2. $u_x(x, y)$ exists in N;
3. $u_y(a, b)$ exists;
4. $u_x(x, y)$ or $u_y(x, y)$ is continuous at (a, b).

4.3 Differentiable Functions of a Complex Variable
4.3.1
Definition $f(z)$ is *differentiable* at z_0 with derivative $f'(z_0)$ iff
$$\frac{f(z) - f(z_0)}{z - z_0} \to l \equiv f'(z_0) \text{ as } z \to z_0.$$

4.3.2
Theorem $f(z) = u(x, y) + iv(x, y)$ *(where u, v are real) is differentiable at a point $c = a + ib$ (where a, b are real) iff*

1. u, v *are differentiable at* (a, b);
2. $u_x(a, b) = v_y(a, b)$ *and* $v_x(a, b) = -u_y(a, b)$.

The last relations (2) are called the *Cauchy–Riemann equations*.*

PROOF

Necessity. Assume that $f'(c)$ exists and $f'(c) = \alpha + i\beta$. Then
$$\frac{f(z) - f(c)}{z - c} = \alpha + i\beta + \epsilon_1 + i\epsilon_2,$$
where ϵ_1, ϵ_2 are real and both $\to 0$ as $z \to c$.

Thus for all $z = x + iy$ in some neighborhood of $c = a + ib$ we have
$$u(x, y) - u(a, b) + i\{v(x, y) - v(a, b)\} = \{(\alpha + \epsilon_1) + i(\beta + \epsilon_2)\}$$
$$\times \{(x - a) + i(y - b)\},$$
that is,
$$u(x, y) - u(a, b) = (\alpha + \epsilon_1)(x - a) - (\beta + \epsilon_2)(y - b),$$
$$v(x, y) - v(a, b) = (\beta + \epsilon_2)(x - a) + (\alpha + \epsilon_1)(y - b).$$
Therefore u, v are differentiable at (a, b) and
$$\frac{\partial u}{\partial x} = \alpha, \quad \frac{\partial u}{\partial y} = -\beta, \quad \frac{\partial v}{\partial x} = \beta, \text{ and } \frac{\partial v}{\partial y} = \alpha.$$
Thus the Cauchy–Riemann equations are satisfied.

Sufficiency. For $z \in N(c, \epsilon)$, that is, for (x, y) near (a, b), we have
$$\frac{\partial u}{\partial x} = \alpha = \frac{\partial v}{\partial y} \quad \text{and} \quad \frac{\partial u}{\partial y} = -\beta = -\frac{\partial v}{\partial x}.$$
Consequently,
$$u(x, y) - u(a, b) = \alpha(x - a) - \beta(y - b) + \epsilon(|x - a| + |y - b|),$$
$$v(x, y) - v(a, b) = \beta(x - a) + \alpha(y - b) + \epsilon'(|x - a| + |y - b|),$$

---

* As given by S. Pollard, in Proc. London Math. Soc., Ser. 2, Vol. 21, pp. 463–466, 1922). Titchmarsh gives these relations as necessary and, together with the continuity of u_x, u_y at (a, b), as sufficient.

where $\epsilon, \epsilon' \to 0$ as $(x, y) \to (a, b)$ (see 4.2.1). Then

$$f(z) - f(c) = (\alpha + i\beta)(x - a) + (i\alpha - \beta)(y - b)$$
$$+ (\epsilon + i\epsilon')(|x - a| + |y - b|)$$
$$= (\alpha + i\beta)(z - c) + (\epsilon + i\epsilon')(|x - a| + |y - b|)$$

and

$$\frac{f(z) - f(c)}{z - c} \to \alpha + i\beta = \left. \begin{array}{l} \dfrac{\partial u}{\partial x} + i \dfrac{\partial v}{\partial x} \\[6pt] = \dfrac{\partial v}{\partial y} - i \dfrac{\partial u}{\partial y} \end{array} \right\} \quad \text{when} \quad z \to c.$$

Observe that

$$\frac{|x - a| + |y - b|}{(x - a) + i(y - b)} = \frac{|\cos \theta| + |\sin \theta|}{\cos \theta + i \sin \theta}$$

where $x - a = r \cos \theta$, $y - b = r \sin \theta$

$$= O(1), \quad \text{when} \quad (x, y) \to (a, b). \qquad \square$$

Note: Differentiability implies continuity but the converse is not necessarily the case.

4.3.3

There is no analogue of Rolle's theorem in the complex plane. For example, if

$$f(z) = 1 - e^{2\pi i z},$$

then

$$f(0) = 0 = f(1),$$

yet no value of z makes $f'(z) = -2\pi i e^{2\pi i z}$ vanish.

EXAMPLE 1. Consider $f(z) = \bar{z} = x - iy$, where $u = x$, $v = -y$. Thus $u_x = 1$, $v_y = -1$, and $u_x \neq v_y$ for any z. Consequently, $f(z)$ is continuous but differentiable nowhere. Alternatively, setting $h = h_1 + ih_2$, where h_1, h_2 are real, we get

$$\frac{f(z + h) - f(z)}{h} = \frac{(x + h_1) - i(y + h_2) - x + iy}{h_1 + ih_2} = \frac{h_1 - ih_2}{h_1 + ih_2},$$

which depends upon the ratio h_1/h_2. Thus the difference quotient does not tend to a definite limit when $h \to 0$.

EXAMPLE 2. Consider $f(z) = |z|^2 = x^2 + y^2$. The function f is clearly continuous everywhere. Since $u = x^2 + y^2$, where $v = 0$, then u, v are both

differentiable at all points (x, y) and thus
$$u_x = 2x, \quad u_y = 2y, \quad v_x = 0, \quad v_y = 0.$$
Thus the Cauchy–Riemann equations are satisfied only at $(0, 0)$, that is, $z = 0$. Consequently, f is continuous everywhere but differentiable only at $z = 0$. Alternatively:
$$\frac{f(h) - f(0)}{h} = \frac{h_1^2 + h_2^2}{h_1 + ih_2} = h_1 - ih_2 \to 0$$
when $(h_1, h_2) \to (0, 0)$, that is, when $h \to 0$.

Take $z_0 = a + ib \neq 0$; then
$$\frac{f(z_0 + h) - f(z_0)}{h} = \frac{(a + h_1)^2 + (b + h_2)^2 - (a^2 + b^2)}{h_1 + ih_2}$$
$$= \frac{h_1^2 + h_2^2}{h_1 + ih_2} + 2\frac{ah_1 + bh_2}{h_1 + ih_2},$$
which clearly depends upon the ratio h_1/h_2 and thus does not tend to a definite limit when $h = h_1 + ih_2 \to 0$.

EXAMPLE 3. Consider $f(z) = R\{z\} = x$; we shall show that this function is continuous everywhere and differentiable nowhere.

Let $u = x$, $v = 0$; thus u, v are differentiable everywhere. However, $u_x = 1$, $u_y = 0$, $v_x = v_y = 0$, and the Cauchy–Riemann equations are satisfied nowhere. Alternatively, with the previous notation,
$$\frac{f(z + h) - f(z)}{h} = \frac{h_1}{h_1 + ih_2},$$
which depends upon the ratio h_1/h_2.

4.4 Analytic Functions

4.4.1

Definition If $f(z)$ is single-valued and differentiable (therefore continuous) throughout some neighborhood $|z - a| < \delta$ ($\delta > 0$) of a point a, we say that $f(z)$ is *analytic* (*regular* or *holomorphic*) at a.

Let us write $f \in A$ or $f(z) \in A$ to mean that f is analytic. If $f \in A$ at every point of a domain D, we say that f is analytic in D. A necessary and sufficient condition for this is that f is single-valued and differentiable at every point of D.

4.4.2

Definition A function $f \in A$ at all points of the finite complex plane is called an *entire* (*integral*) function. We will have more to say on this class of functions later in the text.

4.5

Theorem *If $f'(z) = 0$ everywhere in a domain D, then $f(z) = $ const in D.*

PROOF. Let $a \in D$ be a fixed point and let $\xi \in D$ be a variable point. Clearly a and ξ can be joined by a polygonal line consisting of a finite number of line segments all of whose points belong to D. It is sufficient to show that if $f'(z) = 0$ on a line pq, then $f = $ const on pq. That is, if

$$f'(z) = u_x + iv_x = v_y - iu_y = 0,$$

then

$$u_x = u_y = 0 \quad \text{and} \quad v_x = v_y = 0 \quad \text{on } pq.$$

Consequently u, v are constant on pq. □

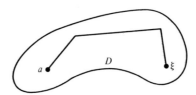

4.5.1

Corollary *If*

1. $f_1(z)$ and $f_2(z) \in A$ in D and
2. $R\{f_1(z)\} = R\{f_2(z)\}$ in D,

then $f_1(z) - f_2(z)$ is constant in D.

PROOF

$$f_1'(z) = u_x + iv_{1x} = u_x - iu_y,$$
$$f_2'(x) = u_x + iv_{2x} = u_x - iu_y.$$

Thus $f_1 - f_2 \in A$ in D and

$$\frac{d}{dz}(f_1 - f_2) = 0 \quad \text{in} \quad D.$$

Consequently $f_1 - f_2 = $ constant in D. □

4.5.2

Corollary *If*

1. $f(z) \in A$ in D and
2. $f(z)$ is real in D,

then $f(z) = $ const in D.

PROOF. We have $f = u + iv$, $v \equiv 0$ in D. Thus

$$f'(z) = u_x + iv_x = v_y + iv_x = 0$$

in D and $f = $ const (real) in D. □

We now return to the problem described in 1.11.1 and discuss the acceptability and consistency of our definition of e^z.

4.6

We shall show that if $f(z) = u(x, y) + iv(x, y)$ satisfies

1. $f(x + i0) = e^x$ and
2. f is entire with $f'(z) = f(z)$,

then the only function with these properties is the function

$$f(z) = e^x \cos y + ie^x \sin y.$$

Since $f'(z) = f(z)$ implies that

$$\frac{\partial u}{\partial x} + i\frac{\partial v}{\partial x} = u + iv$$

then

$$\frac{\partial u(x, y)}{\partial x} = u(x, y) \quad \text{and} \quad \frac{\partial v(x, y)}{\partial x} = v(x, y).$$

By separating the variables we obtain

$$u(x, y) = e^x \varphi(y) \quad \text{and} \quad v(x, y) = e^x \psi(y).$$

Now $f \in A$, thus

$$u_{xx} + u_{yy} = 0 \Rightarrow e^x(\varphi'' + \varphi)y = 0$$
$$\Rightarrow \varphi(y) = a \cos y + b \sin y,$$

where a and b are arbitrary constants; hence,

$$u(x, y) = e^x a \cos y + e^x b \sin y.$$

By the Cauchy–Riemann equations, we have $u_y = -v_x$, thus

$$-e^x a \sin y + e^x b \cos y = -e^x \psi(y)$$
$$\Rightarrow \psi(y) = -b \cos y + a \sin y.$$

We now use property (1) and

$$e^x = u(x, 0) + iv(x, 0)$$
$$= ae^x + ie^x(-b),$$

therefore $a = 1$, $b = 0$. Consequently $f(z) = e^x \cos y + ie^x \sin y$.

EXAMPLE 1. Test the following function for analyticity:

$$f(z) = \frac{z^2}{z + \bar{z}}, \quad x \neq 0.$$

SOLUTION

$$f(z) = \frac{z^2}{2x} = \frac{x^2 - y^2 + 2ixy}{2x} = u + iv,$$

where

$$u = \frac{x^2 - y^2}{2x}, \qquad v = \frac{2xy}{2x} = y,$$

$$u_x = \frac{2x^2 + 2y^2}{4x^2}, \qquad v_x = 0,$$

$$u_y = -\frac{y}{x}, \qquad v_y = 1.$$

Thus $f(z)$ is analytic nowhere.

EXAMPLE 2. Show that

$$f(z) = \frac{xy^2(x + iy)}{x^2 + y^4} \qquad \text{if} \quad z \neq 0,$$

$$f(z) = 0 \qquad \text{if} \quad z = 0$$

is not differentiable at $z = 0$.

SOLUTION

$$f'(0) = \lim_{z \to 0} \frac{f(z) - f(0)}{z} = \lim_{z \to 0} \frac{xy^2}{x^2 + y^4}.$$

Along the curve $y^2 = x$ we have

$$f'(0) = \lim_{z \to 0} \frac{x^2}{x^2 + x^2} = \frac{1}{2};$$

along the line $x = 0$ we have

$$f'(0) = 0.$$

Consequently f is not differentiable at $z = 0$.

EXAMPLE 3. If $f(z) = u + iv \in A$ in D and $u, v \in C^2$ in D, then

a) $\left\{ \dfrac{\partial}{\partial x} |f(z)| \right\}^2 + \left\{ \dfrac{\partial}{\partial y} |f(z)| \right\}^2 = |f'(z)|^2$

b) $\left(\dfrac{\partial^2}{\partial x^2} + \dfrac{\partial^2}{\partial y^2} \right) |f(z)|^2 = 4|f'(z)|^2.$

SOLUTION OF (a). Since $|f| = (u^2 + v^2)^{1/2}$, we have

$$\frac{\partial}{\partial x} |f| = \frac{1}{2(u^2 + v^2)^{1/2}} \cdot (2uu_x + 2vv_x),$$

$$\frac{\partial}{\partial y} |f| = \frac{1}{2(u^2 + v^2)^{1/2}} \cdot (2uu_y + 2vv_y).$$

Functions of a Complex Variable

Consequently,

$$\{|f|_x\}^2 + \{|f|_y\}^2 = \frac{1}{u^2+v^2}\{u^2u_x^2 + v^2v_x^2 + u^2u_y^2 + v^2v_y^2$$
$$+ 2uv(u_xv_x + u_yv_y)\}.$$

Since $f \in A$ in D, then $u_x = v_y$ and $u_y = -v_x$, thus

$$\{|f|_x\}^2 + \{|f|_y\}^2 = \frac{1}{u^2+v^2}\{u^2(u_x^2 + v_x^2) + v^2(u_x^2 + v_x^2)\} = |f'|^2.$$

SOLUTION OF (b). We have, $|f| = (u^2 + v^2)^{1/2}$, thus

$$\frac{\partial}{\partial x}|f| = \frac{2uu_x + 2vv_x}{2(u^2+v^2)^{1/2}}.$$

Also, $|f|^2 = u^2 + v^2$, thus

$$\frac{\partial}{\partial x}|f|^2 = 2uu_x + 2vv_x,$$

$$\frac{\partial^2}{\partial x^2}|f|^2 = 2uu_{xx} + 2u_x^2 + 2vv_{xx} + 2v_x^2.$$

Similarly for $\partial^2|f|^2/\partial y^2$. Thus,

$$\left(\frac{\partial^2}{\partial x^2} + \frac{\partial^2}{\partial y^2}\right)|f|^2 = 2\{uu_{xx} + u_x^2 + vv_{xx} + v_x^2 + uu_{yy} + u_y^2 + vv_{yy} + v_y^2\}$$

$$= 2\{u(u_{xx} + u_{yy}) + (u_x^2 + v_x^2) + (u_y^2 + v_y^2) + v(v_{xx} + v_{yy})\}.$$

Since $f \in A$,

$$u_x = v_y \Rightarrow u_{xx} = v_{xy}$$

and

$$u_y = -v_x \Rightarrow u_{yy} = -v_{yx}.$$

Consequently, $u_{xx} + u_{yy} = v_{xy} - v_{yx} = 0$ since $u, v \in C^2$; also,

$$u_x^2 = v_y^2,$$
$$u_y^2 = v_x^2,$$

and

$$v_{xx} + v_{yy} = 0.$$

Therefore

$$\left(\frac{\partial^2}{\partial x^2} + \frac{\partial^2}{\partial y^2}\right)|f|^2 = 4(u_x^2 + v_x^2) = 4|f'|^2.$$

4.7

At this stage we mention the concept of a harmonic function. A few simple properties will be studied. However harmonic functions are interesting in their own right and several authors spend considerable time studying the deeper structure of these functions.*

4.7.1 Harmonic Functions

Let $f = u + iv \in A$ in D, then $u_x = v_y$ and $u_y = -v_x$. We have essentially anticipated the following property in our previous analysis.

Since $f \in A$, derivatives of all orders exist (and are continuous); thus

$$u_{xx} = v_{xy}; \quad u_{yy} = -v_{yx} \quad \text{and} \quad v_{xy} = v_{yx}.$$

Therefore $u_{xx} + u_{yy} = 0$; similarly, $v_{xx} + v_{yy} = 0$.

A function φ such that $\varphi_{xx} + \varphi_{yy} = 0$ is said to satisfy Laplace's equation in two dimensions. We write Laplace's equation as $\nabla^2 \varphi = 0$, where

$$\nabla^2 \equiv \frac{\partial^2}{\partial x^2} + \frac{\partial^2}{\partial y^2}.$$

The equation has been studied in great detail and is the equation of motion of a vast number of phenomena in the physical sciences.

4.7.2

Definition If $u \in C^2$ and satisfies Laplace's equation in some domain D, then u is called a *harmonic function*.

4.7.3

Definition If $f = u + iv$ is analytic (in some domain D), then u and v are called *conjugate harmonic functions*.

Given one function (say u), we can find the conjugate v by using the Cauchy–Riemann equations.

EXAMPLE. Given $u = \dfrac{y}{x^2 + y^2}$, we get

$$u_x = \frac{-2xy}{(x^2 + y^2)^2}, \quad u_{xx} = \frac{6x^2y - 2y^3}{(x^2 + y^2)^3},$$

$$u_y = \frac{x^2 - y^2}{(x^2 + y^2)^2}, \quad u_{yy} = \frac{-6x^2y + 2y^3}{(x^2 + y^2)^3},$$

then $u_{xx} + u_{yy} = 0$ and u is harmonic.

---

* For example, R. Nevanlinna, and V. Paatero, *Introduction to Complex Analysis*. Reading, Mass.: Addison–Wesley, 1964, Chap. 11).

Now,
$$u_x = -\frac{2xy}{(x^2+y^2)} = v_y,$$

thus
$$v = -2x\int \frac{y\,dy}{(x^2+y^2)^2} + \varphi(x)$$
$$= -2x \cdot \frac{-1}{2(x^2+y^2)} + \varphi(x) = \frac{x}{x^2+y^2} + \varphi(x),$$

whence
$$f = \frac{y+ix}{x^2+y^2} + i\varphi(x).$$

However, $u_y = -v_x$, then
$$\frac{x^2-y^2}{(x^2+y^2)^2} = \frac{x^2-y^2}{(x^2+y^2)^2} + \varphi'(x);$$

consequently $\varphi(x) = k$ (constant), and
$$f = \frac{i}{z} + ik.$$

After we have studied the concept of *analytic continuation*, we shall prove that indeed every harmonic function admits a conjugate function in some suitable simply connected domain.

4.7.4
Exercises

In Exercises 1–8, show that the given function u is harmonic in the indicated domain D and find its harmonic conjugate:

1: $u = \cos x \cosh y$ in $D = \mathcal{C}$.

2: $u = x^2 - 3xy^2$ in $D = \mathcal{C}$.

3: $u = e^x \sin y$ in $D = \mathcal{C}$.

4: $u = x^4 - 6x^2y^2 + y^4 + x^3y - xy^3$ in $D = \mathcal{C}$.

5: $u = \ln(x^2 + y^2)$ in $D = \mathcal{C} - \{[0,0]\}$.

6: $u = e^{x^2-y^2}\cos 2xy$ in $D = \mathcal{C}$.

7: $u = (x-1)^3 - 3xy^2 + 3y^2$ in $D = \mathcal{C}$.

8: $u = \tan^{-1}\left(\dfrac{y - y_0}{x - x_0}\right)$, where x_0, y_0 are constants, in

$$D = \mathcal{C} - \{[x_0, y]\}$$

9: If $f = u + iv \in A$ in a domain D, show that uv is harmonic there, but u^2 need not be.

10: Let $u(x, y)$ be harmonic in a domain D. Show that in D, Laplace's equation in polar coordinates is given by

$$r^2 \frac{\partial^2 u}{\partial r^2} + r \frac{\partial u}{\partial r} + \frac{\partial^2 u}{\partial \theta^2} = 0.$$

11: Show that if $u(x, y)$ is harmonic in a domain D and if partial derivatives of all orders exist and are continuous in D, then the functions $u_x, u_y, u_{xx}, u_{xy}, u_{yy}, u_{xxx}, \ldots$ are all harmonic in D.

12: Prove that Curl Grad $v = 0$ if $I\{v\}$ is harmonic.

13: Prove that a harmonic function $\varphi(x, y)$ remains harmonic under the transformation $w = f(z)$ (that is, $x = x(u, v)$ and $y = y(u, v)$) provided $f(z) \in A$ and $f'(z) \neq 0$.

4.8 Trigonometric Functions and Identities

Since

$$e^{ix} = \cos x + i \sin x \quad \text{and} \quad e^{-ix} = \cos x - i \sin x$$

for real x, then

$$\sin x = \frac{e^{ix} - e^{-ix}}{2i} \quad \text{and} \quad \cos x = \frac{e^{ix} + e^{-ix}}{2}.$$

Thus the following definition is natural.

4.8.1
Definition

$$\sin z = \frac{e^{iz} - e^{-iz}}{2i} \quad \text{and} \quad \cos z = \frac{e^{iz} + e^{-iz}}{2}.$$

Thus $\sin z$ and $\cos z$ are entire, since they are linear combinations of entire functions. Since

$$\frac{d}{dz} e^{iz} = ie^{iz} \quad \text{and} \quad \frac{d}{dz} e^{-iz} = -ie^{-iz},$$

then

$$\frac{d}{dz} \sin z = \cos z \quad \text{and} \quad \frac{d}{dz} \cos z = -\sin z.$$

Functions of a Complex Variable

The functions $\tan z$, $\cot z$, $\sec z$, and $\operatorname{cosec} z$ are defined in the usual sense, as for real trigonometric functions, keeping in mind that $\tan z$ and $\sec z$ are analytic in any domain where $\cos z \neq 0$, while $\cot z$ and $\operatorname{cosec} z$ are analytic in any domain where $\sin z \neq 0$.

From the definition of $\sin z$ and $\cos z$ we have:

$$\sin z = \frac{e^{i(x+iy)} - e^{-i(x+iy)}}{2i} = \frac{e^{ix} \cdot e^{-y} - e^{-ix} \cdot e^{y}}{2i}$$

$$= e^{-y}\frac{(\cos x + i \sin x)}{2i} - e^{y}\frac{(\cos x - i \sin x)}{2i}$$

$$= \frac{\cos x}{i}\left(\frac{e^{-y} - e^{y}}{2}\right) + \sin x\left(\frac{e^{-y} + e^{y}}{2}\right).$$

that is,

$$\sin z = \sin x \cosh y + i \cos x \sinh y$$
$$\cos z = \cos x \cosh y - i \sin x \sinh y.$$

Also, we deduce from these two results that

$$\sinh y = \frac{1}{i}\sin(iy)$$

and

$$\cosh y = \cos iy.$$

Also, $\overline{\sin z} = \sin \bar{z}$ and $\overline{\cos z} = \cos \bar{z}$.

4.8.2

The periodic character of $\sin z$ and $\cos z$ is evident from the expansion of $\sin(x + iy)$ and $\cos(x + iy)$. All trigonometric identities are still valid for complex argument, thus

1. $\sin^2 z + \cos^2 z = 1$,
2. $\sin(z_1 + z_2) = \sin z_1 \cos z_2 + \cos z_1 \sin z_2$,
3. $\sin(-z) = -\sin z$
4. $\sin\left(\frac{\pi}{2} - z\right) = \cos z$
5. $\sin 2z = 2 \sin z \cos z$
6. $\sin z = 0$ for $z = n\pi$, $n = 0, \pm 1, \ldots$

4.8.3
Exercises

1: Show that

$$\sec^2 z = 1 + \tan^2 z,$$

and

$$\operatorname{cosec}^2 z = 1 + \cot^2 z.$$

2: Show that $\cos(i\bar{z}) = \overline{\cos(iz)}$, $\forall z$, and $\sin(i\bar{z}) = \overline{\sin(iz)}$ iff $z = n\pi i$, $n = 0, \pm 1, \ldots$.

3: Show that $|\sinh y| \leq |\sin z| \leq \cosh y$.

4: Prove that $|\sin z| \geq |\sin x|$, and $|\cos z| \geq |\cos x|$.

5: Find all roots of $\cos z = 2$.

6: Show that $\cos \bar{z}$ and $\sin \bar{z}$ are analytic nowhere.

7: Is $\sinh(e^z)$ entire? Why?

8: Prove that $\sin z = 0$ for $z = n\pi$; $n = 0, \pm 1, \ldots$

9*: Solve and discuss the equation $\sin z = \sinh z$.

10: Writing $x = \frac{1}{2}(z + \bar{z})$, $y = \frac{1}{2i}(z - \bar{z})$, introduce the operators
$$\partial = \frac{1}{2}\left(\frac{\partial}{\partial x} - i\frac{\partial}{\partial y}\right) \quad \text{and} \quad \bar{\partial} = \frac{1}{2}\left(\frac{\partial}{\partial x} + i\frac{\partial}{\partial y}\right).$$
Now define $\partial = \partial/\partial z$ and $\bar{\partial} = \partial/\partial \bar{z}$. Show that
$$f = u(x, y) + iv(x, y)$$
is analytic if and only if $\bar{\partial} f = 0$.

5
Curvilinear Integrals

5.1

Let $f(z) = u(x, y) + iv(x, y)$ be defined at all points $z = x + iy$ of a Jordan arc $C(z = \theta(t), a \leqslant t \leqslant b)$ with endpoints α, β. Take $\alpha = z_0$, $z_1, \ldots, z_n = \beta$ to be points in order on C and ξ_r an arbitrary point on the arc $\widehat{z_{r-1}z_r}$. Set $\eta = \max_r |z_r - z_{r-1}|$.

5.1.1
Definition

$$\int_C f(z)\, dz = \lim_{\substack{\eta \to 0 \\ n \to \infty}} \sum_{i=1}^n f(\xi_r)(z_r - z_{r-1}),$$

if the limit on the right-hand side and the value of the right-hand side does not depend upon the partitioning of the arc and the point ξ_r on the arc $\widehat{z_{r-1}z_r}$.

If C' consists of the same points as C but described in the opposite sense, we see that

$$\int_C f(z)\, dz = -\int_{C'} f(z)\, dz,$$

subject to the existence of either side.

Suppose C is given by $z = \theta(t) = \varphi(t) + i\psi(t)$, $a \leqslant t \leqslant b$. (Here C is described in the positive sense as t increases from a to b.) Let t_r, τ_r be particular values corresponding to z_r, ξ_r, respectively. Then we have $a = t_0 < t_1 < \cdots < t_n = b$, $t_{r-1} \leqslant \tau_r \leqslant t_r$ $(r = 1, \ldots, n)$ and

$$\int_C f(z)\, dz = \lim_{\substack{\eta \to 0 \\ n \to \infty}} \sum_{r=1}^n f\{\theta(\tau_r)\}\{\theta(t_r) - \theta(t_{r-1})\}, \quad \tau_r \in [t_{r-1}, t_r],$$

if the limit on the right exists and is independent of the partitioning. Now,

$$\int_{t=a}^{b} f\{\theta(t)\} \, d\theta(t) = \lim_{\substack{\delta \to 0 \\ n \to \infty}} \sum_{r=1}^{n} f\{\theta(\tau_r)\} \{\theta(t_r) - \theta(t_{r-1})\},$$

where $\delta = \max_r |t_r - t_{r-1}|$, if the limit on the right exists. Thus

$$\int_C f(z) \, dz = \int_{t=a}^{b} f\{\theta(t)\} \, d\theta(t)$$

when $\eta \to 0$ iff $\delta \to 0$, and the Stieltjes integral on the right exists.

Since C is a Jordan arc, z is a continuous (hence uniformly continuous) function of t in $[a, b]$, whence, if $\delta \to 0$ then $\eta \to 0$; also, t is a continuous (hence uniformly continuous) function of z on C (by 4.1.4). Thus if $\eta \to 0$ then $\delta \to 0$. Consequently for a Jordan arc we get

$$\int_C f(z) \, dz = \int_{t=a}^{b} f\{\theta(t)\} \, d\theta(t),$$

if the Stieltjes integral on the right exists.

It follows then from the fact that C is a Jordan arc that $\int_C f(z) \, dz$ exists when $f(z)$ is continuous on C and the arc is rectifiable, since $f\{\theta(t)\}$ is then continuous in $a \leq t \leq b$ and $\theta(t) \in B$ and is continuous in $[a, b]$.

Thus $\int_C f(z) \, dz$ exists provided

1. C is a rectifiable Jordan arc,
2. $f(z)$ is continuous on C.

5.1.2

Theorem *If $f(z)$ is continuous on C and $\theta'(t) \in C[a, b]$, where C is a Jordan arc defined by $z = \theta(t)$, $a \leq t \leq b$, then*

$$\int_C f(z) \, dz = \int_a^b f\{\theta(t)\} \theta'(t) \, dt.$$

PROOF. The proof follows from 3.3.3(3). □

5.1.3

Theorem *Let C consist of two rectifiable Jordan arcs C_1 and C_2. Then, if $f(z)$ is continuous on $C = C_1 \cup C_2$,*

$$\int_C f(z) \, dz = \int_{C_1} f(z) \, dz + \int_{C_2} f(z) \, dz.$$

PROOF. The result follows from the limit of the sums representing the two integrals on the right. □

Curvilinear Integrals

5.2 Fundamental Inequality for Integrals

5.2.1

Theorem *If*

1. *C is a rectifiable Jordan arc of length l,*
2. $|f(z)| \leq M$ *on C,*
3. $\int_C f(z)\, dz$ *exists,*

then $|\int f(z)\, dz| \leq Ml.$

PROOF. For all modes of subdivision we have:

$$\left|\sum_{r=1}^{n} f(\xi_r)\{z_r - z_{r-1}\}\right| \leq \sum_{r=1}^{n} |f(\xi_r)|\, |z_r - z_{r-1}|$$

$$\leq M \sum_{r=1}^{n} |z_r - z_{r-1}|$$

$$\leq Ml.$$

Now let $n \to \infty$ and $\eta = \max_r |z_r - z_{r-1}| \to 0.$ □

5.2.2

Theorem *If C is a closed rectifiable Jordan arc, then*

$$\int_C 1\, dz = 0 \quad \text{and} \quad \int_C z\, dz = 0.$$

PROOF. Suppose $C = C_1 \cup C_2$, as shown, and $\eta = \max_r |z_r - z_{r-1}|$. Then

$$\int_{C_1} 1\, dz = \lim_{\substack{\eta \to 0 \\ n \to \infty}} \sum_{r=1}^{n} 1 \cdot (z_r - z_{r-1})$$

$$= \lim_{\substack{\eta \to 0 \\ n \to \infty}} (\beta - \alpha) = \beta - \alpha.$$

Similarly, $\int_{C_2} 1\, dz = \alpha - \beta$. Hence $\int_C 1\, dz = \int_{C_1} 1\, dz + \int_{C_2} 1\, dz = 0.$
Further, $\int_C z\, dz$, $\int_{C_1} z\, dz$, and $\int_{C_2} z\, dz$ all exist, and

$$\int_{C_1} z\, dz = \lim_{\substack{\eta \to 0 \\ n \to \infty}} \sum_{r=1}^{n} z_r(z_r - z_{r-1}) = \lim_{\substack{\eta \to 0 \\ n \to \infty}} \sum_{r=1}^{n} z_{r-1}(z_r - z_{r-1}).$$

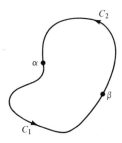

Thus

$$\int_{C_1} z\, dz = \frac{1}{2} \lim_{\substack{\eta \to 0 \\ n \to \infty}} \sum_{r=1}^{n} (z_r^2 - z_{r-1}^2) = \frac{1}{2}(\beta^2 - \alpha^2).$$

Similarly,

$$\int_{C_2} z\, dz = \frac{1}{2}(\alpha^2 - \beta^2);$$

consequently

$$\int_C z\, dz = \int_{C_1} z\, dz + \int_{C_2} z\, dz = 0. \qquad \square$$

5.2.3

Theorem *If C is a circle $|z - a| = r$, $r > 0$, described in the counterclockwise sense and n is an integer, then*

$$\int_C (z - a)^n\, dz = \begin{cases} 2\pi i r^{n+1} & \text{if } n = -1, \\ 0 & \text{if } n \ne -1. \end{cases}$$

PROOF. Let C be $z - a = r \operatorname{cis} \theta = re^{i\theta}$, $0 \le \theta \le 2\pi$. Then

$$\int_C (z - a)^n\, dz = \int_0^{2\pi} r^n e^{in\theta} r(-\sin\theta + i\cos\theta)\, d\theta$$

$$= ir^{n+1} \int_0^{2\pi} \{\cos(n+1)\theta + i\sin(n+1)\theta\}\, d\theta$$

$$= \begin{cases} 2\pi i r^{n+1} & \text{if } n+1 = 0 \quad \text{or } n = -1, \\ 0 & \text{if } n+1 \ne 0 \quad \text{or } n \ne -1.\ (n \text{ integer}). \end{cases}$$

5.3 Integral of a Function around the Boundary of a Domain

Let D be a domain whose boundary C consists of a finite number of rectifiable Jordan arcs $C_1, C_2, \ldots, C_p$, each described in the sense that D is kept on the left. We will sometimes use the symbol ∂D to mean *the boundary of D*. Thus, $C = C_1 \cup C_2 \cup C_3 \cup C_4$.

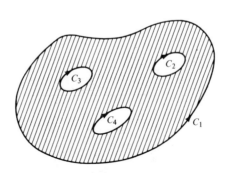

Curvilinear Integrals

Suppose $f(z)$ is continuous on $\overline{D} = D \cup C$. Then we have the following definition.

5.3.1
Definition

$$\int_C f(z)\,dz = \sum_{r=1}^{n} \int_{C_r} f(z)\,dz,$$

when each C_r is described in the above sense.

Let us write "crJc" to mean *closed rectifiable Jordan curve*.

5.3.2

Theorem *Let D be a domain bounded by a crJc. Suppose that the region $\overline{D}$ is divided into a finite number of closed subregions $R_1, R_2, \ldots, R_p$ by a rectifiable Jordan arcs drawn within $\overline{D}$. Let γ_s be the boundary of R_s, $(s = 1, \ldots, p)$ described so that the interior of R_s remains on the left. Then if $f(z)$ is continuous on $\overline{D}$, we have*

$$\int_C f(z)\,dz = \sum_{s=1}^{p} \int_{\gamma_s} f(z)\,dz.$$

PROOF

a. *Case $p = 2$.*

$$\int_{\gamma_1} + \int_{\gamma_2} = \left\{ \int_{ABF} + \int_{FEA} \right\} + \left\{ \int_{AEF} + \int_{FDA} \right\}$$

$$= \int_{ABF} + \int_{FDA} = \int_C.$$

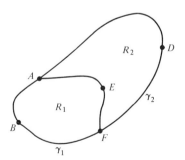

b. *Case $p = k + 1$ $(k \geqslant 2)$.* Suppose that the proposition (π_p) is true for $p = 2, \ldots, k$. Number the R_s so that R_{k+1} has some points in common with C. Then

$$\int_{LQNPL} = \sum_{s=1}^{k} \int_{\gamma_s} \quad \text{and} \quad \int_{LMN} + \int_{NQL} = \int_{\gamma_{k+1}}.$$

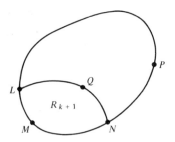

Thus

$$\int_C = \int_{LQNPL} + \int_{\gamma_{k+1}} = \sum_{s=1}^{k+1} \int_{\gamma_s}$$

and π_{k+1} is true. □

We now study Cauchy's theorem in considerable detail. This theorem is perhaps the cornerstone of analytic-function theory and the proof of a very general case will be undertaken.

5.4 Cauchy's Theorem

Let C be a crJc and D the interior domain bounded by C. Cauchy's theorem gives conditions for which

$$\int_C f(z)\,dz = 0.$$

This fact was mentioned by Gauss in a letter to Bessell (1811). The proof was published by Cauchy in 1825. Early attempts at proving the theorem relied on the continuity of $f'(z)$ in a domain containing $\bar{D}$. The following proof based on Green's theorem was suggested by Cauchy in 1846.

5.4.1
Since

$$\int_C (P\,dx + Q\,dy) = \iint_D \left(\frac{\partial Q}{\partial x} - \frac{\partial P}{\partial y} \right) dx\,dy,$$

then

$$\int_C f(z)\,dz = \int_C (u\,dx - v\,dy) + i\int_C (u\,dy + v\,dx)$$

$$= \iint_D \left(-\frac{\partial v}{\partial x} - \frac{\partial u}{\partial y} \right) dx\,dy + i\iint_D \left(\frac{\partial u}{\partial x} - \frac{\partial v}{\partial y} \right) dx\,dy$$

$$= 0 \quad \text{(by the Cauchy–Riemann equations).}$$

Note: Differentiability of $f(z)$ is not sufficient for the above proof.

Curvilinear Integrals

5.4.2

Pollard* proved that $\int_C f(z)\,dz = 0$ for

1. $f(z) \in A$ in D (bounded by C),
2. $f(z)$ is continuous on C.

An elegant proof based on Lebesgue integration was published by Estermann.* The usual proof is that by Goursat (1900)† for $f(z)$ single-valued and differentiable on $\bar{D}$ and C not met by any line (parallel to either axis) in more than a finite number of points.

The following diagram gives an example of a crJc meeting the x-axis in an infinite number of points.

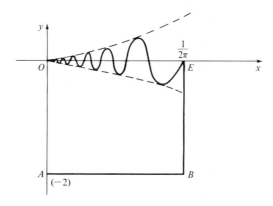

Curve C consists of lines OA, AB, BE and the curve γ:

$$\begin{cases} x = t, & 0 \leqslant t \leqslant \dfrac{1}{2\pi}, \\ y = \begin{cases} t^2 \sin \dfrac{1}{t} & \text{if } 0 < t \leqslant \dfrac{1}{2\pi}, \\ 0 & \text{if } t = 0. \end{cases} \end{cases}$$

Curve γ is rectifiable since dx/dt, dy/dt are bounded on $0 \leqslant t \leqslant 1/2\pi$.

We now need the following lemma proved by Goursat.

5.4.3

Lemma *Let f be single-valued and differentiable at all points of a closed rectangle I, with sides parallel to the axes (regard being paid to values of f at points of I only). Then $\int_C f(z)\,dz = 0$, where C is the boundary of I.*

---

* Proc. London Math. Soc. 21, 1923.
* Math. Zeitschrift, 37, 556–560, 1933.
† Goursat, E., "Cours d'Analyse," Paris, 2 (1918) 74–78.

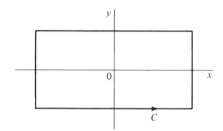

PROOF. Since $f \in C(I)$ then $\int_C f\,dz$ exists. Suppose $\int_C f\,dz = \alpha$. Divide I into four congruent rectangles. Then

$$\int_C = \int_{\gamma_1} + \int_{\gamma_2} + \int_{\gamma_3} + \int_{\gamma_4},$$

where $\gamma_1, \ldots, \gamma_4$ are boundaries of the subrectangles. Thus $|\int_{\gamma_r}| \geq \alpha/4$ for at least one γ_r.

By repeated subdivision, we can find a sequence of nested rectangles $I_1, I_2, \ldots, I_n, \ldots$ (with boundaries $C_1, C_2, \ldots, C_n, \ldots$) such that

1. $I_{n+1} \subset I_n$, $\forall n$;
2. diameter of $I_n \to 0$ as $n \to \infty$;
3. $|\int_{C_n} f(z)\,dz| \geq \alpha/4^n$.

The nest defines a point $c \in I$. Thus given $\epsilon > 0$, $\exists \delta(\epsilon) > 0$ such that

$$\left| \frac{f(z) - f(c)}{z - c} - f'(c) \right| < \epsilon \text{ for } z \in I \text{ and } 0 < |z - c| < \delta,$$

that is

$$f(z) = f(c) + (z - c)\{f'(c) + \eta(z)\},$$

where $|\eta(z)| < \epsilon$ for $z \in I$ and $|z - c| < \delta$.

The neighborhood $|z - c| < \delta(\epsilon)$ contains I_n for $n \geq \nu(\epsilon)$, and then

$$\left| \int_{C_n} f\,dz \right| = \left| \int_{C_n} \{f(c) + (z - c)[f'(c) + \eta(z)]\}\,dz \right|$$

$$= \left| \int_{C_n} (z - c)\eta(z)\,dz \right|$$

since $\int_{C_n} 1\,dz = 0 = \int_{C_n} z\,dz$.

Now, on C_n we have $|(z - c)\eta(z)| < l_n \epsilon$, where l_n is the perimeter of I_n. Thus

$$\left| \int_{C_n} f\,dz \right| \leq l_n^2 \epsilon = \left(\frac{l}{2^n}\right)^2 \epsilon = \frac{l^2}{4^n} \epsilon,$$

where l is the perimeter of I. Hence,

$$\frac{|\alpha|}{4^n} \leq \frac{l^2}{4^n} \epsilon,$$

that is,
$$\alpha = \int_C f\,dz = 0. \qquad \square$$

A similar result for a triangle can be proved (see, for example, the works of Knopp,* Valiron,† and Copson.**

A final lemma attributed to Landau will give us enough leverage to attack the main Cauchy theorem.

5.4.4

Lemma (Landau) *Let a rectifiable curve C of length l be enclosed in a square.*

Suppose the square is subdivided into squares of side δ; all squares are regarded as closed. Then the number of squares containing points of C, cannot exceed
$$4\left(\frac{l}{\delta} + 1\right).$$

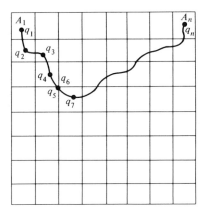

PROOF. Consider a sequence of squares entered by curve C starting at the endpoint q_1 (say, in square A_1). Let the sequence of squares be $A_1, A_2, \ldots, A_n$, no square being counted twice.

The result is clearly true for $n \leq 4$. Suppose $n = 4m + p > 4$, where $m \geq 1$, $p = 1, 2, 3, 4$. Let $q_1, q_2, \ldots, q_n$ be points of C in $A_1, A_2, \ldots, A_n$, respectively. At least *two* of the squares $A_1, A_2, \ldots, A_5$ have no common boundary points; then arc $q_1 q_5 \geq \delta$. Similarly,

$$\text{arc } q_5 q_9 \geq \delta; \quad \text{arc } q_9 q_{13} \geq \delta; \quad \ldots; \quad \text{arc } q_{4m-3} q_{4m+1} \geq \delta.$$

Thus
$$l \geq \text{arc } q_1 q_{4m+1} \geq m\delta$$

---

* Knopp, K., "Theory of Functions," Part I, Dover Publ. (1945) 49.
† Valiron, G., "Cours d'analyse mathématiques," 2 vols. Masson, Paris, (1954).
** Copson, E. T., "Introduction to the Theory of Functions of a Complex Variable," Clarendon Press, Oxford (1960) 61.

and

$$n = 4m + p \leq 4m + 4 \leq 4\frac{l}{\delta} + 4.$$ □

5.4.5

Theorem (Strong Cauchy) *If*

1. *C is a closed rectifiable Jordan curve, D is the interior domain bounded by C (therefore D is simply connected), and C is of length l;*
2. *$f(z) \in A$ in D;*
3. *$f(z)$ is continuous on $\overline{D}$;*
4. *C is not met by a line (parallel to either axis) in more than a finite number of points, then $\int_C f(z)\,dz = 0$.*

PROOF. Since f is continuous, f is uniformly continuous on $\overline{D}$; thus for ϵ, $k > 0$, $\exists \delta(\epsilon, k) > 0$ such that $|f(z_1) - f(z_2)| < \epsilon/k$ for $z_1 \in \overline{D}$, $z_2 \in \overline{D}$, $|z_1 - z_2| < 2\delta$. We may choose $\delta < 1$.

Enclose $\overline{D}$ in a square Q with sides parallel to the axes; let the length of the sides be an integral multiple of δ. Divide Q into squares of side δ by lines parallel to the axes. According to (4) this resolves D into a finite number of open regions $R_1, R_2, \ldots, R_m, R_{m+1}, \ldots, R_p$ with boundaries $\gamma_1, \gamma_2, \ldots, \gamma_p$, each $\overline{R}_s$ being contained in $\overline{D}$. We assume that γ_s contains no points of C for $s = m+1, \ldots, p$ and that γ_s contains at least one point of C for $s = 1, 2, \ldots, m$.

Then $\int_C f\,dz$ exists and

$$\int_C f\,dz = \sum_{s=1}^{p} \int_{\gamma_s} f\,dz = \sum_{s=1}^{m} \int_{\gamma_s} f\,dz,$$

since $\int_{\gamma_s} f\,dz = 0$ for $s = m+1, \ldots, p$.

Let α_s be a fixed point of R_s, where $s = 1, \ldots, m$; then

$$|z - \alpha_s| \leq \sqrt{2}\,\delta < 2\delta \qquad \text{for } z \in \gamma_s$$

and

$$|f(z) - f(\alpha_s)| < \epsilon/k.$$

Thus

$$\left| \int_{\gamma_s} f(z)\,dz \right| = \left| \int_{\gamma_s} \{f(z) - f(\alpha_s)\}\,dz \right| \leq \frac{\epsilon}{k} l_s,$$

where l_s is equal to the length of γ_s and

$$\left| \int_C f(z)\,dz \right| \leq \sum_{s=1}^{m} \left| \int_{\gamma_s} f(z)\,dz \right| \leq \frac{\epsilon}{k} \left\{ \sum_{s=1}^{m} l_s \right\},$$

Curvilinear Integrals

that is,

$$\left| \int_C f(z)\,dz \right| \leq \frac{\epsilon}{k}\{l + 4\delta\}$$

[Number of closed subsquares with at least 1 point of C]}

$$\leq \frac{\epsilon}{k}\left\{l + 4\delta\left[4\frac{l}{\delta} + 4\right]\right\}$$

$$< \frac{\epsilon}{k}\{17l + 16\}, \quad \text{since } 0 < \delta < 1.$$

Choosing $k = 17l + 16$, we have

$$\left| \int_C f(z)\,dz \right| < \epsilon, \text{ or } \int_C f(z)\,dz = 0. \qquad \square$$

5.4.6

We can extend Cauchy's theorem to multiply-connected domains in the following way.

Let C be a crJc and C' another crJc inside C, such that $C \cap C' = \phi$, and let C and C' satisfy condition (4) of Cauchy's theorem. Let $f(z) \in A$ in the domain D consisting of points interior to C but exterior to C', and let f be continuous on C and C'. We create two domains D_1 and D_2, where $D_1 \cup D_2 = D$, as follows. Join the curve C to C' by a straight line l_1 from a point P on C to Q on C'; move partway around C' and again join C' to C by a straight line l_2 from a point R on C' to S on C, where $l_1 \cap l_2 = \phi$.

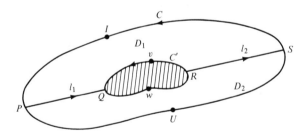

Then, $f(z) \in A$ in D_1 bounded by the crJc made up of:
1. an arc $\widehat{STP}$,
2. the line l_1 from P to Q,
3. an arc $\widehat{QVR}$, and
4. the line l_1 from R to S.

Similarly, $f(z) \in A$ in D_2 bounded by the crJc made up of:
1. an arc $\widehat{PUS}$,
2. the line l_2 from S to R,

3. an arc $R\widehat{W}Q$, and
4. the line l_1 from Q to P.

Note that $(S\widehat{T}P \cup P\widehat{U}S) = C$ and $(Q\widehat{V}R \cup R\widehat{W}Q) = C'$. Thus, by the Cauchy theorem,

$$\left\{\int_{S\widehat{T}P} + \int_{PQ} + \int_{Q\widehat{V}R} + \int_{RS}\right\} f(z)\,dz = 0$$

and also

$$\left\{\int_{P\widehat{U}S} + \int_{SR} + \int_{R\widehat{W}Q} + \int_{QP}\right\} f(z)\,dz = 0.$$

Adding these two results and observing that

$$\left\{\int_{PQ} + \int_{QP}\right\} f(z)\,dz = 0$$

and

$$\left\{\int_{RS} + \int_{SR}\right\} f(z)\,dz = 0,$$

we get

$$\int_C f(z)\,dz - \int_{C'} f(z)\,dz = 0$$

or

$$\int_C f(z)\,dz = \int_{C'} f(z)\,dz.$$

Similarly, we can establish the following.

5.4.7

Theorem Let $f(z) \in A$ in a domain D consisting of points interior to a crJc C but exterior to n disjoint curves $C_1, C_2, \ldots, C_n$, all of which are interior to C; also, let f be continuous on the curves $C, C_1, C_2, \ldots, C_n$, all of which satisfy condition (4) of Cauchy's theorem. Then

$$\int_C f(z)\,dz = \sum_{i=1}^{n} \int_{C_i} f(z)\,dz\ f(z)\,dz.$$

5.5

We will demonstrate that it is possible to approximate (arbitrarily closely) the integral of a continuous function $f(z)$ around a crJc by integrating the function along an appropriate sequence of line segments. However, although the construction of the line segments may help in obtaining numerical results for the integral, it does not help in proving the above form of Cauchy's theorem.

Curvilinear Integrals

5.5.1

Definition We say that a polygon is a *Jordon polygon* and is closed if it has no multiple points except the initial point $z_1(=z_n$, the last point).

5.5.2

Theorem *If $f(z)$ is continuous in a domain D and C is a rectifiable Jordan arc lying in D, then for every $\epsilon > 0$ there exists a polygon $\Gamma \subset D$ and joining the endpoints α, β of C, such that*

$$\left| \int_C f(z)\,dz - \int_\Gamma f(z)\,dz \right| < \epsilon.$$

PROOF. By the definition of $\int_C f(z)\,dz$, for a given $\epsilon > 0$, $\exists \eta_1 > 0$ and an integer N such that

$$\left| \int_C f(z)\,dz - \sum_{r=1}^n f(z_r)(z_r - z_{r-1}) \right| < \frac{\epsilon}{2},$$

where $\alpha = z_0, z_1, \ldots, z_n = \beta$ are points in order on C and

$$\delta = \max_r |z_r - z_{r-1}| < \eta_1 \text{ and } n > N.$$

Let Γ be a polygon with vertices $z_0, z_1, \ldots, z_n$. If we take the set of points $\{z_r\}$ such that $\delta < \frac{1}{2}\rho\{C, \partial D\}$, then $\Gamma \subset F = \{z \mid \rho(C, z) \leq \eta_2\}$, where $\eta_2 = \frac{1}{2}\rho\{C, \partial D\}$.

Now, $F \subset D$, thus $f(z)$ is uniformly continuous on F and $\exists \eta_3 > 0$ such that

$$|f(z+h) - f(z)| < \frac{\epsilon}{2l}$$

whenever $z \in F$, $z + h \in F$, and $|h| < \eta_3$ (l is the length of C). Construct Γ so that $0 < \delta < \eta = \min(\eta_1, \eta_2, \eta_3)$; then $\Gamma \subset F \subset D$ and

$$\left| \int_\Gamma f(z)\,dz - \sum_{r=1}^n f(z_r)(z_r - z_{r-1}) \right| = \left| \sum_{r=1}^n \int_{\gamma_r} \{f(z) - f(z_r)\}\,dz \right|, \quad (5.1)$$

where γ_r is the line segment joining z_{r+1} to z_r. Thus, (5.1) does not exceed

$$\sum_{r=1}^n \left| \int_{\gamma_r} \{f(z) - f(z_r)\}\,dz \right| \leq \sum_{r=1}^n \frac{\epsilon}{2l} |z_r - z_{r-1}| \leq \frac{\epsilon}{2}$$

and the result follows. □

Note, that the theorem in 5.5.2 is of no use for the purpose of proving the strong form of Cauchy's theorem, since in the polygon constructed around the crJc some line segments may lie outside the region $D \cup C = \overline{D}$, and we ostensibly have no information about $f(z)$ outside of $\overline{D}$. However we could use this result to prove a weaker form of Cauchy's theorem, namely:

If $f(z) \in A$ in a simply-connected domain D and C is a crJc interior to

D, then $\int_C f(z)\,dz = 0$. Cauchy's theorem has remarkable consequences and we now establish Cauchy's integral formula that essentially evaluates a function at a point (in terms of an integral).

5.6 Cauchy's Integral Formula

5.6.1

Theorem *If*

1. *D is a bounded domain whose boundary C consists of a finite number of crJc,*
2. *$f \in A$ in D,*
3. *f is continuous on $\bar{D} = D \cup C$,*

then

$$f(a) = \frac{1}{2\pi i} \int_C \frac{f(z)}{z-a}\,dz, \qquad \forall a \in D,$$

the boundary being traversed in the positive sense so as to keep the interior of D on the left.

Note: In the statement of the theorem and in its proof we use Cauchy's theorem under the Pollard conditions; i.e., we have not postulated that C should be met in at most a finite number of points by any line parallel to either axis.

PROOF. Suppose $a \in D$. Then there exists $\delta > 0$ such that $z \in D$ whenever $|z - a| < \delta$.

Let C_ρ be the circle $|z - a| = \rho$, where $0 < \rho < \delta$. Let D_1 be the domain bounded by C and C_ρ. Then $f(z)/(z - a)$ is analytic in D_1 and continuous on $\bar{D}_1$. Thus

$$\int_C \frac{f(z)}{z-a}\,dz - \int_{C_\rho} \frac{f(z)}{z-a} = 0,$$

that is,

$$\int_C \frac{f(z)}{z-a}\,dz = \int_{C_\rho} \frac{f(z)-f(a)}{z-a}\,dz + f(a)\int_{C_\rho} \frac{dz}{z-a}$$

$$= \int_{C_\rho} \frac{f(z)-f(a)}{z-a}\,dz + 2\pi i \cdot f(a).$$

Now

$$\left|\int_{C_\rho} \frac{f(z)-f(a)}{z-a}\,dz\right| \leq \frac{M_\rho}{\rho} 2\pi\rho = 2\pi M_\rho,$$

where $M_\rho = \max_{|z-a|=\rho}\{|f(z)-f(a)|\}$. Thus,

$$\int_{C_\rho} \frac{f(z)-f(a)}{z-a}\,dz \to 0 \qquad \text{as} \qquad \rho \to 0,$$

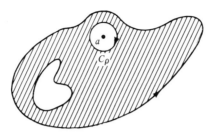

by continuity of $f(z)$ at $z = a$. Letting $\rho \to 0$, we have

$$\int_C \frac{f(z)}{z-a}\,dz = 2\pi i \cdot f(a).$$

□

5.6.2

Corollary *If $f, g \in A$ in D and are continuous on $\overline{D}$, while $f = g$ everywhere on C, then $f = g$, $\forall z \in \overline{D}$. In particular, if $f' \in A$ in D and continuous on $\overline{D}$, while $f(z) = k$ (constant) on C, then $f(z) = k$ in $\overline{D}$.*

The next result generalizes Cauchy's integral formula and is of considerable importance.

5.6.3

Theorem *If $f(z) \in A$ in D and continuous on $\overline{D}$ (where D is a bounded domain whose boundary C consists of a finite number of rectifiable Jordan curves), then for all $a \in D$,*

$$f^{(n)}(a) = \frac{n!}{2\pi i} \int_C \frac{f(z)}{(z-a)^{n+1}}\,dz, \quad \text{where } n = 0, 1, \ldots, \quad (5.2)$$

the boundary being traversed in the positive sense.

The following diagram illustrates the situation.

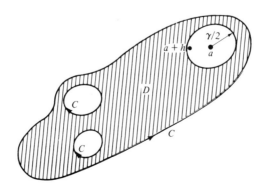

PROOF. Denote expression (5.2) by π_n. According to 5.6.1, expression π_0 is true. Suppose π_k is true, where k is zero or a positive integer. Let a be any fixed point of D; then C is a closed set, thus $\gamma = \rho(a, C) > 0$. Let δ be the diameter of $\overline{D}$. Choose h such that $0 < |h| < \gamma/2$. Then $(a + h) \in D$ and

$$\frac{f^{(k)}(a+h) - f^{(k)}(a)}{h} - \frac{(k+1)!}{2\pi i} \int_C \frac{f(z)}{(z-a)^{k+2}} \, dz$$

$$= \frac{k!}{2\pi i h} \int_C \left\{ \frac{f(z)}{(z-a-h)^{k+1}} - \frac{f(z)}{(z-a)^{k+1}} - \frac{h(k+1)f(z)}{(z-a)^{k+2}} \right\} dz$$

$$= \frac{k!}{2\pi i h} \int_C \frac{f(z)g(z)}{(z-a-h)^{k+1}(z-a)^{k+2}} \, dz,$$

where

$$g(z) = (z-a)^{k+2} - (z-a)(z-a-h)^{k+1} - h(k+1)(z-a-h)^{k+1}$$

$$= (z-a)^{k+2} - \left\{ (z-a)^{k+2} - (k+1)(z-a)^{k+1} h \right.$$

$$+ \binom{k+1}{2}(z-a)^k h^2 - \cdots \right\}$$

$$- h(k+1)\{(z-a)^{k+1} - (k+1)(z-a)^k h + \cdots \}$$

$$= \sum_{r=2}^{k+2} b_r (z-a)^{k+2-r} h^r$$

and b_r is independent of $z, a,$ and h. On C,

$$|g(z)| \le \sum_{r=2}^{k+2} |b_r| \delta^{k+2-r} \left(\frac{\gamma}{2}\right)^{r-2} |h|^2 = A|h|^2,$$

where A is independent of z and h. Hence if $z \in C$, then

$$\left| \frac{f(z)g(z)}{(z-a-h)^{k+1}(z-a)^{k+2}} \right| \le \frac{MA|h|^2}{\left(\frac{\gamma}{2}\right)^{k+1} \gamma^{k+2}} = K|h|^2.$$

Thus,

$$\left| \frac{k!}{2\pi i h} \int_C \frac{f(z)g(z)}{(z-a-h)^{k+1}(z-a)^{k+2}} \, dz \right| \le \frac{k!}{2\pi |h|} \cdot K|h|^2 \cdot l = \frac{k! K l}{2\pi} |h|,$$

where l is the length of C. Whence it follows that

$$\lim_{h \to 0} \frac{f^{(k)}(a+h) - f^{(k)}(a)}{h} = f^{(k+1)}(a) = \frac{(k+1)!}{2\pi i} \int_C \frac{f(z)}{(z-a)^{k+2}} \, dz,$$

that is, π_{k+1} is true. The result follows by induction. $\square$

Curvilinear Integrals

Note: If $g(z)$ is continuous on C and we define a function f over D such that if $a \in D$, then

$$f(a) = \frac{1}{2\pi i} \int_C \frac{g(z)}{z - a} \, dz,$$

it follows from the above arguments that

$$f^{(n)}(a) = \frac{n!}{2\pi i} \int_C \frac{g(z)}{(z - a)^{n+1}} \, dz.$$

5.6.4

Corollary *If $f(z) \in A$ in D, then $f(z)$ has derivatives of all orders in D.*

PROOF. Suppose $a \in D$. Then there exists $\rho > 0$ such that $\overline{N(a, \rho)} \subseteq D$. Taking the circle $|z - a| = \rho$ (described positively) for C, we get

$$f^{(n)}(a) = \frac{n!}{2\pi i} \int_C \frac{f(z)}{(z - a)^{n+1}} \, dz,$$

that is, $f^{(n)}(a)$ exists for $n = 1, 2, \ldots$ and all $a \in D$. □

5.6.5

Corollary 2 *If $f(z) \in A$ in D, then $f'(z) \in A$ in D.*

We now prove a lemma that is related to the previous result and which will be useful later in determining the analyticity of a function.

5.7

Lemma *Let $\varphi(\xi)$ be continuous on a rectifiable Jordan arc γ. Then the function*

$$F_n(z) = \int_\gamma \frac{\varphi(\xi)}{(\xi - z)^n} \, d\xi$$

is analytic in each of the regions determined by γ and its derivative is $F_n'(z) = n F_{n+1}(z)$.

PROOF. We first show that $F_1(z)$ is continuous. Let $z_0 \notin \gamma$; choose the neighborhood $\{|z - z_0| < \delta\} \cap \gamma = \emptyset$. Restricting z to the smaller neighborhood $|z - z_0| < \delta/2$, we have that $|\xi - z| > \delta/2$ for $\xi \in \gamma$. Since

$$F_1(z) - F_1(z_0) = (z - z_0) \int_\gamma \frac{\varphi(\xi) \, d\xi}{(\xi - z)(\xi - z_0)},$$

then

$$|F_1(z) - F_1(z_0)| < |z - z_0| \frac{2}{\delta^2} \int_\gamma |\varphi| \, |d\xi|,$$

φ is continuous and therefore bounded on γ, consequently $F_1(z)$ is continuous at z_0.

Applying this part of the lemma to $\varphi(\xi)/(\xi - z_0)$, we see that the difference quotient

$$\frac{F_1(z) - F_1(z_0)}{z - z_0} = \int_\gamma \frac{\varphi(\xi)\,d\xi}{(\xi - z)(\xi - z_0)}$$

tends to the limit $F_2(z_0)$ as $z \to z_0$. Thus it is clear that

$$F_1'(z) = F_2(z).$$

The general case is proved by induction. (Note that we only require the continuity of φ on γ.) □

One or two more properties emerge that will be useful in the evaluation of integrals.

5.8

Theorem *Let $f(z) \in A$ and be single-valued in a simply connected domain D; let C be a rectifiable Jordan curve lying in D; let α, β be the initial and final points of C. Then $\int_C f(z)\,dz$ depends only upon α and β and is thus independent of the path C.*

PROOF. Let C' be a rectifiable Jordan curve drawn from β to α such that $\Gamma = C \cup C'$ is simple and rectifiable. Then

$$\int_\Gamma f(z)\,dz = \int_C + \int_{C'} = 0,$$

that is,

$$\int_C f(z)\,dz = -\int_{C'} = \int_{-C'} f(z)\,dz.$$

It can be shown that the above analysis follows through even if Γ is not simple (but still rectifiable). Thus the integral is independent of the curve C. □

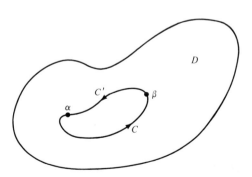

5.8.1

Theorem *Let $f(z) \in A$ and be single-valued in a simply connected domain D. Let C be a rectifiable Jordan curve in D with initial point α and terminal point β.*

If $G(z)$ is any single-valued function such that $G'(z) = f(z)$, then

$$\int_C f(z)\,dz = G(\beta) - G(\alpha).$$

PROOF. Let z be any point on C. Let $C(z)$ represent a rectifiable Jordan curve drawn from α to z. Then

$$F(z) = \int_{C(z)} f(\xi)\,d\xi = \int_\alpha^z f(\xi)\,d\xi$$

is independent of the path from α to z. Choose any point $z + h$ interior to a circle centered at z and contained in D.

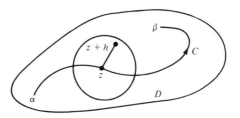

Take a straight-line path from z to $z + h$ and let $F(z + h)$ be the integral of f from α to z along $C(z)$ and then from z to $z + h$ along the straight line. Thus,

$$F(z + h) - F(z) = \left\{ \int_\alpha^{z+h} - \int_\alpha^z \right\} f(\xi)\,d\xi$$

$$= \int_z^{z+h} f(\xi)\,d\xi.$$

Now $f(z)$ is continuous, therefore

$$\lim_{\xi \to z} f(\xi) = f(z)$$

that is,

$$f(\xi) = f(z) + \eta(\xi)$$

where $\eta \to 0$ as $\xi \to z$.

Thus for a given $\epsilon > 0$ there exists $\delta > 0$ such that $|\eta(\xi)| < \epsilon$ for $|\xi - z| < \delta$ and

$$F(z + h) - F(z) = \int_z^{z+h} f(z)\,d\xi + \int_z^{z+h} \eta(\xi)\,d\xi.$$

But
$$\int_z^{z+h} f(z)\,d\xi = f(z)h,$$

Consequently,
$$\left|\frac{F(z+h)-F(z)}{h} - f(z)\right| < \epsilon \quad \text{when} \quad 0 < |h| < \delta.$$

Hence $F(z)$ is differentiable and $F'(z) = f(z)$. Also,
$$F(\beta) = \int_\alpha^\beta f(\xi)\,d\xi.$$

If $G(z)$ is single-valued and has the property that
$$G'(z) = f(z),$$

since
$$G'(z) - F'(z) = 0 = \frac{d}{dz}\{G(z) - F(z)\},$$

then by our previous results,
$$G(z) = F(z) + \kappa, \quad \kappa \text{ constant}$$

and
$$G(\alpha) = F(\alpha) + \kappa, \quad \kappa \text{ constant}.$$

But
$$F(\alpha) = 0,$$

thus
$$G(\beta) - G(\alpha) = F(\beta) = \int_\alpha^\beta f(\xi)\,d\xi = \int_C f(z)\,dz. \quad \square$$

We are now able to establish the same results for functions of a complex variable as we did for real functions; for example,
$$\int_\alpha^\beta z^n\,dz = \frac{\beta^{n+1} - \alpha^{n+1}}{n+1}, \quad \text{where } n \neq -1,$$

$$\int_\alpha^\beta e^{az}\,dz = \frac{e^{a\beta} - e^{a\alpha}}{a}, \quad \text{etc.}$$

Note: If the function is not analytic in a simply connected domain (bounded by a crJc), the following can happen.

Consider $\int_C dz/z$; if C is any closed rectifiable Jordan curve encircling the origin, we might expect $\log z \big|_\alpha^\alpha = 0$.

However for a circle of radius r (in particular $z = re^{i\theta}$),
$$dz = ire^{i\theta}\,d\theta \quad \text{and} \quad \int_0^{2\pi} i\,d\theta = 2\pi i.$$

5.8.2

We will now conclude the chapter by working several examples using the apparatus we have developed.

EXAMPLE 1. Evaluate $\int_{-i}^{i} (x^2 + iy^2)\, dz$ on the right half of the unit circle described counterclockwise.

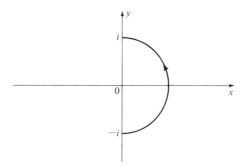

SOLUTION. Use the transform $x = r \cos \theta$, $y = r \sin \theta$, $r = 1$ on $|z| = 1$:
$$dz = dx + i\, dy = -(\sin \theta - i \cos \theta)\, d\theta.$$

We will use $s \equiv \sin \theta$ and $c \equiv \cos \theta$ when it is clear that the angle is just θ and not, for example, 2θ, $\theta/2$, etc. Thus,

$$I = \int_{-\pi/2}^{\pi/2} (c^2 + is^2)(-s + ic)\, d\theta$$

$$= \int_{-\pi/2}^{\pi/2} (-sc^2 + ic^3 - is^3 - s^2 c)\, d\theta.$$

Since s^3 is an odd function of θ, then

$$\int_{-\pi/2}^{\pi/2} s^3\, d\theta = 0;$$

Since $-sc^2$ is an odd function of θ, then

$$\int_{-\pi/2}^{\pi/2} (-sc^2)\, d\theta = 0;$$

Since c^3 is an even function of θ, then

$$\int_{-\pi/2}^{\pi/2} c^3\, d\theta = 2\int_{0}^{\pi/2} c^3\, d\theta;$$

Since $-s^2 c$ is an even function of θ, then

$$\int_{-\pi/2}^{\pi/2} (-s^2 c)\, d\theta = 2\int_{0}^{\pi/2} (-s^2 c)\, d\theta.$$

Thus

$$I = 2\int_0^{\pi/2}(ic^3 - s^2c)\,d\theta = 2\left[i\frac{2}{3}\right] - 2\left[\frac{\sin^3\theta}{3}\right]_0^{\pi/2}$$
$$= \frac{2}{3}[2i - 1].$$

EXAMPLE 2. Show that $|\int_C(x^2 + iy^2)\,dz| \leq \pi$, where C is the semicircle in the previous example.

SOLUTION. If $I = \int_C(x^2 + iy^2)\,dz$, then

$$|I| \leq \int_C |x^2 + iy^2|\,|dz|.$$

Now, $|x^2 + iy^2| = \sqrt{x^4 + y^4} \leq \sqrt{x^4 + 2x^2y^2 + y^4} = x^2 + y^2 = 1$. Consequently,

$$|I| \leq \int_C |x^2 + iy^2|\,|dz| \leq \int_C |dz| = \pi. \qquad \square$$

EXAMPLE 3. Evaluate $I = \int_C \pi \exp(\pi\bar{z})\,dz$, where C is a square with vertices at $z = 0, 1, 1 + i, i$.

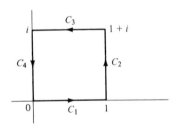

SOLUTION

$$\exp(\pi\bar{z}) = e^{\pi(x - iy)} = e^{\pi x - i\pi y}$$

and

$$dz = dx + i\,dy.$$

Thus

$$I = \pi\sum_r \int_{C_r} e^{\pi x - i\pi y}(dx + i\,dy), \quad \text{where } r = 1, 2, 3, 4.$$

For C_1 we have $y = 0$, $dy = 0$; thus,

$$\pi\int_0^1 e^{\pi x}\,dx = e^{\pi x}\big|_0^1 = e^\pi - 1.$$

Curvilinear Integrals

For C_2 we have $x = 1$, $dx = 0$; thus,

$$\pi \int_0^1 e^{\pi - i\pi y} i \, dy = \pi e^{\pi} i \int_0^1 e^{-i\pi y} \, dy$$

$$= \frac{i\pi e^{\pi}}{-i\pi} e^{i\pi y}\big|_0^1$$

$$= -e^{\pi}\{e^{i\pi} - 1\} = 2e^{\pi}.$$

For C_3 we have $y = 1$, $dy = 0$; thus,

$$\pi \int_1^0 e^{\pi x - i\pi} \, dx = -\pi e^{-i\pi} \frac{1}{\pi} e^{\pi x}\big|_0^1$$

$$= -e^{-i\pi}(e^{\pi} - 1) = e^{\pi} - 1.$$

For C_4 we have $x = 0$, $dx = 0$; thus,

$$\pi \int_1^0 e^{-i\pi y} i \, dy = i\pi(-1) \frac{1}{-i\pi} e^{-i\pi y}\big|_0^1$$

$$= e^{-i\pi} - 1 = -2.$$

Adding the four integrals, we get

$$\int_C \pi \exp(\pi \bar{z}) \, dz = 4(e^{\pi} - 1).$$

For applications of Cauchy's formula consider the following examples.

EXAMPLE 4. Evaluate

$$\int_C \frac{\cosh z + \cos z}{(z + 3i)(z^2 + 16)} \, dz, \quad \text{where } C : |z| = 2.$$

SOLUTION. Take

$$f(z) = \frac{\cosh z + \cos z}{(z + 3i)(z^2 + 16)}.$$

Clearly f is not analytic at $z = -3i$ and $\pm 4i$, however these three points are outside $|z| = 2$. Thus f is analytic in D, that is, in $|z| < 2$, and certainly continuous on $|z| = 2$, thus $\int_C f \, dz = 0$.

EXAMPLE 5. Evaluate

$$\int_C \frac{-z - 7}{z^2 - 1} \, dz, \quad \text{where } C : |z| = 2.$$

SOLUTION. Take

$$\frac{-z - 7}{z^2 - 1} = \frac{3}{z + 1} - \frac{4}{z - 1}.$$

Now using Cauchy's integral formula

$$f(a) = \frac{1}{2\pi i} \int_C \frac{f(z)}{z-a} \, dz$$

we get

$$\int_{|z|=2} \frac{3}{z+1} \, dz = 2\pi i \cdot 3 \quad \text{since} \quad -1 \in D$$

and

$$\int_{|z|=2} \frac{-4}{z-1} \, dz = 2\pi i(-4) \quad \text{since} \quad 1 \in D.$$

Adding these results, we obtain

$$\int_{|z|=2} \frac{-z-7}{z^2-1} \, dz = 6\pi i - 8\pi i = -2\pi i.$$

EXAMPLE 6. Show that for $f \in A$ in D (and continuous on $\bar{D}$) bounded by a crJc C and for $z_0 \in D$ and a positive integer n we have

$$\frac{1}{2\pi i} \int_C \frac{\{f(z)\}^n}{z-z_0} \, dz = \left\{ \frac{1}{2\pi i} \int_C \frac{f(z)}{z-z_0} \, dz \right\}^n.$$

SOLUTION. Since

$$f(z_0) = \frac{1}{2\pi i} \int_C \frac{f(z)}{(z-z_0)} \, dz,$$

then

$$\{f(z_0)\}^n = \left\{ \frac{1}{2\pi i} \int_C \frac{f(z)}{z-z_0} \, dz \right\}^n.$$

Now consider $\varphi(z) = \{f(z)\}^n$, where n is a positive integer. Then since $\varphi \in A$ in D (and continuous on $\bar{D}$) we have

$$\frac{1}{2\pi i} \int_C \frac{\varphi(z)}{z-z_0} \, dz = \varphi(z_0) = \{f(z_0)\}^n.$$

Thus

$$\frac{1}{2\pi i} \int_C \frac{\{f(z)\}^n}{z-z_0} \, dz = \left\{ \frac{1}{2\pi i} \int_C \frac{f(z)}{z-z_0} \, dz \right\}^n.$$

EXAMPLE 7. For n a positive integer, show that

$$\frac{1}{2\pi} \int_0^{2\pi} (2\cos\theta)^{2n} \, d\theta = \frac{(2n)!}{(n!)(n!)}.$$

Curvilinear Integrals

SOLUTION. If $z = e^{i\theta}$, then $z + 1/z = 2\cos\theta$; thus

$$\int_{|z|=1}\left(z + \frac{1}{z}\right)^{2n}\frac{dz}{z} = \int_0^{2\pi}(2\cos\theta)^{2n}\cdot\frac{iz\,d\theta}{z}$$

$$= i\int_0^{2\pi}(2\cos\theta)^{2n}d\theta.$$

Also,

$$\int_{|z|=1}\left(z + \frac{1}{z}\right)^{2n}\frac{dz}{z} = \int_{|z|=1}\frac{(z^2+1)^{2n}}{z^{2n+1}}dz$$

$$= \int_{|z|=1}\left\{\frac{1}{z^{2n+1}}\sum_{r=0}^{2n}\binom{2n}{r}z^{2r}\right\}dz$$

$$= \sum_{r=0}^{2n}\binom{2n}{r}\int_0^{2\pi}e^{i(2r-2n-1)\theta}ie^{i\theta}d\theta$$

$$= i\sum_{r=0}^{2n}\binom{2n}{r}\int_0^{2\pi}\{\cos(2r-2n)\theta + i\sin(2r-2n)\theta\}d\theta$$

$$= i\sum_{r=0}^{2n}\binom{2n}{r}\delta_n^r 2\pi$$

$$= 2\pi i\binom{2n}{n}$$

since $\delta_n^r = 1$ if $n = r$ and $\delta_n^r = 0$ otherwise, and the result follows.

5.8.3
Exercises

1: Show that $\left|\int_C \frac{dz}{z^2+1}\right| \leq \frac{\pi}{3}$; C: 1st quadrant of $|z| = 2$.

2: Evaluate $\int_C \frac{dz}{z^4(z^2+16)}$; C: $1 < |z| < 3$.

3: Evaluate $\int_C \frac{\sinh^2 z + \cos z}{z - \pi i}dz$; C: $|z| = 4$.

4: Evaluate $\int_C \frac{e^z - 1}{z}dz$; C: $|z| = 1$.

5: Show that $\int_C \frac{\cos z}{z(z^2+8)}dz = \frac{\pi i}{4}$; C: a square with vertices at $(2, 0)$, $(-2, 0)$, $(0, 2)$, and $(0, -2)$.

6: Evaluate $\int_C \dfrac{dz}{z^4 - 1}$; $C: |z| = 2$.

7: Evaluate $\int_C \dfrac{z \sin^2 z}{z^2 - 7z + 7} \, dz$; $C: |z| = 1$.

8: Evaluate $\int_C \dfrac{3z^2 + 4z - 1}{(z^2 + 4)(z^2 + 1)} \, dz$; $C: |z| = 3$.

9: Given $C: z = \exp i\theta$ $(-\pi \leqslant \theta \leqslant \pi)$ and $k = \text{const}$, show that

$$\int_C \frac{e^{kz}}{z} \, dz = 2\pi i$$

and hence deduce that

$$\int_0^{\pi} e^{k \cos \theta} \cos(k \sin \theta) \, d\theta = \pi.$$

10: Prove that if $f \in A$ in D (and continuous on D) bounded by a crJc C, then for $z_0 \in D$ and n a positive integer we have

$$\frac{n!}{2\pi i} \int_C \frac{f(z)}{z - z_0} \, dz = \frac{1}{2\pi i} \int_C \frac{f^{(n)}(z)}{z - z_0} \, dz.$$

11: Evaluate $\int_C \dfrac{\tan(z/2)}{[z - (\pi/4)]^2} \, dz$; C: square of Exercise 5.

12: If $f(z)$ and $g(z)$ are entire functions, show that the integral

$$\frac{1}{2\pi i} \int_C \left\{ \frac{f(w)}{w - z} + z \frac{g(1/w)}{zw - w^2} \right\} dw$$

represents $f(z)$ inside the unit circle C and $g(1/z)$ outside it.

We now study a kind of converse of Cauchy's theorem, which is attributed to Morera.

5.9

Theorem (Morera's) *Let f be single-valued and continuous in a domain D, of the z-plane. Let $\int_C f(z) \, dz = 0$ for every crJc C in D. Then $f \in A$ in D.*

PROOF. We have that $F(z) = \int_\alpha^z f(z) \, dz$ is independent of the path of integration ($\alpha \in D$). Hence $F'(z)$ exists and equals $f(z)$, $\forall z \in D$.

Note: all that is needed is the continuity of f.

Curvilinear Integrals

Thus $F(z)$ is an analytic function. However, the derivative of an analytic function is analytic and since $F'(z) = f(z)$, then $f(z) \in A$ in D. □

Several important consequences of Cauchy's integral formula now emerge and are studied in the subsequent theorems.

5.10

Theorem (Cauchy's Inequality for $f^{(n)}(z_0)$) Let $f \in A$ in a circle C (and continuous on $\overline{C}$) with center z_0 and radius r; let $M = \max_{\overline{C}} |f|$. Then

$$|f^{(n)}(z_0)| \leq \frac{Mn!}{r^n} \; ; \quad n = 0, 1, \ldots$$

PROOF. We have $C: |z - z_0| = r$ and by Cauchy's formula we obtain

$$|f^{(n)}(z_0)| = \left| \frac{n!}{2\pi i} \int_C \frac{f(z)\, dz}{(z - z_0)^{n+1}} \right| \leq \frac{n!}{2\pi} \frac{M}{r^{n+1}} \cdot 2\pi r = \frac{Mn!}{r^n}. \quad □$$

5.11

Theorem (Liouville) If $f \in A$ and is bounded for all z in the finite complex plane, then f is a constant.

PROOF. We have $|f(z)| \leq M$ for all z. Take $n = 1$ in Cauchy's inequality, then

$$|f'(z_0)| \leq \frac{M}{r}, \quad \forall r > 0.$$

Thus for r sufficiently large, $|f'|$ becomes arbitrarily small, that is, $f'(z_0) = 0$. Since z_0 is any point in the complex plane, we have $f'(z) = 0$ for all z. Thus f is a constant (by 4.5). □

6

Series of Complex Functions, Taylor's Theorem, Uniform Convergence

6.1
Definition An infinite series of the form

$$\sum_{n=0}^{\infty} a_n(z - z_0)^n = a_0 + a_1(z - z_0) + \cdots + a_n(z - z_0)^n + \cdots,$$

where z is a complex variable and $z_0, a_0, a_1, \ldots$ are fixed complex numbers, is called a *power series*, in powers of $(z - z_0)$.

6.1.1
Definition The *circle of convergence* C, of a power series $\sum_{n=0}^{\infty} a_n(z - z_0)^n$, is a circular set of points such that the series converges for all z interior to C, diverges for all z outside C, and may or may not converge at none, all, or some of the points on the boundary.

The following theorem pertains to the radius of this circle of convergence.

6.1.2
Theorem Every power series $\sum_{n=0}^{\infty} a_n(z - z_0)^n$ has a *radius of convergence* R, $0 < R < \infty$, such that the series converges absolutely when $|z - z_0| < R$ and diverges when $|z - z_0| > R$.

For $R = 0$, the series converges for $z = z_0$ only, while for $R = \infty$ the series converges for all z.

The number R is given by

$$(a) \quad R = \frac{1}{\lim_{n \to \infty} \sqrt[n]{|a_n|}}$$

or

$$(b) \quad R = \overline{\lim_{n \to \infty}} \left| \frac{a_n}{a_{n+1}} \right|.$$

PROOF. Let $u_n(z) = a_n(z - z_0)^n$. Since

$$|u_n|^{1/n} = |a_n|^{1/n} |z - z_0|,$$

by using Cauchy's root test we get

$$\overline{\lim_{n \to \infty}} |u_n|^{1/n} = \overline{\lim_{n \to \infty}} |a_n|^{1/n} |z - z_0| = \frac{|z - z_0|}{R} \quad \text{(by hypothesis)}.$$

Thus if $|z - z_0| < R$, the series converges absolutely for all z in the open disk of radius R centered at z_0.

Further, if $|z - z_0| > R$, then the series diverges for all z outside the circle. For $|z - z_0| = R$ we have no information and require to use more erudite techniques to determine the behavior of the series on the boundary.

Note:

i. $\sum_{n=0}^{\infty} z^n$ does not converge for any point on its circle of convergence $|z| = 1$ since $\lim_{n \to \infty} e^{in\theta}$ does not exist.
ii. $\sum_{n=1}^{\infty} z^n/n^2$ converges for all points on its circle of convergence $|z| = 1$ since $|z^n/n^2| \leq 1/n^2$ on $|z| = 1$.

Using d'Alembert's ratio test, it is easy to see that the radius of convergence may be given by (b). Clearly, if $R = 0$, then we have convergence for $z = z_0$ only, however for $R = \infty$ we have convergence for all z. □

6.2 Uniform Convergence

6.2.1

Definition If $\{f_n(z)\}$ is a sequence of complex-valued functions defined over a set S, then we say that $f_n(z) \to f(z)$ *uniformly* on S if for every $\epsilon > 0$ there exists $\delta(\epsilon) > 0$ and independent of z, such that

$$|f_n(z) - f(z)| < \epsilon, \quad \forall n > \delta(\epsilon) \quad \text{and} \quad \forall z \in S.$$

Several consequences of the concept of uniform convergence will be listed and a few proofs given, however, the standard theorems from real-variable theory carry over directly to the complex-function theory (with few exceptions).

6.2.2

Theorem *If* (i) $f_n(z) \to f(z)$ *uniformly on a set S and* (ii) *each $f_n(z)$ is continuous on S, then $f(z)$ is continuous on S.*

PROOF. Given $\epsilon, k > 0$, there exists $\nu(\epsilon) > 0$ such that
$$|f_n(z) - f(z)| < \frac{\epsilon}{k} \quad \text{for} \quad n > \nu(\epsilon) \quad \text{and} \quad \forall z \in S.$$
Take any $z_0 \in S$. Then, given any fixed $n > \nu(\epsilon)$, there exists $\delta(\epsilon) > 0$ (δ depends upon z_0 and n) such that
$$|f_n(z) - f_n(z_0)| < \frac{\epsilon}{k} \quad \text{for} \quad |z - z_0| < \delta(\epsilon) \quad \text{when} \quad z \in S.$$
For such n and z we then have
$$|f(z) - f(z_0)| = |f(z) - f_n(z) + f_n(z) - f_n(z_0) + f_n(z_0) - f(z_0)|$$
$$\leqslant \frac{\epsilon}{k} + \frac{\epsilon}{k} + \frac{\epsilon}{k} = \frac{3\epsilon}{k}.$$
Choosing $k = 3$, we see that
$$|f(z) - f(z_0)| < \epsilon \quad \text{for} \quad |z - z_0| < \delta(\epsilon) \quad \text{and} \quad z \in S. \quad \square$$

6.2.3

Theorem *If (i) $f_n(z) \to f(z)$ uniformly on a rectifiable Jordan arc C of length l and (ii) each $f_n(z)$ is continuous on C, then*
$$\int_C f_n(z) \, dz \to \int_C f(z) \, dz \quad \text{as} \quad n \to \infty.$$

PROOF. Continuity of $f(z)$ follows from 6.2.2, consequently $\int_C f_n(z) \, dz$ and $\int_C f(z) \, dz$ exist. By the first condition, given $\epsilon, k > 0$, there exists $\nu(\epsilon) > 0$ such that
$$|f_n(z) - f(z)| < \frac{\epsilon}{k} \quad \text{for} \quad n > \nu(\epsilon) \quad \text{and} \quad z \in C.$$
For this n we have
$$\left| \int_C \{f_n(z) - f(z)\} \, dz \right| < \frac{\epsilon}{k} l.$$
Choosing $k = l + 1$, we get
$$\left| \int_C \{f_n(z) - f(z)\} \, dz \right| < \epsilon \quad \text{for} \quad n > \nu(\epsilon),$$
whence
$$\lim_{n \to \infty} \int_C f_n(z) \, dz = \int_C f(z) \, dz. \quad \square$$

6.2.4 Corresponding Results for Series

Write $S(z) = \sum_{r=0}^{\infty} u_r(z)$ and $S_n(z) = \sum_{r=0}^{n} u_r(z)$. If $S_n(z) \to S(z)$ uniformly on a set S as $n \to \infty$, then we say that the *series converges uniformly* on S.

The proofs of the following theorems are virtually the same as for real variables.

6.2.5

Theorem *If (i) $u_r(z)$ is continuous on a set D, where $r = 0, 1, \ldots,$ and (ii) $\sum_r u_r(z)$ converges uniformly to $S(z)$ on the set D, then $S(z)$ is continuous on D.*

6.2.6

Theorem *If (i) $u_r(z)$ is continuous on a rectifiable Jordan curve C and (ii) $\sum_r u_r(z)$ converges uniformly to $S(z)$ on C, then*

$$\int_C S(z)\,dz = \sum_{r=0}^{\infty} \left\{ \int_C u_r(z)\,dz \right\}.$$

6.2.7

Theorem (Weierstrass M-test) *If (i) $|u_n(z)| \leq M_n$, $\forall z \in D$, where M_n is independent of z and (ii) $\sum_{n=0}^{\infty} M_n < \infty$, then $\sum_{n=0}^{\infty} u_n(z)$ converges uniformly on D.*

The following theorem will be particularly useful in later analysis. It essentially states that *Every power series is uniformly convergent within its circle of convergence.*

6.2.8

Theorem *If $\sum_{n=0}^{\infty} a_n(z - z_0)^n$ has a radius of convergence $R > 0$ and if $0 \leq R_1 < R$, then $\sum_{n=0}^{\infty} a_n(z - z_0)^n$ is uniformly convergent in $|z - z_0| \leq R_1$.*

PROOF. Let z_1 be a point in $R_1 < |z - z_0| < R$, then $\sum_{n=0}^{\infty} a_n(z_1 - z_0)^n < \infty$ and thus $|a_n(z_1 - z_0)^n| < 1$ for $n > N$. Since

$$|a_n(z - z_0)^n| = |a_n(z_1 - z_0)^n| \frac{|a_n(z - z_0)^n|}{|a_n(z_1 - z_0)^n|} < \left|\frac{z - z_0}{z_1 - z_0}\right|^n = M^n,$$

where $M < 1$, provided $z \in \{|z - z_0| \leq R_1\}$, then by the M-test we have uniform convergence. □

6.2.9

Exercises

Find the radius of convergence for the following series:

1: $\sum_{n=1}^{\infty} \frac{z^n}{n(n+1)}$

2: $\sum_{n=1}^{\infty} (1 - \frac{1}{n})^{n^2} z^n$

3: $\displaystyle\sum_{n=2}^{\infty} \frac{\log n^n}{n!} z^n$

4: Show that $\sum_n a_n z^n$ and $\sum_n a_n \dfrac{z^{n+1}}{n+1}$ have the same circle of convergence.

We now study Taylor's theorem for analytic functions, which in a sense turns out to have far less pathology associated with it than that for the case of a real variable.

6.3
Theorem (Taylor) *If $f(z) \in A$ in $|z - a| < R$, where $R > 0$, then*

$$f(z) = \sum_{n=0}^{\infty} \frac{f^{(n)}(a)}{n!} (z - a)^n \quad \text{for} \quad |z - a| < R.$$

PROOF. Take h such that $|h| < R$ and ρ such that $|h| < \rho < R$. Let C be the circle $|z - a| = \rho$. Then $f(z) \in A$ in and on C. Thus by Cauchy's integral formula we have

$$f(a + h) = \frac{1}{2\pi i} \int_C \frac{f(z)}{z - a - h} dz,$$

that is,

$$f(a + h) = \frac{1}{2\pi i} \int_C \frac{f(z)}{z - a} \left\{1 - \frac{h}{z - a}\right\}^{-1} dz$$

$$= \frac{1}{2\pi i} \int_C \sum_0^{\infty} \left\{\frac{f(z) h^r}{(z - a)^{r+1}}\right\} dz,$$

$$\left(\text{since } \left|\frac{h}{z - a}\right| = \frac{|h|}{\rho} < 1 \text{ on } C\right).$$

Now, the integrand is a convergent series of terms each of which is continuous on C, and the series is also uniformly convergent on C because (M-test) it is majorized by $\sum_0^{\infty} M |h|^r / \rho^{r+1}$, a convergent series of nonnegative terms, where M is the upper bound for $|f(z)|$ on C. Thus,

$$f(a + h) = \sum_0^{\infty} \left\{\frac{1}{2\pi i} \int_C \frac{f(z)}{(z - a)^{r+1}} dz\right\} h^r$$

$$= \sum_0^{\infty} \frac{f^{(r)}(a)}{r!} h^r.$$

This holds for all h such that $|h| < R$. $\square$

Series of Complex Functions, Taylor's Theorem, Uniform Convergence

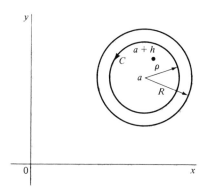

An important point is that if $f \in A$ in some circle C, then its Taylor series converges in C and no test is required.

6.3.1

Consequences

1. If $f(z), g(z) \in A$ for $|z - a| < R$, $R > 0$, and $f^{(n)}(a) = g^{(n)}(a)$ for $n = 0, 1, 2, \ldots$, then $f(z) = g(z)$ for $|z - a| < R$.
2. If $f(z), g(z) \in A$ for $|z - a| < R$, $R > 0$, and $f(z) = g(z)$ at all points on a closed rectifiable Jordan curve C containing a in its interior domain and lying in the domain $|z - a| < R$, then $f(z) = g(z)$ for $|z - a| < R$.

PROOF. Curve C is assumed to be described positively. For $n = 0, 1, 2, \ldots$, we have

$$f^{(n)}(a) = \frac{n!}{2\pi i} \int_C \frac{f(z)}{(z-a)^{n+1}} \, dz = \frac{n!}{2\pi i} \int_C \frac{g(z)}{(z-a)^{n+1}} \, dz = g^{(n)}(a). \quad \square$$

3. If $f(z) \in A$ for $|z - a| < R$, $R > 0$, and $f'(z) = 0$ for $|z - a| < r$, $0 < r < R$, then $f(z) = f(a)$ for $|z - a| < R$.

PROOF. Let C be the circle $|z - a| = r/2$. Then, for $n = 2, 3, \ldots$, we have

$$f^{(n)}(a) = \frac{(n-1)!}{2\pi i} \int_C \frac{f'(z)}{(z-a)^n} \, dz = 0,$$

since $f'(z) = 0$ on C; thus,

$$0 = f^{(1)}(a) = f^{(2)}(a) = \cdots = f^{(n)}(a) = \cdots$$

and $f(z) = f(a)$ for $|z - a| < R$. $\quad \square$

Specific examples of Taylor's theorem follow immediately: e.g.,

1. $e^z = \sum_{n=0}^{\infty} \frac{z^n}{n!}$, $|z| < \infty$;
2. $\cos z = 1 + \sum_{n=1}^{\infty} (-1)^n \frac{z^{2n}}{(2n)!}$, $|z| < \infty$;

3. $\dfrac{1}{1+z} = \sum_{n=0}^{\infty}(-1)^n z^n, \quad |z| < 1;$

4. If we expand $f(z) = 1/z$ about $z = 1$, we obtain

$$\frac{1}{z} = \sum_{n=0}^{\infty}(-1)^n (z-1)^n, \quad |z-1| < 1.$$

The function $f(z) = 1/z$ is clearly analytic at all points except $z = 0$.

6.4 Zeros of Analytic Functions

If $f(z) \in A$ in a domain D and $f(z) \not\equiv 0$ in D, we call a a *zero* of $f(z)$ if $a \in D$ and $f(a) = 0$.

We will discuss the concept of a *limit point* of zeros after we have studied analytic continuation.

6.4.1 Order of a zero

By Taylor's theorem, if $f(z) \in A$ in a domain D containing the point $z = a$, then the representation

$$f(z) = \sum_{n=0}^{\infty} \frac{f^{(n)}(a)}{n!}(z-a)^n$$

holds in all disks $C: |z - a| < R$ contained in D. Let us write

$$a_n = \frac{f^{(n)}(a)}{n!}, \quad n = 0, 1, \ldots.$$

Then if $a_0 = a_1 = a_2 = \cdots = a_{m-1} = 0$ and $a_m \neq 0$ we have

$$f(z) = a_m(z-a)^m + a_{m+1}(z-a)^{m+1} + \cdots$$

$$= \sum_{n=m}^{\infty} a_n(z-a)^n,$$

and in this case we say that f has a *zero of order m* at $z = a$. Clearly, $f(z) = (z-a)^m g(z)$, where $g(z) \in A$ in D and $g(a) \neq 0$.

Alternatively, we can define $z = a$ to be a zero of order m of $f(z)$ if $f(a) = 0$ and its derivatives up to the $(m-1)$th order vanish at $z = a$.

A function may have infinitely many zeros in a domain D in which it is analytic, however, unless $f(z) \equiv 0$, the limit points of these zeros must lie on the boundary of D.

For example, $f(z) = \sin(1/z)$ has an infinite number of zeros in D: $0 < |z| < 1$. The limit point of the set of zeros $\{z = 1/n\pi, n = \pm 1, \pm 2, \ldots\}$, namely $z = 0$, clearly lies on the boundary of D.

Some useful applications and ramifications of Taylor's theorem are exhibited in the next section.

6.4.2 Some Properties of Entire Functions

If $f(z)$ is entire, then by Taylor's theorem, we can write

$$f(z) = \sum_{n=0}^{\infty} a_n z^n, \quad \forall z.$$

We have already shown in 5.11 that if a function is analytic and bounded for all z, then it must be a constant. We now extend this result of Liouville as follows.

6.4.3

Theorem *If (i) $f(z)$ is entire and (ii) $f(z) = O(|z|^k)$ as $z \to \infty$, where k is real, then $f(z)$ is a polynomial of degree not exceeding k.*

PROOF. We have $f(z) = \sum_{n=0}^{\infty} a_n z^n$ for all z and $a_n = f^{(n)}(0)/n!$. For any $r > 0$, Cauchy's inequalities give

$$|a_n| \leqslant \frac{M(r)}{r^n}; \quad \text{where } n = 0, 1, \ldots$$

and there exist $K > 0$ and $r_0 > 0$ such that $|f(z)|/|z|^k \leqslant K$ for $|z| \geqslant r_0$.

Thus $|a_n| \leqslant Kr^k/r^n$ for $r \geqslant r_0$. Letting $r \to \infty$, it follows that $a_n = 0$ if $n > k$ and we have the result. □

6.4.4 The Case of a Polynomial

If $f(z) = a_0 + a_1 z + \cdots + a_n z^n$, where $a_n \neq 0$, then for $z \neq 0$ we get

$$\frac{f(z)}{z^n} = \frac{a_0}{z^n} + \frac{a_1}{z^{n-1}} + \cdots + a_n \to a_n \neq 0 \quad \text{as} \quad |z| \to \infty.$$

Thus $|f(z)| \to \infty$ as $|z| \to \infty$. This result leads immediately to the fundamental theorem of algebra.

6.4.5

Theorem *If $f(z)$ is a polynomial of degree $k \geqslant 1$, then $f(z)$ has at least one zero.*

By the remainder (factor) theorem and induction it follows that $f(z)$ has exactly k zeros.

PROOF. Suppose $f(z)$ has no zeros, then $1/f(z)$ is entire. Thus, since $|f(z)| \to \infty$ as $|z| \to \infty$, then $|1/f(z)| \to 0$ as $|z| \to \infty$. Consequently, $1/f(z) = O(1)$ as $|z| \to \infty$, and by Liouville's theorem it follows that $1/f(z) = \text{const}$. This contradiction proves the result. □

6.4.6

Corollary *A polynomial $f(z)$ of degree $k \geqslant 1$ takes every complex value c exactly k times (multiplicity of solutions being taken into account).*

6.5 Cauchy Product of Two Series

Given two power series, under certain conditions, we can multiply the series and write the coefficient of the nth term in a closed form.

6.5.1
Definition Given two series $\sum_{n=0}^{\infty} a_n$ and $\sum_{n=0}^{\infty} b_n$, define

$$c_n = \sum_{k=0}^{n} a_k b_{n-k}, \, n = 0, 1, 2, \ldots.$$

The series $\sum_{n=0}^{\infty} c_n$ is called the Cauchy product of $\sum_n a_n$ and $\sum_n b_n$.

6.5.2
Theorem Suppose that $\sum_n a_n z^n$ and $\sum_n b_n z^n$ have radii of convergence not less than $r_0 > 0$; define $c_n = \sum_{k=0}^{n} a_k b_{n-k}$. Then $\sum_n c_n z^n$ has radius of convergence not less than r_0 and inside this circle of radius r_0 we have

$$\sum_{n=0}^{\infty} c_n z^n = \left(\sum_{n=0}^{\infty} a_n z^n \right)\left(\sum_{n=0}^{\infty} b_n z^n \right).$$

PROOF. The method is essentially a generalization of the manner in which polynomials are multiplied. Let $K = \{z \mid |z| < r_0\}$ and

$$f(z) = \sum_{n=0}^{\infty} a_n z^n \quad \text{and} \quad g(z) = \sum_{n=0}^{\infty} b_n z^n.$$

Then $f, g \in A$ in K and consequently $fg \in A$ in K. By Taylor's theorem, we have

$$(f \cdot g)(z) = \sum_{n=0}^{\infty} \frac{(f \cdot g)^{(n)}(0) z^n}{n!}, \quad \forall z \in K.$$

It is not difficult to show that the nth derivative of the product $f(z)g(z)$ is given by

$$(f \cdot g)^{(n)}(z) = \sum_{k=0}^{n} \binom{n}{k} f^{(k)}(z) g^{(n-k)}(z).$$

Thus,

$$\frac{(f \cdot g)^{(n)}(0)}{n!} = \sum_{k=0}^{n} \frac{1}{k!(n-k)!} f^{(k)}(0) g^{(n-k)}(0)$$

$$= \sum_{k=0}^{n} a_k b_{n-k}$$

and $\sum_{n=0}^{\infty} c_n z^n$ converges on K. The radius of convergence is not less than r_0 (by Taylor's theorem) and consequently on K:

$$\sum_{n=0}^{\infty} c_n z^n = \left(\sum_{n=0}^{\infty} a_n z^n \right)\left(\sum_{n=0}^{\infty} b_n z^n \right). \qquad \square$$

Actually, we can do a lot better than this, since this result is quite restrictive. The more general result is attributed to Mertens.

6.5.3

Theorem (Mertens) *Let $\sum_{n=0}^{\infty} a_n$ converge absolutely to A and let $\sum_{n=0}^{\infty} b_n$ converge to the sum B. Then the Cauchy product of these two series converges to AB.*

PROOF. Define $A_n = \sum_{k=0}^{n} a_k$, $B_n = \sum_{k=0}^{n} b_k$, $C_n = \sum_{k=0}^{n} c_k$, where $c_k = \sum_{r=0}^{k} a_r b_{k-r}$ and $k = 0, 1, \ldots$. Let $d_n = B - B_n$ and $e_n = \sum_{k=0}^{n} a_k d_{n-k}$. Then

$$C_p = \sum_{n=0}^{p} \sum_{k=0}^{n} a_k b_{n-k} = \sum_{n=0}^{p} \sum_{k=0}^{p} f_n(k),$$

where

$$f_n(k) = \begin{cases} a_k b_{n-k} & \text{if } n \geq k, \\ 0 & \text{if } n < k. \end{cases}$$

Thus,

$$C_p = \sum_{k=0}^{p} \sum_{n=0}^{p} f_n(k) = \sum_{k=0}^{p} \sum_{n=k}^{p} a_k b_{n-k}$$

$$= \sum_{k=0}^{p} a_k \sum_{m=0}^{p-k} b_m$$

$$= \sum_{k=0}^{p} a_k B_{p-k}$$

$$= \sum_{k=0}^{p} a_k (B - d_{p-k})$$

$$= A_p B - e_p.$$

It now suffices to show that $e_p \to 0$ as $p \to \infty$. The sequence $\{d_n\}$ converges to 0 since $B = \sum_{n=0}^{\infty} b_n$. Now choose $M > 0$ such that $|d_n| \leq M$ for all n and let $K = \sum_{n=0}^{\infty} |a_n|$. Given $\epsilon > 0$, choose N such that for $n > N$ we get $|d_n| < \epsilon/2K$ and also $\sum_{n=N+1}^{\infty} |a_n| < \epsilon/2M$. Then for $p > 2N$, we have

$$|e_p| \leq \sum_{k=0}^{N} |a_k d_{p-k}| + \sum_{k=N+1}^{p} |a_k d_{p-k}| \leq \frac{\epsilon}{2K} \sum_{k=0}^{N} |a_k| + M \sum_{k=N+1}^{p} |a_k|$$

$$\leq \frac{\epsilon}{2K} \sum_{k=0}^{\infty} |a_k| + M \sum_{k=N+1}^{\infty} |a_k|$$

$$< \frac{\epsilon}{2} + \frac{\epsilon}{2} = \epsilon.$$

Thus $e_p \to 0$ as $p \to \infty$ and hence $C_p \to AB$ as $p \to \infty$. $\square$

6.5.4

Corollary *If two power-series expansions about the origin are*

$$f(z) = \sum_{n=0}^{\infty} a_n z^n \quad \text{and} \quad g(z) = \sum_{n=0}^{\infty} b_n z^n, \quad |z| < R,$$

then the product $f(z)g(z)$ is given by

$$f(z)g(z) = \sum_{n=0}^{\infty} c_n z^n, \quad z \in \{|z| < r\} \cap \{|z| < R\},$$

where

$$c_n = \sum_{k=0}^{n} a_k b_{n-k}, \quad n = 0, 1, \ldots.$$

PROOF. The proof follows from Mertens theorem since

$$\sum_{n=0}^{\infty} \left(\sum_{k=0}^{n} a_k z^k b_{n-k} z^{n-k} \right) = \sum_{n=0}^{\infty} c_n z^n.$$

$\square$

EXAMPLES

1. Show that the Cauchy product of $\sum_{n=0}^{\infty} (-1)^{n+1}/\sqrt{n+1}$ with itself is a divergent series.
2. Show that the Cauchy product of $\sum_{n=0}^{\infty} (-1)^{n+1}/(n+1)$ with itself, is given by

$$2 \sum_{n=1}^{\infty} \frac{(-1)^{n+1}}{n+1} \left(1 + \frac{1}{2} + \cdots + \frac{1}{n} \right).$$

Does this converge? Why?

SOLUTION 1. Let $A = \sum_{n=0}^{\infty} (-1)^{n+1}/\sqrt{n+1}$, which clearly converges conditionally. Then $A^2 = \sum_{n=0}^{\infty} c_n$, where

$$c_n = \sum_{k=0}^{n} \frac{(-1)^{k+1}}{\sqrt{k+1}} \frac{(-1)^{n-k+1}}{\sqrt{n-k+1}} = \sum_{k=1}^{n} \frac{(-1)^n}{\sqrt{k(n-k+2)}}.$$

Now, $0 < k(n-k+2) < (k+n)^2$ if $n \geq 2$; thus $c_n > \sum_{k=1}^{n} (-1)^n/(n+k)$ and

$$\sum_{k=1}^{n} \frac{(-1)^n}{n+k} = (-1)^n \left[\frac{1}{n+1} + \cdots + \frac{1}{2n} \right] \approx (-1)^n [\log_e 2 + \gamma_n],$$

where γ_n tends to a finite constant as $n \to \infty$. (Why?). Thus

$$\sum_{n=0}^{\infty} (-1)^n [\log_e 2 + \gamma_n]$$

diverges, which proves part (1).

2. Let $A = \sum_{n=0}^{\infty} (-1)^{n+1}/(n+1)$, which clearly converges conditionally.

Series of Complex Functions, Taylor's Theorem, Uniform Convergence   93

Then $A^2 = \sum_{n=0}^{\infty} c_n$, where

$$c_n = \sum_{k=0}^{n} \frac{(-1)^{k+1}}{k+1} \frac{(-1)^{n-k+1}}{n-k+1}$$

$$= (-1)^n \sum_{k=0}^{n} \frac{1}{(k+1)(n-k+1)}$$

$$= \frac{(-1)^n}{n+2} \sum_{k=0}^{n} \left\{ \frac{1}{k+1} + \frac{1}{n-k+1} \right\}.$$

We can write

$$\sum_{k=0}^{n} \left\{ \frac{1}{k+1} + \frac{1}{n-k+1} \right\} = 2\left(1 + \frac{1}{2} + \cdots + \frac{1}{n+1}\right),$$

thus

$$\sum_{n=0}^{\infty} c_n = \sum_{n=1}^{\infty} \frac{(-1)^{n-1}}{n+1}\left(1 + \frac{1}{2} + \cdots + \frac{1}{n}\right).$$

By Euler's formula we have for γ:

$$1 + \frac{1}{2} + \cdots + \frac{1}{n} = \log_e n + \gamma_n,$$

where γ_n tends to a finite limit as $n \to \infty$; thus

$$\sum_{n=1}^{\infty} \frac{(-1)^{n-1}}{n+1}\left(1 + \cdots + \frac{1}{n}\right) = \sum_{n=1}^{\infty} \frac{(-1)^n}{n+1}(\log_e n + \gamma_n),$$

which converges. (Why?)

These examples show that the Cauchy product of two conditionally convergent series may not converge. However, if the Cauchy product of two conditionally convergent series converges, the question arises as to whether we can equate these two numbers, i.e., if $\sum a_n = A$, $\sum b_n = B$, and $\sum c_n = C$, where $c_n = \sum_k a_k b_{n-k}$, then does $AB = C$? The answer is yes, and to prove this, we require the following theorem.

6.5.5

Theorem (Abel) Let $f(x) = \sum_{n=0}^{\infty} a_n x^n$ for $-r < x < r$. If the series converges at $x = r$, then $\lim_{x \to r^-} f(x)$ exists and we have

$$\lim_{x \to r^-} f(x) = \sum_{n=0}^{\infty} a_n r^n.$$

PROOF. We can assume without loss of generality that $r = 1$. Thus we have $f(x) = \sum_{n=0}^{\infty} a_n x^n$ for $-1 < x < 1$ and $\sum_{n=0}^{\infty} a_n$ converges.

We write $f(1) = \sum_{n=0}^{\infty} a_n$ and wish to prove that $\lim_{x \to 1^-} f(x) = f(1)$, that is, f is left-continuous at $x = 1$. Multiplying $f(x)$ by the geometric series (about $x = 0$) for $1/(1-x)$, we obtain (by the corollary to Mertens theorem) that

$$\frac{1}{1-x} f(x) = \sum_{n=0}^{\infty} c_n x^n, \quad \text{where} \quad c_n = \sum_{k=0}^{n} a_k.$$

Thus, $f(x) - f(1) = (1-x)\sum_{n=0}^{\infty} [c_n - f(1)] x^n$ if $-1 < x < 1$.

By hypothesis, $\lim_{n \to \infty} c_n = f(1)$. Therefore, given $\epsilon > 0$, there exists N such that $n \geq N$ implies $|c_n - f(1)| < \epsilon/2$. Now we can write

$$f(x) - f(1) = (1-x) \sum_{n=0}^{N-1} [c_n - f(1)] x^n + (1-x) \sum_{n=N}^{\infty} [c_n - f(1)] x^n.$$

Let M be the largest of the N numbers $[c_n - f(1)]$, $n = 0, 1, \ldots, N-1$. If $0 < x < 1$, then we have

$$|f(x) - f(1)| \leq (1-x)NM + (1-x) \frac{\epsilon}{2} \sum_{n=N}^{\infty} x^n$$

$$= (1-x)NM + (1-x) \frac{\epsilon}{2} \frac{x^N}{1-x} < (1-x)NM + \frac{\epsilon}{2}.$$

Now write $\delta = \epsilon/2NM$; then $0 < (1-x) < \delta$ implies that

$$|f(x) - f(1)| < \epsilon,$$

which means that $\lim_{x \to 1^-} f(x) = f(1)$. □

Finally we can prove our previous assertion.

6.5.6

Theorem *Let $\sum_{n=0}^{\infty} a_n$ and $\sum_{n=0}^{\infty} b_n$ be two convergent series and let $\sum_{n=0}^{\infty} c_n$ be their Cauchy product. If $\sum_{n=0}^{\infty} c_n$ converges, then*

$$\sum_{n=0}^{\infty} c_n = \left(\sum_{n=0}^{\infty} a_n \right) \left(\sum_{n=0}^{\infty} b_n \right).$$

PROOF. The two power series $\sum_{n=0}^{\infty} a_n x^n$ and $\sum_{n=0}^{\infty} b_n x^n$ converge for $x = 1$, thus they converge in $|x| < 1$. With $|x| < 1$, write

$$\sum_{n=0}^{\infty} c_n x^n = \left(\sum_{n=0}^{\infty} a_n x^n \right) \left(\sum_{n=0}^{\infty} b_n x^n \right)$$

by the corollary to Mertens theorem. Now let $x \to 1^-$ and apply Abel's theorem. □

Note: The converse of Abel's theorem is false in general, that is, $\lim_{x \to r^-} f(r)$ may exist but the series $\sum_{n=0}^{\infty} a_n r^n$ may fail to converge.

Consider, for example, $f(x) = 1/(1+x)$, $-1 < x < 1$. We then have $a_n = (-1)^n$ and $f(x) \to \frac{1}{2}$ as $x \to 1^-$, but $\sum_{n=0}^{\infty} (-1)^n$ diverges..

Tauber (1897)* showed that if we place further restrictions on a_n, namely, $a_n = o(1/n)$, then a converse to Abel's theorem is possible. Actually, it is possible to replace $o(1/n)$ by the more general $O(1/n)$, however the proof of the case $a_n = O(1/n)$ is reasonably difficult.[†] We will demonstrate the proof for the case $a_n = o(1/n)$ after we study Abel's theorem for $f(z) = \sum_{n=0}^{\infty} a_n z^n$.

The transition to the complex case is not too difficult. We demonstrate a theorem showing that in the complex plane $z = 1$ must be approached along a path with considerable constraints placed upon it.

6.5.7

Theorem *Let $f(z) = \sum_{n=0}^{\infty} a_n z^n$ wherever the series converges. If this includes $z = 1$, then $\lim_{z \to 1^*}$ exists and we have $\lim_{z \to 1^*} f(z) = f(1)$, where $z \to 1^*$ means that $z \to 1$ along any path lying between two chords of the unit circle that pass through $z = 1$.*

PROOF. We prove uniform convergence of the power series in a domain bounded by the two chords through $z = 1$ and a sufficiently small circle centered at $z = 1$.

Let $S_{n,p} = a_n + a_{n+1} + \cdots + a_p$, so that $|S_{n,p}| < \epsilon$ when $n_0 \leq n < p$. Then

$$\sum_{\nu=n}^{m} a_\nu z^\nu = S_{n,n} z^n + (S_{n,n+1} - S_{n,n})z^{n+1} + \cdots + (S_{n,m} - S_{n,m-1})z^m$$

$$= S_{n,n}(z^n - z^{n+1}) + \cdots + S_{n,m-1}(z^{m-1} - z^m) + S_{n,m} z^m.$$

Consequently, for $n \geq n_0$ we have

$$\left| \sum_{\nu=n}^{m} a_\nu z^\nu \right| \leq \epsilon \left\{ \sum_{\nu=n}^{m-1} |z^\nu - z^{\nu-1}| + |z|^m \right\}$$

$$< \epsilon \left\{ |1 - z| \sum_{\nu=0}^{\infty} |z|^\nu + 1 \right\}$$

$$= \epsilon \left\{ \frac{|1 - z|}{1 - |z|} + 1 \right\}.$$

Thus we have uniform convergence provided that $|1 - z|/(1 - |z|)$ is bounded as $z \to 1$ along the path considered, and then $f(z)$ is a continuous function of z (in the previously mentioned domain). We need to restrict the path, since $|1 - z|/(1 - |z|)$ can be made arbitrarily large by taking z close to 1 but even closer to the circumference.

Suppose that $|1 - z| \leq k(1 - |z|)$, where $k > 1$. This inequality is satisfied in a region bounded by the curve $|1 - z| = k(1 - |z|)$.

---

* Tauber, A., "Ein Satz aus der Theorie der unendlichen Reihen," Monatschefte f. Math. u. Phys., Vol. 8, (1897) 273–277.
[†] Titchmarsh, E. C., "Theory of Functions," Oxford University Press (1939) 231.

Write $1 - z = \rho e^{i\varphi}$, then

$$\rho = k - k|1 - \rho e^{i\varphi}|$$

that is,

$$(\rho - k)^2 = k^2(1 - 2\rho \cos \varphi + \rho^2)$$

or

$$\rho = 2 \frac{k^2 \cos \varphi - k}{k^2 - 1}.$$

This equation represents two curves (one using φ and one using $-\varphi$) passing through $z = 1$ and making an angle of arc $\cos(1/k)$ with the real axis. With k large enough, the curve can be made to include any region of the required type. The theorem now follows since $|1 - z|/(1 - |z|) \leq k$ inside the curve. □

The converse of this result, whilst not being true in general, can be demonstrated, provided the coefficients are sufficiently well behaved.

PROOF. A simple lemma is first required. □

Lemma *If $b_n \to 0$ as $n \to \infty$, then*

$$\frac{b_0 + b_1 + \cdots + b_n}{n + 1} \to 0.$$

PROOF. If $|b_n| < k$ for all n and $|b_n| < \epsilon$ for $n > n_0$, then

$$\left| \frac{b_0 + b_1 + \cdots + b_n}{n + 1} \right| \leq \left| \frac{b_0 + \cdots + b_{n_0}}{n + 1} \right| + \left| \frac{b_{n_0+1} + \cdots + b_n}{n + 1} \right|$$

$$\leq \frac{(n_0 + 1)k}{n + 1} + \frac{(n - n_0)\epsilon}{n + 1} < 2\epsilon,$$

when $n > (n_0 + 1)k/\epsilon$. This proves the lemma. □

We have the following theorem.

6.5.8

Theorem *Let $f(z) = \sum_{n=0}^{\infty} a_n z^n$ for $|z| < 1$. If $\lim_{z \to 1^*} f(z) = s$ and $a_n = o(1/n)$ for $n = 1, 2, \ldots$, then $f(1) = s$.*

PROOF. We have to show that as $z \to 1$,

$$\sum_{n=0}^{\infty} a_n z^n - \sum_{n=0}^{N} a_n = 0, \quad \text{where} \quad N = \left[\frac{1}{1 - |z|} \right].$$

Equivalently, we have to show that

$$\sum_{n=N+1}^{\infty} a_n z^n - \sum_{n=0}^{N} a_n(1 - z^n) \to 0.$$

Series of Complex Functions, Taylor's Theorem, Uniform Convergence 97

Let us call these sums S_1 and S_2, respectively. If $|na_n| < \epsilon$ for $n > N$, then

$$|S_1| = \left| \sum_{n=N+1}^{\infty} na_n \cdot \frac{z^n}{n} \right| < \frac{\epsilon}{N+1} \sum_{n=N+1}^{\infty} |z|^n < \frac{\epsilon}{(N+1)(1-|z|)} < \epsilon.$$

Also, $|1 - z^n| = |(1-z)(1 + z + \cdots + z^{n-1})| \leq |1-z|n$. Thus, if $|1-z| \leq k(1-|z|)$, $k > 1$, is satisfied (as in the previous theorem), then

$$|S_2| \leq \sum_{n=0}^{N} |na_n(1-z)| \leq k(1-|z|) \sum_{n=0}^{N} n|a_n| \leq \frac{k}{N} \sum_{n=0}^{N} n|a_n|$$

and $\sum_{n=0}^{N} \frac{n|a_n|}{N} \to 0$ by the lemma. This proves the theorem. □

6.6 Convergence of Series and Analyticity of Functions

In general, it is not possible to use convergence or divergence of the series expansion of a function to determine whether or not it is analytic at certain points. The trouble is that all possible relations can occur. Thus, if

$$f(z) = \sum_{n=1}^{\infty} \frac{(-1)^n z^n}{n} = \log \frac{1}{1+z},$$

the series converges at $z = 1$ and the function is analytic at $z = 1$. The series diverges at $z = -1$ and the function is not analytic at $z = -1$.
If

$$f(z) = \sum_{n=0}^{\infty} (-1)^n z^n = \frac{1}{1+z},$$

the series diverges at $z = 1$ but the function is analytic there.
If

$$f(z) = \sum_{n=1}^{\infty} \frac{z^n}{n^2} = \int_0^z \frac{1}{w} \log \frac{1}{1-w} dw,$$

the series converges at $z = 1$ but $f(z)$ is not analytic at $z = 1$. However, there is one case that can be resolved, and we study it in detail in the following theorem.

6.6.1

Theorem *If $f(z) = \sum_0^\infty a_n z^n$ and $a_n \to 0$ as $n \to \infty$, the series converges at every point of the boundary of the unit circle where the function is analytic.*

PROOF. Without loss of generality, we may take the point in question at $z = 1$. We may suppose further, that $f(1) = 0$. Thus we are required to prove that $\sum_0^\infty a_n$ converges, i.e., that $S_n \to 0$. Since

$$S_n = a_0 + a_1 + \cdots + a_n \quad \text{and} \quad a_k = \frac{1}{2\pi i} \int_{|z|=r<1} \frac{f(z)}{z^{k+1}} dz,$$

then

$$\frac{f(z)}{1-z} = \sum_{n=0}^{\infty} a_n z^n \sum_{n=0}^{\infty} z^n = \sum_{n=0}^{\infty} S_n z^n$$

and therefore
$$S_n = \frac{1}{2\pi i} \int_{|z|=r<1} \frac{f(z)}{(1-z)z^{n+1}} dz.$$
Transforming $z = re^{i\theta}$, $0 < r < 1$, we get
$$S_n = \frac{1}{2\pi r^n} \int_{-\pi}^{\pi} \frac{f(re^{i\theta})}{1 - re^{i\theta}} e^{-in\theta} d\theta.$$
We now choose a function $\varphi(\theta) = \varphi(\theta, \delta, r)$, where $0 < \delta \leq \pi$, such that:
1. $\varphi(\theta) = 1/(1 - re^{i\theta})$ for $\delta < |\theta| < \pi$,
2. $\varphi(\theta)$, $\varphi'(\theta)$, and $\varphi''(\theta)$ are continuous for $-\pi < \theta < \pi$,
3. $|\varphi(\theta)| < K$, $|\varphi'(\theta)| < K$, $|\varphi''(\theta)| < K$ for $-\pi < \theta < \pi$, where $K = K(\theta)$, but K does not depend on r.

For example, if
$$\varphi(\theta) = b_0\theta^5 + b_1\theta^4 + b_2\theta^3 + b_3\theta^2 + b_4\theta + b^5, \text{ where } -\pi - \delta < \theta \leq \delta,$$
then we can determine the six coefficients, so that
$$\varphi(\pm\delta) = \frac{1}{1 - re^{\pm i\delta}}, \qquad \varphi'(\pm\delta) = \frac{ire^{\pm i\delta}}{(1 - re^{\pm i\delta})^2},$$
$$\varphi''(\pm\delta) = \frac{re^{\pm i\delta}(1 + re^{\pm i\delta})}{(1 - re^{\pm i\delta})^3}.$$
Consequently, condition (2) is satisfied and b_i, $1 \leq i \leq 5$, are linear function of $\varphi(\pm\delta)$, $\varphi'(\pm\delta)$, $\varphi''(\pm\delta)$. Also,
$$|\varphi(\pm\delta)|^2 = \frac{1}{|1 + r^2 - 2r\cos\theta|^2} = \frac{1}{(1 + r^2)^2} \frac{1}{\left(1 - \frac{2r}{1 + r^2}\cos\theta\right)^2}$$
and since $1 + r^2 > 2r$ (when $r > 0$),
$$|\varphi(\pm\delta)| \leq \frac{1}{2}\operatorname{cosec}\frac{\delta}{2}, \qquad |\varphi'(\pm\delta)| \leq \frac{1}{4}\operatorname{cosec}^2\frac{\delta}{2},$$
$$|\varphi''(\pm\delta)| \leq \frac{1}{4}\operatorname{cosec}^3\frac{\delta}{2},$$
thus condition (3) is satisfied. We can now write
$$2\pi r^n S_n = \int_{-\delta}^{\delta} \frac{f(re^{i\theta})}{(1 - re^{i\theta})} e^{-in\theta} d\theta + \int_{-\pi}^{\pi} f(re^{i\theta}) \varphi(\theta) e^{-in\theta} d\theta$$
$$- \int_{-\delta}^{\delta} f(re^{i\theta}) \varphi(\theta) e^{-in\theta} d\theta$$
$$= I_1 + I_2 - I_3.$$

To study the bounds of these three integrals, we invoke the analyticity of $f(z)$ at $z = 1$ and the fact that $f(1) = 0$. Clearly the Taylor series for f in the neighborhood of $z = 1$ gives

$$f(z) = f(1) + (z - 1)f'(1) + \cdots$$

and thus

$$f(z) = O\{|1 - z|\}$$

or, more precisely,

$$f(re^{i\theta}) = O\{1 - re^{i\theta}\}$$

for $|\theta| \leq \theta_0$ and uniformly for $r_0 \leq r \leq 1$.

Consequently,

$$I_1 = \int_{-\delta}^{\delta} O(1)\, d\theta = O(\delta).$$

Now fix δ; then we have

$$I_2 = \int_{-\pi}^{\pi} \sum_{m=0}^{\infty} a_m r^m e^{i(m-n)\theta} \varphi(\theta)\, d\theta.$$

Since the series converges uniformly for $0 < r < 1$, we have

$$I_2 = \sum_{m=0}^{\infty} a_m r^m \int_{-\pi}^{\pi} e^{i(m-n)\theta} \varphi(\theta)\, d\theta$$

$$= a_n r^n \int_{-\pi}^{\pi} \varphi(\theta)\, d\theta + \sum_{m \neq n} a_m r^m \int_{-\pi}^{\pi} e^{i(m-n)\theta} \varphi(\theta)\, d\theta.$$

Now by integrating the second integral by parts twice, we get

$$I_2 = a_n r^n \int_{-\pi}^{\pi} \varphi(\theta)\, d\theta - \sum_{m \neq n} \frac{a_m r^m}{(m-n)^2} \int_{-\pi}^{\pi} e^{i(m-n)\theta} \varphi''(\theta)\, d\theta.$$

Let $\epsilon_\nu = \max_{m \geq \nu}(|a_m|)$, then $\epsilon_\nu \to 0$ as $\nu \to \infty$ and

$$|I_2| \leq 2\pi K \left\{ \epsilon_n + \epsilon_0 \sum_{m \leq n/2} \frac{1}{(m-n)^2} + \epsilon_{n/2} \sum_{m > n/2} \frac{1}{(m-n)^2} \right\}$$

$$= O\{\epsilon_n\} + O\left\{\frac{1}{n}\right\} + O\{\epsilon_{n/2}\}.$$

Finally, consider I_3 and integrate by parts:

$$I_3 = \left[\frac{f\varphi e^{-in\theta}}{-in} \right]_{-\delta}^{\delta} + \frac{1}{in} \int_{-\delta}^{\delta} (f'\varphi + f\varphi') e^{-in\theta}\, d\theta = O\left\{\frac{1}{n}\right\},$$

since by condition (3) function φ is bounded and for fixed r, $-\delta < \theta < \delta$, the functions f and f' are bounded.

Thus, given ϵ, we can choose δ so that $|I_1| < \epsilon/3$ for all n, and with δ now fixed, we choose $n_0 = n_0(\epsilon)$ so large that $|I_2| < \epsilon/3$ and $|I_3| < \epsilon/3$ for

$n > n_0$. Hence
$$2\pi r^n |S_n| < \epsilon \text{ for } n > n_0.$$
The right-hand side is clearly independent of r, therefore
$$2\pi |S_n| < \epsilon \text{ as } r \to 1.$$
and thus
$$S_n \to 0, \quad n > n_0. \qquad \square$$

Note*:

1. We need to get a bound for S_n independent of r; thus we construct a function φ such that φ, φ', and φ'' are bounded by a constant that does not depend on r.
2. $\varphi''(\theta)$ need not be continuous for the proof to carry through, however we clearly require $|\varphi''| < K(\delta)$. Thus we choose φ a polynomial in θ of degree five in order to have six equations to determine $\varphi(\pm\delta)$, $\varphi'(\pm\delta)$, and $\varphi''(\pm\delta)$.
3. The hypothesis that $f(z) \in A$ at $z = 1$, could be replaced by $f(z) = O\{|1 - z|^\alpha\}$ for $\alpha > 0$.

6.7

We now study several further properties of uniformly convergent series of analytic functions and deduce some results that will be of use in later analysis.

6.7.1

Theorem *If (i) $\sum_{n=1}^{\infty} u_n(z)$ converges to $S(z)$, $\forall z \in D$ (simply connected), (ii) $\sum_{n=1}^{\infty} u_n(z)$ converges uniformly in every closed region $R \subset D$, and (iii) $u_n \in A$ in D, for $n = 1, 2, \ldots$, then $S(z) \in A$ in D and*
$$S^{(r)}(z) = \sum_{n=1}^{\infty} u_n^{(r)}(z), \quad \forall z \in D.$$

PROOF. Take a point $\xi \in D$. Then there exists $\eta > 0$ such that $z \in D$ whenever $|z - \xi| \leq \eta$. By the second assumption, $\sum_{n=1}^{\infty} u_n(z)$ is uniformly convergent for $|z - \xi| \leq \eta$. Let C be a crJc contained in the region $E = \{z | |z - \xi| \leq \eta\}$. Then $\sum_{n=1}^{\infty} u_n(z)$ is a uniformly convergent series of continuous functions on C; whence $S(z)$ is continuous and
$$\int_C S(z)\,dz = \sum_{n=1}^{\infty} \int_C u_n(z)\,dz = 0,$$
since $u_n(z) \in A$ for $n = 1, 2, \ldots$. Thus by Morera's theorem, $S(z) \in A$ in

---

* For further discussion see W. H. Young, *On Restricted Fourier Series and the Convergence of Power Series*. Proc. London Math. Soc., 2, 12 pp. 71–88, 1913.

E, that is, $S(z) \in A$ at ξ. Now ξ was any point in D, hence $S(z) \in A$ in D. Take C to be the circle $|z - \xi| = \eta/2$, then

$$\frac{S(z)}{(z - \xi)^{r+1}} \quad \text{and} \quad \frac{u_n(z)}{(z - \xi)^{r+1}}, \quad \text{where } r = 1, 2, \ldots$$

are continuous on C and $\sum_{n=1}^{\infty} u_n(z)/(z - \xi)^{r+1}$ converges uniformly to $S(z)/(z - \xi)^{r+1}$ on C, where $r = 1, 2, \ldots$. Further, $S(z)$ and $u_n(z) \in A$ within (and on) C; thus from Cauchy's integral formula we get

$$S^{(r)}(\xi) = \frac{r!}{2\pi i} \int_C \frac{S(z)}{(z - \xi)^{r+1}} dz = \sum_{n=1}^{\infty} \frac{r!}{2\pi i} \int_C \frac{u_n(z)}{(z - \xi)^{r+1}} dz = \sum_{n=1}^{\infty} u_n^{(r)}(z).$$

Since $\xi \in D$ is arbitrary, then

$$S^{(r)}(z) = \sum_{n=1}^{\infty} u_n^{(r)}(z), \quad \forall z \in D. \qquad \square$$

Note that since we have assumed $u_n(z) \in A$, we have in effect assumed a lot more than mere differentiability, and the hypothesis required for the theorem in the real-variable case, (namely, that the derived series must converge uniformly) does not arise here.

The next theorem follows naturally.

6.7.2

Theorem *If $\sum_{n=0}^{\infty} a_n z^n$ has radius of convergence $R > 0$ and*

$$f(z) = \sum_{n=0}^{\infty} a_n z^n$$

in $|z| < R$, then $f(z) \in A$ in $|z| < R$ and

$$f^{(r)}(z) = \sum_{n=r}^{\infty} n(n-1) \ldots (n - r + 1) a_n z^{n-r}$$

for $|z| < R$, where $r = 1, 2 \ldots$.

PROOF. $\sum_{n=0}^{\infty} a_n z^n$ is a uniformly convergent series of analytic functions in every closed set interior to $|z| < R$; thus by the previous theorem

$$f(z) = \sum_{n=0}^{\infty} a_n z^n \in A \quad \text{and} \quad f^{(r)}(z) = \sum_{n=r}^{\infty} n(n-1) \cdots (n - r + 1) a_n z^{n-r}.$$

$\square$

6.7.3

Theorem *If (i) $f(z) \in A$ in $|z| < R$, so that $f(z) = \sum_{n=0}^{\infty} a_n z^n$ for $|z| < R$, where*

$$a_n = \frac{f^{(n)}(0)}{n!}, \quad \text{and} \quad (n = 0, 1, \ldots),$$

and (ii) $0 \leq r < R$, then
$$\int_0^{2\pi} |f(re^{i\theta})|^2 \, d\theta = \sum_{n=0}^{\infty} |a_n|^2 r^{2n}.$$

PROOF

$$\frac{1}{2\pi} \int_0^{2\pi} |f(re^{i\theta})|^2 \, d\theta = \frac{1}{2\pi} \int_0^{2\pi} f(re^{i\theta}) \left\{ \sum_{n=0}^{\infty} \bar{a}_n r^n e^{-in\theta} \right\} d\theta$$

$$= \sum_{n=0}^{\infty} \frac{1}{2\pi} \bar{a}_n r^n \int_0^{2\pi} f(re^{i\theta}) e^{-in\theta} \, d\theta$$

$$= \sum_{n=0}^{\infty} \bar{a}_n r^{2n} \frac{1}{2\pi i} \int_0^{2\pi} \frac{f(re^{i\theta})}{(re^{i\theta})^{n+1}} \, d(re^{i\theta})$$

$$= \sum_{n=0}^{\infty} \bar{a}_n r^{2n} a_n = \sum_{n=0}^{\infty} |a_n|^2 r^{2n}.$$

The interchange of $\sum$ and $\int$ is justified since $\sum_{n=0}^{\infty} \bar{a}_n r^n e^{-in\theta}$ is uniformly convergent $(0 \leq \theta \leq 2\pi)$. □

This theorem can be used, for example, to prove Cauchy's inequality which we restate as follows:

If (i) $f(z) \in A$ in $|z| < R$, so that $f(z) = \sum_{n=0}^{\infty} a_n z^n$ for $|z| < R$, where
$$a_n = \frac{f^{(n)}(0)}{n!} \quad \text{and} \quad n = 0, 1, \ldots,$$
and (ii) $0 \leq r < R$, then $|a_n r^n| \leq M(r)$, where $n = 0, 1, \ldots$ and $M(r) = \max_{|z|=r} |f(z)|$. If $0 < r < R$, then we can say $|a_n| \leq M(r)/r^n$. Further, if $|a_n| r^n = M(r)$ for some n, then $f(z) = a_n z^n$.

PROOF. From 6.7.3 and with $0 \leq r < R$, we have
$$\sum_{n=0}^{\infty} |a_n|^2 r^{2n} = \frac{1}{2\pi} \int_0^{2\pi} |f(re^{i\theta})|^2 \, d\theta \leq \{M(r)\}^2.$$

Thus
$$|a_n|^2 r^{2n} \leq \{M(r)\}^2, \quad \text{or} \quad |a_n| r^n \leq M(r) \quad \text{for} \quad n = 0, 1, \ldots.$$
Further, if $|a_n| r^n = M(r)$, then $a_\nu = 0$ for $\nu \neq n$, and $f(z) = a_n z^n$. □

We now examine a theorem similar to Cauchy's inequality but involving the upper bound of $R\{f(z)\}$ on $|z| = r$ rather than $M(r)$. Let $A(r)$ be the upper bound of the real part of $f(z)$ on $|z| = r$.

6.7.4

Theorem If $f(z) = \sum_{n=0}^{\infty} a_n z^n$, then for all $n > 0$ and r,
$$|a_n| r^n \leq \max\{4A(r), 0\} - 2R\{f(0)\}.$$

Series of Complex Functions, Taylor's Theorem, Uniform Convergence

PROOF. Let $z = re^{i\theta}$ and $f(z) = \sum_{n=0}^{\infty} a_n z^n = U(r, \theta) + iV(r, \theta)$; also let $a_n = \alpha_n + i\beta_n$. Then

$$U(r, \theta) = \sum_{n=0}^{\infty} (\alpha_n \cos n\theta - \beta_n \sin n\theta) r^n.$$

The series converges uniformly with respect to θ. Thus we may multiply by $\cos n\theta$ or $\sin n\theta$ and integrate term by term. Thus

$$\int_0^{2\pi} U(r, \theta) \cos n\theta \, d\theta = \int_0^{2\pi} \alpha_n r^n \cos^2 n\theta \, d\theta = \pi \alpha_n r^n.$$

Similarly,

$$\int_0^{2\pi} U(r, \theta) \sin n\theta \, d\theta = -\pi \beta_n r^n, \quad n > 0;$$

also,

$$\int_0^{2\pi} U(r, \theta) \, d\theta = 2\pi \alpha_0.$$

Hence

$$a_n r^n = (\alpha_n + i\beta_n) r^n = \frac{1}{\pi} \int_0^{2\pi} U(r, \theta) e^{-in\theta} \, d\theta, \quad n > 0$$

and

$$|a_n| r^n \leq \frac{1}{\pi} \int_0^{2\pi} |U(r, \theta)| \, d\theta.$$

Thus,

$$|a_n| r^n + 2\alpha_0 \leq \frac{1}{\pi} \int_0^{2\pi} \{|U(r, \theta)| + U(r, \theta)\} \, d\theta.$$

Note that $|u| + u = 0$ for $u < 0$. Hence if $A(r) < 0$, the right-hand side of the latter expression becomes zero, and if $A(r) \geq 0$, the right-hand side does not exceed

$$\frac{1}{\pi} \int_0^{2\pi} 2A(r) \, d\theta = 4A(r).$$

Thus

$$|a_n| r^n \leq \max\{4A(r), 0\} - 2R\{f(0)\}. \qquad \square$$

This may be improved as follows. Write $a_n = \alpha_n + i\beta_n = |a_n| e^{i\theta_n}$, then

$$|a_n| r^n = (\alpha_n + i\beta_n) r^n e^{-i\theta_n}$$

$$= \frac{1}{\pi} \int_0^{2\pi} U(r, \theta) e^{-in\theta} e^{-i\theta_n} \, d\theta$$

$$= \frac{1}{\pi} \int_0^{2\pi} U(r, \theta) \cos(n\theta + \theta_n) \, d\theta, \quad |a_n| r^n \text{ real}.$$

Thus

$$|a_n|r^n + 2\alpha_0 = \frac{1}{\pi} \int_0^{2\pi} U(r,\theta)\{1 + \cos(n\theta + \theta_n)\}\, d\theta$$

$$\leqslant \frac{1}{\pi} \int_0^{2\pi} A(r)\{1 + \cos(n\theta + \theta_n)\}\, d\theta$$

$$= \frac{A(r)}{\pi} \int_0^{2\pi} \{1 + \cos(n\theta + \theta_n)\}\, d\theta$$

$$= 2A(r)$$

and

$$|a_n|r^n \leqslant 2[A(r) - R\{f(0)\}].$$

6.7.5
Exercises

1: If $|z| < 1$, prove that

$$\tan^{-1} z = z - \frac{z^3}{3} + \frac{z^5}{5} - \cdots, \text{ (Gregory's series)}.$$

Take the principal value of $\tan^{-1} z$.

2: Show that the series

$$\text{(i) } 1 - \frac{z^2}{2!} + \frac{z^4}{4!} - \cdots$$

$$\text{(ii) } z - \frac{z^3}{3!} + \frac{z^5}{5!} - \cdots$$

are absolutely convergent for all values of z.

3: Show that the series

$$1 + \frac{z}{1+z} + \left(\frac{z}{1+z}\right)^2 + \cdots$$

is convergent for $R(z) > -\frac{1}{2}$ and find its sum.

4: Show that the series

$$1 + \frac{z}{1!} + \frac{2m+2}{2!} z^2 + \cdots + \frac{(nm+2)(nm+3)\cdots(nm+n)}{n!} z^n + \cdots,$$

is absolutely convergent if

$$|z| < m^m/(m+1)^{m+1}, \text{ where } m \text{ a positive integer.}$$

Series of Complex Functions, Taylor's Theorem, Uniform Convergence

5: Show that e^{-1}/b is the radius of convergence of the series

$$1 + az + \frac{a(a-2b)}{2!}z^2 + a\frac{(a-3b)^2}{3!}z^3 + \cdots.$$

6: Show that the series

$$\frac{z}{1-z^2} + \frac{z^2}{1-z^4} + \frac{z^4}{1-z^8} + \cdots.$$

is convergent for $|z| < 1$ and also if $|z| > 1$. Find the respective sums.

7: Show that the series

$$\sum_{n=1}^{\infty} \frac{z^{n+1}}{n(n+1)}$$

is absolutely convergent at all points on its circle of convergence.

8: Show that the series

$$1 - z + \frac{z(z-1)}{2!} - \frac{z(z-1)(z-2)}{3!} + \cdots$$

is convergent for $R(z) > 0$ and divergent for $R(z) < 0$.

9: Show that the series $\sum_{n=0}^{\infty} ze^{-nz}$ converges absolutely but not uniformly in the sector $|z| \leqslant R$, $|\arg z| \leqslant \delta$, where $0 < \delta < \pi/2$. Is the convergence uniform when $r \leqslant |z| \leqslant R$, $|\arg z| \leqslant \delta$?

10: Show that if $\sum_{n=1}^{\infty} a_n$ is convergent, the series $\sum_{n=1}^{\infty} a_n n^{-z}$ converges when $R(z) > 0$ but is not necessarily convergent when $R(z) = 0$. Prove that the convergence is uniform in any bounded closed region for which $R(z) \geqslant \delta > 0$.

11: Show that each of the following functions is an entire function of z:

(i) $\sum_{n=1}^{\infty} \frac{z^n}{(n!)^{1/2}}$ (ii) $\sum_{n=1}^{\infty} \frac{1}{2^n} \frac{1}{n^z}$ (iii) $\sum_{n=1}^{\infty} \frac{\sin nz}{n!}$

12: Show that if $a > 0$, each of the following functions represents an analytic function in $R(z) > 0$:

$$\sum_{n=1}^{\infty} e^{-n^2 az} \qquad \sum_{n=1}^{\infty} \frac{e^{-anz}}{(a+n)^z}$$

13: We define the Riemann zeta-function by

$$\zeta(z) = \sum_{n=1}^{\infty} \frac{1}{n^z}, \quad R(z) > 1.$$

Show that $\zeta(z)$ is analytic for $R(z) > 1$.

14: Suppose that sequences $s_n(z) \to s(z)$ and $t_n(z) \to t(z)$ uniformly in a domain D and suppose that $s(z)$ and $t(z)$ are bounded in D. Prove that

$$s_n(z)t_n(z) \to s(z)t(z).$$

uniformly in D.

[Hint: Study $s_n(t_n - t) + t(s_n - s)$.]

7
Maximum Modulus

7.1 Maximum Modulus

We now study the maximum modulus $M(r) = \max|f(z)|$ of a function $f(z)$ on the disk $|z| \leq r$. The *maximum principle* illustrated by the *maximum-modulus theorem*, a theorem central to all analytic-function theory, demonstrates the remarkable property of these functions, namely, that if a function $f(z)$ is analytic in an open connected set K and is not constant in K, then $f(z)$ has no maximum value in K. Further, if $f(z)$ is continuous on a closed connected set (region) D and analytic in the interior of D, then $f(z)$ attains its maximum value on the boundary of D.

The problem of finding the exact position or positions of these maxima can offer considerable difficulty.

We prove the maximum-modulus theorem for a function analytic in and continuous on the boundary of a domain D.

7.1.1
Lemma Let $f(z) \in A$ for $|z - a| < R$, $R > 0$, and
$$M(r) = \max_{|z-a|=r} |f(z)|, \quad 0 \leq r < R.$$
Then $|f(a)| \leq M(r)$, $0 \leq r < R$, and $|f(z)| = M(r)$, for $0 < r < R$, if and only if $f(z) = \text{const} = M(r)e^{i\alpha}$, α real and $|z - a| < R$.

PROOF. Since we may write $f(z) = \sum_0^\infty a_n(z - a)^n$, $|z - a| < R$, we have
$$\sum_{n=0}^\infty |a_n|^2 r^{2n} = \frac{1}{2\pi} \int_0^{2\pi} |f(a + re^{i\theta})|^2 \, d\theta, \quad 0 \leq r < R,$$

valid at $r = 0$, since $|a_0|^2 = |f(a)|^2$. Thus

$$\sum_0^\infty |a_n|^2 r^{2n} \leqslant \{M(r)\}^2 \text{ and } |a_0|^2 \leqslant \{M(r)\}^2, \text{ or } |a_0| \leqslant M(r).$$

Also, since $|a_0|^2 + |a_1|^2 r^2 + \cdots + |a_n|^2 r^{2n} \leqslant \{M(r)\}^2$, we get $|a_0| < M(r)$, $0 < r < R$, unless $a_n = 0$; where $n = 1, 2, \ldots$. Whence $f(z) = \text{const} = M(r)$, $0 < r < R$.

Thus, either $f(z)$ is constant for $|z - a| < R$ or $|f(z)|$ takes values greater than $|f(a)|$ in every neighborhood of $z = a$. Alternatively, if $f(z)$ is analytic and not constant in a domain D, then $|f(z)|$ cannot take a maximum value at any point $a \in D$. □

7.1.2

The Maximum-Modulus Theorem *If $f(z) \in A$ in a bounded domain D and continuous on $\overline{D}$, and $M = \max_{z \in C} |f(z)|$, where C is the boundary of D, then*

$$|f(z)| \leqslant M \text{ in } D.$$

Further, $|f(z)| < M$ in D except in the case where $f(z)$ is a constant.

PROOF. Clearly, if $f(z) = K$ (const), $\forall z$, $z \in D$, then since f is continuous on $\overline{D}$, $f(z) = K$, $\forall z$, $z \in \overline{D}$, and hence $f(z) = K$, $\forall z$, $z \in C$, so that $|K| = M$ and $|f(z)| = M$, $\forall z, z \in \overline{D}$.

Assume that $f(z)$ is not constant in D. Write $K = \max_{z \in \overline{D}} |f(z)|$. Since $\overline{D}$ is compact, K is finite and $|f(z)|$ attains the value K at least once in $\overline{D}$. By the previous lemma, $|f(z)| \neq K$, $\forall z = a \in D$ and $|f(z)| > |f(a)|$ in every neighborhood of a. Thus the max K of $|f(z)|$ on $\overline{D}$ is attained at some $z \in C$. Accordingly, $K = M$ and $|f(z)| < M$ in D. □

7.2 The Minimum Principle

If $f(z) \in A$ in D and is never zero there, then $[f(z)]^{-1} \in A$ in D. Since the minimum of $|f(z)|$, that is, $m(r)$, is attained at the same points as the maximum of $|f(z)|^{-1}$, it follows from the maximum principle that $|f(z)|$ cannot attain its minimum in the interior of D. Further, since the inability of a function to attain its maximum at an interior point is a local property, the maximum principle is valid for analytic functions not single-valued in a multiply connected domain.

Thus we have the minimum principle: if $f \neq 0$ in D, then $|f(z)|$ attains its minimum value on the boundary of a domain D, and if the minimum value is found inside D, then $f(z)$ is necessarily a constant.

We cannot expect $m(r)$ to behave as simply as $M(r)$, since it vanishes whenever r is the modulus of a zero of $f(z)$. Except in the immediate neighborhood of these exceptional points, a lower limit can be set for $m(r)$. Generally $m(r) \to 0$ in somewhat the same way as $1/M(r)$.

We shall not pursue the subject further since theorems concerning the connection between $M(r)$ and $m(r)$ are of considerable difficulty.*

EXAMPLE. Given $f(z) \in A$ in $|z| < 1$ and $|f(z)| \leq 1/(1 - |z|)$, find a good estimate for $|f^{(n)}(0)|$.

SOLUTION. $|f^{(n)}(0)| \leq M(r) n!/r^n$ where $M(r) = \max_{|z|=r} |f(z)|$.
Our task is to minimize $M(r) r^{-n}$. Since

$$\frac{M(r)}{r^n} \leq \frac{1}{(1-r)r^n},$$

we maximize $(1 - r)r^n$. By differentiation, $\max\{r^n - r^{n+1}\}$ occurs for $r = n/(n + 1)$, thus

$$\min\left\{ \frac{1}{r^n(1-r)} \right\} = \frac{(1+n)^n (n+1)}{n^n} = \frac{(1+n)^{n+1}}{n^n}.$$

Consequently,

$$|f^{(n)}(0)| \leq n! \frac{(n+1)^{n+1}}{n^n} < e(n+1)! \, .$$

We now study some applications of the maximum principle. For functions $f(z) \in A$ in an open disk $|z| < R$ and continuous on the closed disk $|z| \leq R$, we have that $|f| \leq M$ on $|z| = R$ implies that $|f| \leq M$ on $|z| \leq R$ and that equality holds only for constant functions. Consequently, if we know that $|f| < M$, we may expect that under suitable conditions we could improve this inequality. The following theorem, known as Schwarz's lemma, moves a little in this direction.

7.3

Theorem (Schwarz's lemma) *If $f(z) \in A$ for $|z| < R$ and $|f(z)| \leq M$ and $f(0) = 0$, then $|f(z)| \leq |z|M/R$, $0 \leq |z| < R$. Further, $|f'(0)| \leq M/R$. Equality holds for $f(z) = zMe^{i\gamma}/R$.*

PROOF. Let $g(z) = f(z)/z$ for $0 < |z| < R$ and

$$g(0) = f'(0).$$

On $|z| = r < R$, $|g(z)| \leq M/r$. By the maximum principle, since $g(z) \in A$ for $0 \leq |z| \leq r$, then $|g(z)| \leq M/r$ for $|z| \leq r$. We now let $r \to R$ and hence we have the result that $|f(z)| \leq |z|M/R$ for $0 \leq |z| < R$.

---

*See for example A.S.B. Holland, *Introduction to the Theory of Entire Functions*. New York: Academic Press, 1973.

Clearly,
$$|f'(0)| = |g(0)| \leq M/R.$$

Also, if equality holds at some point, then $|g(z)|$ attains its maximum and thus $g(z)$ must reduce to a constant; hence, $f(z) = zMe^{i\gamma}/R$, γ real. $\square$

Before proceeding to applications of Schwarz's lemma, it will be convenient to study the problem of finding a function $f(z)$ when its boundary values are given. In particular, in the case of a closed disk $|z| \leq R$ we have the following formula.

7.4 Poisson's Integral Formula

Theorem Let $f(z) \in A$ in a region D including $|z| \leq R$ and let $U(r, \theta)$ be $R\{f(z)\}$. Then for $0 \leq r < R$ we have

$$U(r, \theta) = \frac{1}{2\pi} \int_0^{2\pi} \frac{R^2 - r^2}{R^2 - 2Rr\cos(\theta - \varphi) + r^2} U(R, \varphi) \, d\varphi.$$

Similarly for $V(r, \theta) = I\{f(z)\}$.

PROOF. Let $f(z) = \sum_{n=0}^{\infty}(\alpha_n + i\beta_n)r^n e^{in\theta}$, $r \leq R$. We have

$$U(r, \theta) = \sum_{n=0}^{\infty}(\alpha_n \cos n\theta - \beta_n \sin n\theta)r^n,$$

where

$$\alpha_n R^n = \frac{1}{\pi} \int_0^{2\pi} U(R, \varphi)\cos n\varphi \, d\varphi, \quad n > 0,$$

$$\beta_n R^n = -\frac{1}{\pi} \int_0^{2\pi} U(R, \varphi)\sin n\varphi \, d\varphi, \quad n > 0,$$

Thus,
$$\alpha_0 = \frac{1}{2\pi} \int_0^{2\pi} U(R, \varphi) \, d\varphi.$$

$$U(r, \theta) = \frac{1}{2\pi} \int_0^{2\pi} U(R, \varphi) \, d\varphi$$
$$+ \frac{1}{\pi} \sum_{n=1}^{\infty} \frac{r^n}{R^n} \int_0^{2\pi} U(R, \varphi)\{\cos n\theta \cos n\varphi + \sin n\theta \sin n\varphi\} \, d\varphi$$
$$= \frac{1}{\pi} \int_0^{2\pi} U(R, \varphi)\left\{\frac{1}{2} + \sum_{n=1}^{\infty} \cos n(\theta - \varphi)\left(\frac{r}{R}\right)^n\right\} d\varphi. \quad \square$$

The inversion of the summation and integral processes is justified by uniform convergence. To sum the series $\sum_{n=1}^{\infty} \cos n(\theta - \varphi)(r/R)^n$, use

$$C = \sum_{n=1}^{\infty} \cos n\alpha \cdot x^n \quad \text{and} \quad iS = i \sum_{n=1}^{\infty} \sin n\alpha \cdot x^n.$$

Thus
$$C = \frac{x\cos\alpha - x^2}{1 - 2x\cos\alpha + x^2}$$
and adding $\frac{1}{2}$ we obtain Poisson's formula. Thus we have for a fixed $z_0 = r_0 e^{i\theta_0}$ interior to $\Gamma:[z], |z| \leq R]$:

$$f(z_0) = \frac{1}{2\pi} \int_0^{2\pi} \frac{R^2 - r_0^2}{R^2 - 2Rr_0\cos(\theta - \theta_0) + r_0^2} f(Re^{i\theta})\, d\theta;$$

for $z_0 = 0$ we have

$$f(0) = \frac{1}{2\pi} \int_0^{2\pi} f(Re^{i\theta})\, d\theta.$$

This basically says that *the value of an analytic function at the center of a circle equals the arithmetic mean of the values on the circumference.*

7.5 Theorem of Borel and Carathéodory

The following result enables us to deduce an upper bound for the modulus of a function on $|z| = r$ from bounds for its real or imaginary parts on a larger concentric circle $|z| = R$.

7.5.1

Theorem *Let $f(z) \in A$ for $|z| \leq R$ and let $M(r)$ and $A(r)$ denote the $\max |f(z)|$ and $\max R\{f(z)\}$, respectively, on $|z| = r$. Then for $0 < r < R$ we have*

$$M(r) \leq \frac{2r}{R-r} A(R) + \frac{R+r}{R-r} |f(0)| < \frac{R+r}{R-r} \{A(R) + |f(0)|\}.$$

PROOF. The result is clearly true for $f(z) = \text{const}$. For $f(z)$ nonconstant, suppose $f(0) = 0$. Then $A(R) > A(0) = 0$. Since by Poisson's integral formula

$$U(0) = \frac{1}{2\pi} \int_0^{2\pi} U(R, \varphi)\, d\varphi,$$

and if $A(R) = \max_\varphi U(R, \varphi)$, then $U(R, \varphi) \leq A(R), \forall \varphi$, and

$$U(0) = \frac{1}{2\pi} \int_0^{2\pi} U(R, \varphi)\, d\varphi \leq \frac{1}{2\pi} \int_0^{2\pi} A(R)\, d\varphi = A(R).$$

Thus $U(0) = A(R)$ implies $U(R, \varphi) = A(R), \forall \varphi$ and $A(0) \leq U(0) \leq A(R)$, unless the function is a constant.

Let

$$\varphi(z) = \frac{f(z)}{2A(R) - f(z)}.$$

Then $\varphi(z) \in A$ for $|z| \leq R$ since the real part of the denominator does not vanish. Now $\varphi(0) = 0$ and if $f(z) = u + iv$, then

$$|\varphi(z)|^2 = \frac{u^2 + v^2}{\{2A(R) - u\}^2 + v^2} \leq 1$$

since $-2A(R) + u \leq u \leq 2A(R) - u$.

Also, Schwarz's lemma gives $|\varphi(z)| \leq r/R$, since $|\varphi(z)|$ is bounded by 1. Hence

$$|f(z)| = \frac{|2A(R)\varphi(z)|}{|1 + \varphi(z)|} \leq \frac{2A(R) \cdot r}{R - r}$$

(use $|R + R\varphi| \geq R - R|\varphi|$). Thus we obtain the result for $f(0) = 0$.
If $f(0) \neq 0$, apply the result to $f(z) - f(0)$; then

$$|f(z) - f(0)| \leq \frac{2r}{R - r} \cdot \max_{|z|=R} \{R|f(z) - f(0)|\}$$

$$\leq \frac{2r}{R - r} \{A(R) + |f(0)|\}$$

since $f(0)$ may be negative. Adding $|f(0)|$ to both sides, we have that

$$|f(z)| \leq \frac{2r}{R - r} A(R) + \frac{R + r}{R - r} |f(0)|. \qquad \square$$

We also show that

$$\max_{|z|=r} |f^{(n)}(z)| \leq \frac{2^{n+2} n! R}{(R - r)^{n+1}} \{A(R) + |f(0)|\},$$

since

$$f^{(n)}(z) = \frac{n!}{2\pi i} \int_C \frac{f(w) - f(0)}{(w - z)^{n+1}} \, dw \qquad (7.1)$$

where C is the circle

$$|w - z| = \delta = \tfrac{1}{2}(R - r)$$

and

$$|w| \leq r + \tfrac{1}{2}(R - r) = \tfrac{1}{2}(R + r),$$

which ensures that C is inside $|z| = R$. The previous theorem gives

$$\max |f(z) - f(0)| \leq \frac{R + \tfrac{1}{2}(R + r)}{R - \tfrac{1}{2}(R + r)} \{A(R) + |f(0)|\},$$

Maximum Modulus

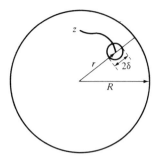

and since $r < R$, we have

$$\max|f(z) - f(0)| < \frac{4R}{R-r}\{A(R) + |f(0)|\}.$$

Thus from Equation (7.1) we get

$$|f^{(n)}(z)| \leq \frac{n!\,4R}{\delta^n(R-r)}\{A(R) + |f(0)|\} = \frac{2^{n+2}n!\,R}{(R-r)^{n+1}}\{A(R) + |f(0)|\}.$$

7.5.2
Exercises

1: Let $U(r,\theta)$ be continuous for $r \leq R$ and harmonic for $r < R$. Show by Poisson's formula that

$$U(0, R) = \frac{1}{2\pi}\int_0^{2\pi} U(R,\theta)\,d\theta.$$

2: If $U(r,\theta)$ of Exercise 1 satisfies $U(R,\phi) \equiv U(\phi) \geq 0$ and if $U(r,\theta) \not\equiv 0$, show that

$$\frac{R-r}{R+r} \leq \frac{U(r,\theta)}{U(0,\theta)} \leq \frac{R+r}{R-r}, \quad 0 \leq r < R,$$

This is called Harnack's inequality.

[Hint: Note that

$$\frac{R^2 - r^2}{R^2 - 2Rr\cos(\theta - \phi) + r^2} U(\phi)$$

lies between

$$\frac{R^2 - r^2}{(R+r)^2} U(\phi) \quad \text{and} \quad \frac{R^2 - r^2}{(R-r)^2} U(\phi).$$

3: By assuming $z = re^{i\theta}$ in the power series for $1/(1-z)$, deduce that

$$\frac{1-r^2}{1-2r\cos\theta + r^2} = 1 + r\cos\theta + r^2\cos 2\theta + \ldots, \qquad 0 \leq r < 1.$$

Use this result to show that

$$\int_0^{2\pi} \frac{1-r^2}{1-2r\cos\theta + r^2} \cos n\theta \, d\theta = r^n, \quad n = 1, 2, 3, \ldots.$$

4: Formulate Exercise 3 for e^z instead of $1/(1-z)$.

Let us reconsider the remark made in 7.2, that the inability of a function to attain its maximum at a point interior to its domain of definition is a *local* property and that the maximum principle is valid for analytic functions not single-valued in a multiply-connected domain. We may now study an important application of this statement.

7.6

Theorem (Hadamard's Three-Circle) *If $f(z) \in A$ and is single-valued in $\rho < |z| < R$ and continuous on $|z| = \rho$, $|z| = R$ and if $M(r)$ denotes the maximum of $|f(z)|$ on $|z| = r$, $\rho < r < R$, then $\log M(r)$ is a convex function of $\log r$, that is, for $\rho < r_1 < r_2 < r_3 < R$, we have*

$$\log M(r_2) \leq \log M(r_1) \frac{\log r_3 - \log r_2}{\log r_3 - \log r_1} + \log M(r_3) \frac{\log r_2 - \log r_1}{\log r_3 - \log r_1}.$$

The above equation is equivalent to:

$$M(r_2)^{\log(r_3/r_1)} \leq M(r_1)^{\log(r_3/r_2)} M(r_3)^{\log(r_2/r_1)}.$$

We recall that a function is convex (downward) in the following sense: $y = \varphi(x)$ is convex if the curve $\varphi(x)$ between x_1 and x_2 is always below the chord joining $(x_1, \varphi(x_1))$ and $(x_2, \varphi(x_2))$, that is,

$$\varphi(x) < \frac{x_2 - x}{x_2 - x_1} \varphi(x_1) + \frac{x - x_1}{x_2 - x_1} \varphi(x_2).$$

PROOF. Let $\varphi(z) = z^\lambda f(z)$, where λ is to be a *determined constant* (real). Since the function $z^\lambda f(z)$ is not in general single-valued in $r_1 \leq |z| \leq r_3$, we cut the annulus along the negative part of the real axis, obtaining a domain in which the principal branch of this function is analytic. The maximum modulus of this branch of the function in the cut annulus is attained on the boundary of the domain. Since λ is real, all the branches of $z^\lambda f(z)$ have the same modulus. By considering a branch of the function analytic in that part of the annulus for which $\pi/2 \leq \arg z \leq 3\pi/2$, we can see clearly that the principal value cannot attain its maximum modulus on the cut and therefore must attain it on one of the boundary circles of the annulus.

Thus the maximum of $|\varphi(z)|$ occurs on one of the bounding circles, that is,
$$|\varphi(z)| \leq \max\{r_1^\lambda M(r_1), r_3^\lambda M(r_3)\}.$$
Hence, on $|z| = r_2$,
$$|f(z)| \leq \max\{r_1^\lambda r_2^{-\lambda} M(r_1), r_3^\lambda r_2^{-\lambda} M(r_3)\}.$$
We choose λ such that $r_1^\lambda M(r_1) = r_3^\lambda M(r_3)$ and
$$\lambda = -\frac{\{\log[M(r_3)/M(r_1)]\}}{\{\log(r_3/r_1)\}}.$$
With this λ we get $M(r_2) \leq (r_2/r_1)^{-\lambda} M(r_1)$ and
$$M(r_2)^{\log(r_3/r_1)} \leq (r_2/r_1)^{\log\{M(r_3)/M(r_1)\}} M(r_1)^{\log(r_3/r_1)};$$
taking the logarithms, we obtain
$$M(r_2)^{\log(r_3/r_1)} \leq M(r_1)^{\log(r_3/r_2)} M(r_3)^{\log(r_2/r_1)}.$$

Clearly equality is achieved when $\varphi(z)$ is constant, that is, $f(z)$ is of the form Cz^u for some real u.

We may express the conclusion of the three-circle theorem in the form:
$$\begin{vmatrix} \log r_1 & \log M(r_1) & 1 \\ \log r_2 & \log M(r_2) & 1 \\ \log r_3 & \log M(r_3) & 1 \end{vmatrix} \geq 0. \qquad \square$$

7.6.1

Let φ be a function such that φ'' is nondecreasing; then we say that φ is convex. Thus for $x_1 < x < x_2$, we have
$$\frac{1}{x - x_1} \int_{x_1}^x \varphi'(t)\, dt \leq \varphi'(x) \leq \frac{1}{x_2 - x} \int_x^{x_2} \varphi'(t)\, dt,$$
integration of which gives the previous definition of convexity.

Alternatively, putting $x = \frac{1}{2}(x_1 + x_2)$, we obtain another more general definition of convexity, namely,
$$\varphi\{\tfrac{1}{2}x_1 + \tfrac{1}{2}x_2\} \leq \tfrac{1}{2}\{\varphi(x_1) + \varphi(x_2)\}.$$

Actually a lot more can be said about convex functions which are of particular interest in the theory of harmonic functions. We state one or two results that are not difficult to prove.*

1. A function $f(x)$ continuous in an open interval I is convex if and only if for every pair of points $x_1, x_2 \in I$ and every pair of nonnegative

---

*See S. Saks and A. Zygmund, *Analytic Functions*. Monographie Matematyczne, Vol. 28, Polska Akademia Nauk, Warsaw, 1965.

numbers p_1, p_2, where $p_1 + p_2 > 0$, we have

$$f\left(\frac{p_1 x_1 + p_2 x_2}{p_1 + p_2}\right) \leq \frac{p_1 f(x_1) + p_2 f(x_2)}{p_1 + p_2}.$$

This clearly reduces to our previous definition if $p_1 = p_2 = 1$.

2. If $g(x)$ is continuous in a closed interval $[a, b]$ and $f(x)$ is convex in an open interval containing all the values of $g(x)$, then

$$f\left\{\frac{1}{b-a}\int_a^b g(x)\,dx\right\} \leq \frac{1}{b-a}\int_a^b f\{g(x)\}\,dx.$$

We continue the chapter with a theorem related to the previous material which has far-reaching consequences in the study of entire and meromorphic functions.

7.7

Theorem (Jensen's) Let $f(z) \in A$ for $|z| < R$. Suppose $f(0) \neq 0$ and let $r_1, r_2, \ldots, r_n, \ldots$ be the moduli of the zeros of $f(z)$ in $|z| < R$ arranged as a nondecreasing sequence. If $r_n \leq r \leq r_{n+1}$, then

$$\log \frac{r^n |f(0)|}{r_1 r_2 \cdots r_n} = \frac{1}{2\pi}\int_0^{2\pi} \log|f(re^{i\theta})|\,d\theta,$$

where a zero of order p is counted p times. (This formula connects the modulus of a function with the moduli of the zeros.)

PROOF. First write the formula in another way. If $r_n \leq r \leq r_{n+1}$,

$$\log \frac{r^n}{r_1 r_2 \cdots r_n} = n \log r - \sum_{m=1}^{n} \log r_m$$

$$= \sum_{m=1}^{n-1} m(\log r_{m+1} - \log r_m) + n(\log r - \log r_n)$$

$$= \sum_{m=1}^{n-1} m \int_{r_m}^{r_{m+1}} \frac{dx}{x} + n \int_{r_n}^{r} \frac{dx}{x}.$$

Let $n(x)$ denote the number of zeros of $f(z)$ in $|z| \leq x$. We have $m = n(x)$ for $r_m \leq x < r_{m+1}$, and $n = n(x)$ for $r_n \leq x < r$. Thus

$$\log \frac{r^n}{r_1 \cdots r_n} = \int_0^r \frac{n(x)}{x}\,dx = \sum_{m=1}^{n-1} m \int_{r_m}^{r_{m+1}} \frac{dx}{x} + n \int_{r_n}^{r} \frac{dx}{x}.$$

[Check:

$$\int_0^r \frac{n(x)}{x}\,dx = \int_0^{r_1} 0 \cdot \frac{dx}{x} + \int_{r_1}^{r_2} 1 \cdot \frac{dx}{x} + \cdots + \int_{r_n}^{r} n \cdot \frac{dx}{x}$$

$$= 1 \cdot \int_{r_1}^{r_2} \frac{dx}{x} + \cdots + (n-1)\int_{r_{n-1}}^{r_n} \frac{dx}{x} + n \int_{r_n}^{r} \frac{dx}{x}.$$

Thus it is necessary to prove that
$$\int_0^r \frac{n(x)}{x} dx = \frac{1}{2\pi} \int_0^{2\pi} \log|f(re^{i\theta})| d\theta - \log|f(0)|.$$

Clearly both sides of the formula are equal for $r = 0$.

If $f(z)$ has no zero on $|z| = r$, then

$$n(r) = \frac{1}{2\pi i} \int_{|z|=r} \frac{f'(z)}{f(z)} dz = \frac{1}{2\pi} \int_0^{2\pi} \frac{f'(re^{i\theta})}{f(re^{i\theta})} re^{i\theta} d\theta. \qquad (7.2)$$

Obviously, we cannot divide by r, integrate with respect to r, or take real parts (by virtue of infinities of the integrand). In an interval between the moduli of two zeros r_n, r_{n+1} each side of Jensen's formula has a continuous derivative. We examine the assertion of the theorem. The derivative of the left-hand side is

$$\frac{d}{dr}\left\{\log \frac{r^n |f(0)|}{r_1 \ldots r_n}\right\} = \frac{n}{r}.$$

The derivative of the right-hand side is

$$\frac{1}{2\pi} \int_0^{2\pi} \frac{d}{dr}\{\log|f(re^{i\theta})|\} d\theta = \frac{1}{4\pi} \int_0^{2\pi} \frac{d}{dr}\{\log f(re^{i\theta}) + \log \bar{f}(re^{-i\theta})\} d\theta$$

(since $\log|f|^2 = 2\log|f| = \log f \cdot \bar{f}$)

$$= \frac{1}{4\pi} \int_0^{2\pi} \left\{\frac{f'(re^{i\theta})}{f(re^{i\theta})} e^{i\theta} + \frac{\bar{f}'(re^{-i\theta})}{\bar{f}(re^{-i\theta})} e^{-i\theta}\right\} d\theta$$

$$= R\left\{\frac{1}{2\pi} \int_0^{2\pi} \frac{f'(re^{i\theta})}{r(re^{i\theta})} e^{i\theta} d\theta\right\}.$$

This result equals $n(r)/r = n/r$ by (7.2) also.

Hence the two derivatives are equal in any such interval; thus the two sides of Jensen's formula differ by a constant in any such interval. Since both sides of Jensen's formula are equal for $r = 0$, the constant is 0 (in that particular interval). Therefore it is sufficient to prove that each side (of Jensen's formula) is continuous when $r = r_n$. This is clearly true for the left-hand side. Next we consider the right-hand side. It is sufficient to assume that there is one zero of modulus r_n and amplitude zero.

Thus $z_n = r_n$ and $f(z) = (z_n - z)\varphi(z)$, where $\varphi(z)$ is analytic and nonzero in the neighborhood of $z = z_n$. Consequently

$$\log|f(z)| = \log|r_n - re^{i\theta}| + \log|\varphi(z)|$$

$$= \log\left|1 - \frac{r}{r_n} e^{i\theta}\right| + \psi(r, \theta)$$

where $\psi(r,\theta) \neq 0$ and continuous in the neighborhood of $z = z_n$. Since we are considering a neighborhood of $z = z_n$, that is, $r = r_n$, let $r/r_n < 2$ and $|\theta| < \pi$. Then

$$\left|1 - \frac{r}{r_n} e^{i\theta}\right|^2 < 9, \quad \text{since} \quad \left|1 - \frac{r}{r_n} e^{i\theta}\right| \leq 1 + \left|\frac{r}{r_n}\right| < 3$$

and therefore

$$9 > \left|1 - \frac{r}{r_n} e^{i\theta}\right|^2 = 1 - \frac{2r}{r_n} \cos\theta + \left(\frac{r}{r_n}\right)^2$$

$$= \sin^2\theta + \left(\cos\theta - \frac{r}{r_n}\right)^2 \geq \sin^2\theta.$$

Further, in the neighborhood of the real zero $z = r_n$ we must consider a range of integration about $\theta = 0$, namely, $[-\delta, \delta]$ as $\delta \to 0$. Thus

$$\left|\int_{-\delta}^{\delta} \log\left|1 - \frac{r}{r_n} e^{i\theta}\right| d\theta\right| \leq \int_{-\delta}^{\delta} \left|\log\left|1 - \frac{r}{r_n} e^{i\theta}\right|\right| d\theta$$

$$< \int_{-\delta}^{\delta} \{\log 3 + |\log|\sin\theta||\} d\theta.$$

By writing

$$\left(1 - \frac{r}{r_n} e^{i\theta}\right) = \alpha,$$

we get $|\log|\alpha|| < \log 3$ and $\log|\alpha| \geq \log|\sin\theta|$; thus

$$|\log|\alpha|| \leq |\log|\sin\theta||$$

and hence

$$|\log|\alpha|| < \log 3 + |\log|\sin\theta||.$$

Also, for $0 < \theta < \pi/2$, we have $\sin\theta > 2\theta/\pi$. Thus

$$\log\sin\theta = \log|\sin\theta| > B + \log\theta$$

(since $\sin\theta$ is positive) and

$$|\log|\sin\theta|| < |B + \log\theta| \leq C + |\log\theta|, \quad \text{since} \quad \log|\sin\theta| < 0.$$

Hence

$$\left|\int_{-\delta}^{\delta} \log\left|1 - \frac{r}{r_n} e^{i\theta}\right| d\theta\right| < \int_{-\delta}^{\delta} \{A + |\log|\theta||\} d\theta$$

$$= 2A\delta - 2 \int_{\epsilon}^{\delta} \log\theta \, d\theta \quad (\text{since } \epsilon \to 0)$$

$$= 2A\delta - 2\{\theta\log\theta - \theta\}_{\epsilon}^{\delta}$$

$$= 2A\delta - 2\delta\log\delta + 2\delta \quad (\text{since } \epsilon\log\epsilon \to 0)$$

$$\to 0 \quad (\text{since } \delta\log\delta \to 0).$$

Maximum Modulus

Thus a contribution to the integral in the neighborhood of zero is arbitrarily small; therefore the whole integral is continuous and Jensen's theorem is established. □

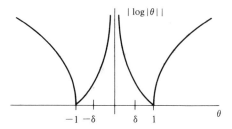

The theorem may be extended to functions with poles as well as zeros.

Let $f(z)$ satisfy the same conditions of the original theorem and let it have zeros $a_1, \ldots, a_m$ and poles $b_1, \ldots, b_n$ with moduli less than or equal to r. Then

$$\log\left\{\left|\frac{b_1 \ldots b_n}{a_1 \ldots a_m} f(0)\right| r^{m-n}\right\} = \frac{1}{2\pi} \int_0^{2\pi} \log|f(re^{i\theta})| \, d\theta,$$

Let

$$f(z) = g(z) \div \left(1 - \frac{z}{b_1}\right)\left(1 - \frac{z}{b_2}\right) \cdots \left(1 - \frac{z}{b_n}\right) = \frac{g(z)}{h(z)}.$$

Since $g(z)$ has zeros $a_1, \ldots, a_m$, and $f(0) = g(0)$,

$$\log \frac{r^m g(0)}{|a_1 a_2 \ldots a_m|} = \frac{1}{2\pi} \int_0^{2\pi} \log|g(re^{i\theta})| \, d\theta$$

and

$$\log \frac{r^n h(0)}{|b_1 b_2 \ldots b_n|} = \log \frac{r^n}{|b_1 b_2 \ldots b_n|} = \frac{1}{2\pi} \int_0^{2\pi} \log|h(re^{i\theta})| \, d\theta$$

The result follows by subtracting the latter equation from the former one.

7.7.1 The Poisson–Jensen Formula

Let $f(z)$ have zeros at $a_1, a_2, \ldots, a_m$ and poles at $b_1, b_2, \ldots, b_n$ inside the circle $|z| \leq R$ and let $f(z)$ be analytic elsewhere inside and on the circle. Then

$$\log|f(re^{i\theta})| = \frac{1}{2\pi} \int_0^{2\pi} \frac{R^2 - r^2}{R^2 - 2Rr\cos(\theta - \varphi) + r^2} \log|f(Re^{i\varphi})| \, d\varphi$$
$$- \sum_{\mu=1}^{m} \log\left|\frac{R^2 - \bar{a}_\mu re^{i\theta}}{R(re^{i\theta} - a_\mu)}\right| + \sum_{\nu=1}^{n} \log\left|\frac{R^2 - \bar{b}_\nu re^{i\theta}}{R(re^{i\theta} - b_\nu)}\right|.$$

This contains Poisson's and Jensen's formulas as special cases. For the case when there are no zeros or poles, we have Poisson's formula for the real part of log $f(z)$, namely,

$$\log|f(re^{i\theta})| = \frac{1}{2\pi} \int_0^{2\pi} \frac{R^2 - r^2}{R^2 - 2Rr\cos(\theta - \varphi) + r^2} \log|f(Re^{i\varphi})| \, d\varphi.$$

For $r = 0$,

$$\log|f(0)| = \frac{1}{2\pi} \int_0^{2\pi} \log|f(Re^{i\varphi})| \, d\varphi - \log\left\{ \left| \frac{b_1 b_2 \ldots b_n}{a_1 a_2 \ldots a_m} \right| R^{m-n} \right\}.$$

1. Let $f(z) = z - a$, $|a| < R$. We have to prove that

$$\log|re^{i\theta} - a| = \frac{1}{2\pi} \int_0^{2\pi} \frac{R^2 - r^2}{R^2 - 2Rr\cos(\theta - \varphi) + r^2} \log|Re^{i\varphi} - a| \, d\varphi$$

$$- \log \left| \frac{R^2 - \bar{a}re^{i\theta}}{R(re^{i\theta} - a)} \right|,$$

that is,

$$\log\left| R - \frac{\bar{a}re^{i\theta}}{R} \right| = \frac{1}{2\pi} \int_0^{2\pi} \frac{R^2 - r^2}{R^2 - 2Rr\cos(\theta - \varphi) + r^2} \log|Re^{i\varphi} - a| \, d\varphi.$$

This follows from the application of Poisson's formula to the real part of $\log(R - (\bar{a}z/R))$ which is analytic for $|z| \leq R$. Since $U(r, \theta)$ corresponds to $\log|R - (\bar{a}re^{i\theta}/R)|$, $U(R, \varphi)$ corresponds to $\log|R - \bar{a}e^{i\varphi}|$, $a = \alpha + i\beta$, and

$$U(R, \varphi) = \frac{1}{2} \log\{(R - \alpha\cos\varphi - \beta\sin\varphi)^2 + (\beta\cos\varphi - \alpha\sin\varphi)^2\},$$

$$\log|Re^{i\varphi} - a| = \frac{1}{2} \log\{(R\cos\varphi - \alpha)^2 + (R\sin\varphi - \beta)^2\};$$

thus

$$\log|R - \bar{a}e^{i\varphi}| = \log|Re^{i\varphi} - a|.$$

2. If $f(z) = 1/(z - b)$, the Poisson–Jensen formula is equivalent to Poisson's formula for the real part of $\log(R - (\bar{b}z/R))$.
3. If $f(z)$ is analytic with no poles or zeros in $|z| \leq R$, the formula is Poisson's formula for the real part of $\log f(z)$.

We now add all these cases to obtain the Poisson–Jensen formula.

A much more elegant proof of Jensen's theorem is available, however it involves the theory of residues and as such is not quite so elementary.*

---

*See, Ahlfors, L.V., "Complex Analysis," McGraw-Hill Book Company, 2nd Ed., (1966), 205.

8
Analytic Continuation

8.1

There are several ways to initiate a study of this subject and we choose a very classical approach. The issue to be discussed can arise as follows. Consider a real polynomial of degree four:

$$f(x) = a_0 + a_1 x + a_2 x^2 + a_3 x^3 + a_4 x^4.$$

It is a remarkable fact that if $f(x)$ is known for five distinct values of x, no matter how close these values of x may be, then the polynomial with all its properties (maxima, minima, zeros, points of inflection) is completely known. We ask, what constraints must be placed upon a function $f(z)$ so that from its behavior in a domain (however small) of the z-plane one can deduce its behavior in the remaining part of the entire plane? We show that if $f(z) \in A$ in a domain, then it is fully determined if the values of $f(z)$ are known along any small arc or infinite set of points (with the limit point in D).* The classical approach is by Taylor series expansions, however we illustrate some other methods.

We require three basic results that we prove immediately.

8.2 Identity Theorems for Analytic Functions

8.2.1

Theorem *If two power series $\sum_{n=0}^{\infty} a_n (z - z_0)^n$ and $\sum_{n=0}^{\infty} b_n (z - z_0)^n$ defined in a domain D have a positive radius of convergence and if their sums*

---

*An illuminating discussion of problems arising in this connection is to be found in the *Theory of Functions* by K. Knopp, Part I. New York: Dover Publications, 1945.

coincide at an infinite number of points with the limit point z_0, then the two series are identical.

PROOF. Assume that the first m coefficients have been shown to be identical respectively, then

$$a_{m+1} + (z - z_0)a_{m+2} + \cdots = b_{m+1} + (z - z_0)b_{m+2} + \cdots$$

for all these infinitely many points. Let $z \to z_0$ by means of these points, since the power series are continuous functions. Thus $b_{m+1} = a_{m+1}$ and the series are identical. □

We develop a lemma required in the subsequent theorem that generalizes Theorem 8.2.1.

8.2.2

Lemma *If a path K (or closed point set) lies within a domain D, then there exists $\rho > 0$, such that the distance of all points on the path from the boundary of D exceeds ρ (that is, the path K does not come arbitrarily close to the boundary).*

PROOF. Since every point $z \in K$ lies in D, a circular neighborhood with center z and radius ρ lies within D.

Let there be a correspondence between each point z and a circle with center z and radius ρ_z/z. By the Heine–Borel theorem, a finite number of these circles are needed to cover K. Let ρ be the radius of the smallest of these circles, then ρ satisfies the theorem, since the circle of radius ρ lies entirely within D. □

8.2.3

Theorem *If two functions $f_1(z)$, $f_2(z)$ are analytic in a domain D and if they coincide at an infinite number of distinct points with limit point $z_0 \in D$, then $f_1(z) \equiv f_2(z)$.*

PROOF. Let K_0 be the largest circle with center z_0, that lies entirely within D. Both functions may be developed in unique power series convergent in K_0, and by 8.2.1 and the assumption of 8.2.3, the two series are identical in K_0. Thus, $f_1 \equiv f_2$ everywhere in K_0. Let ξ be an arbitrary point on D. We require to show that $f_1(\xi) = f_2(\xi)$. Connect z_0 and ξ by a path K lying entirely within D. Let ρ be the positive number whose existence was proved in 8.2.2.

Divide K arbitrarily by points $z_0, z_1, \ldots, z_m = \xi$ into subpaths whose lengths are less than ρ. About each center z_i describe the largest circle K_i lying entirely within D. All radii $r_i \geq \rho$ and each circle contains the center of the next.

Expand f_1 and f_2 in power series about each center z_i, as was done for $i = 0$. All expansions converge at least in K_i. Since the functions f_1 and f_2 are identical in K_0, they coincide at z_1 and in the neighborhood thereof. Thus by 8.2.1 they coincide in K_1, and the functions are equal at and in a neighborhood of z_2, etc.

Thus they coincide at $z_n = \xi$ (and in a neighborhood thereof). □

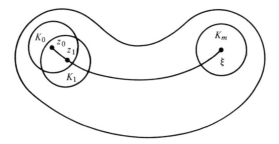

We are now in a position to give a brief survey of the concept of analytic continuation; a thorough analysis requires the concept of Riemann surfaces and will not be included here.

8.3

If we consider the two functions

$$1 + z + z^2 + \cdots, \quad \text{where } |z| < 1$$

and

$$\int_0^\infty e^{-t(1-z)}\, dt, \quad \text{where } R(z) < 1,$$

then their values coincide for certain values of z although they appear to be different. In fact, they coincide with another function $f(z) = 1/(1 - z)$ defined for all z except $z = 1$. We are dealing here with essentially single-valued functions, i.e., for each value of $z(\neq 1)$ there exists one and only one value of $f(z)$. Our task is to develop a theory that will allow us to generate a process for extending our definition of a function beyond a limited region, in which it is already defined. The process is called *analytic continuation* and has no counterpart in the theory of functions of a real variable.

8.3.1

Definition Suppose f_1 and $f_2 \in A$ in domains D_1 and D_2, respectively, and $D_1 \cap D_2 = S \neq \emptyset$. Assume that $f_1 = f_2$ in S. We consider all the values of f_1 and f_2 at points interior to D_1 and D_2 as a single analytic function $f(z)$. Thus,

$$f(z) \in A \text{ in } T = D_1 \cup D_2,$$
$$f(z) = f_1(z) \text{ in } D_1,$$
$$f(z) = f_2(z) \text{ in } D_2$$

We call $f_2(z)$ an analytic continuation of $f_1(z)$ (or f_1 an analytic continuation of f_2).*

---

* For an example of the difficulties encountered in trying to extend the definition of a real function $f(x)$ to values of x different from the original defining set see *Theory of Functions* by E. C. Titchmarsh, Oxford, 1960, p. 139.

8.3.2

By the identity theorem, once an analytic continuation is made from D_1 to D_2 (where $D_1 \cap D_2 \neq \emptyset$), then it is unique. For if there exists a function $F_2(z)$ in D_2 different from $f_2(z)$, we would have $f_1(z)$ and $F_2(z) \in A$ in $D_1 \cup D_2$ giving rise to a function $F(z)$ such that both $F(z)$ and $f(z) \in A$ in $D_1 \cup D_2$. However the identity theorem ensures that $f(z) = F(z)$. If we analytically continue a function in different ways, there is no reason to assume that the final functions are the same. Thus, if $f_1 \in D_1$ is analytically continued to $f_2 \in D_2$ $(D_1 \cap D_2 \neq \emptyset)$ and $f_1 \in D_1$ is analytically continued to $f_3 \in D_3$ $(D_1 \cap D_3 = \emptyset)$, and if $D_2 \cap D_3 \neq \emptyset$, then it may be that $f_2 \neq f_3$, in which case we say that

$$f = f_1 \in D_1$$
$$= f_2 \in D_2$$
$$= f_3 \in D_3$$

is *many-valued*.

We shall elaborate upon the subject of many-valued functions in a later section.

EXAMPLE 1. Suppose we wish to find a function that is the analytic continuation of $f_1(z) = \sum_{n=0}^{\infty} z^n$ defined in the circle $w_1 = \{z \mid |z| < 1\}$ to the circle $w_2 = \{z \mid |z - i| < \sqrt{2}\}$.

If we expand $F(z) = 1/(1 - z)$ about $z = i$, then the function

$$f_2(z) = \frac{1}{1-i} \sum_{n=0}^{\infty} \left(\frac{z-i}{1-i}\right)^n$$

converges in w_2, and since $f_1 = f_2$ on $w_1 \cap w_2 \neq \emptyset$, then f_2 is the required unique analytic continuation of f_1 (to w_2).

Also, if $f_1 = \sum_{n=0}^{\infty} z^n$, which converges if and only if $|z| < 1$, then it is the Maclaurin expansion of $f(z) = 1/(1 - z)$. Thus $f_1(z) = 1/(1 - z)$, where $|z| < 1$, but it is not defined for $|z| \geq 1$. The function $F(z) = 1/(1 - z) \in A$ in $\{C \setminus (1)\}$. It is identical to f_1 interior to $|z| = 1$ and is the only possible analytic continuation of f_1 beyond $|z| = 1$, that is, $\{C \setminus (|z| \leq 1)\}$. We call f_1 and f_2 function elements of F. If we begin with $\sum_0^{\infty} z^n$ which, we are told, converges and represents an analytic function in $|z| < 1$ and its sum is $1/(1 - x)$ when $z = x$ (real), we can conclude that its sum is $1/(1 - z)$ whenever $|z| < 1$, since $1/(1 - z)$ is the analytic function interior to the circle that assumes the values $1/(1 - x)$ along the segment of the x-axis inside the circle.

EXAMPLE 2. Consider $g_1(z) = \int_0^{\infty} e^{-zt} \, dt$:

$$g_1(z) = -\left.\frac{e^{-zt}}{z}\right|_{t=0}^{\infty} = \frac{1}{z} \quad \text{when } x > 0.$$

Thus g_1 is defined only in the domain $x > 0$: D_1 in the diagram, further, $g_1 \in A$ in D_1.

Analytic Continuation

Define
$$g_2(z) = i \sum_{n=0}^{\infty} \left(\frac{z+i}{i}\right)^n, \quad \text{when } |z+i| < 1.$$

This series converges to $1/z$ within the above circle (Taylor expansion about $z = -i$).

Note: $g_2(z) = i/[1 - (z+i)/i] = 1/z$ for $z \in D_2$ namely, $|z+i| < 1$. Thus $g_1 = g_2$ in $D_1 \cap D_2$ and g_2 is the analytic continuation of g_1 into D_2. The function $G(z) = 1/z$, $z \neq 0$, is the analytic continuation of both g_1 and g_2 into D_3 consisting of all points in the plane except the origin, $\{C \setminus 0\}$, and g_1, g_2 are elements of G.

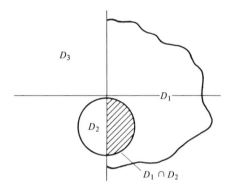

EXAMPLE 3. Consider the function $\exp z$. This is the only (entire) function that can assume the value e^x along a segment of the real axis. Also, since $e^x \cdot e^{-x} = 1$ for all real x and since $\exp z$ and $\exp(-z)$ are entire functions, $\{\exp z\}\{\exp(-z)\} - 1$ is an entire function that vanishes along the real axis, that is, therefore, everywhere.

Thus it follows from the identity for real functions and the analyticity of the function that $\exp(-z) = 1/\exp z$ for all complex z.

Similarly, $\sin^2 z + \cos^2 z = 1$ follows from the identity $\sin^2 x + \cos^2 x = 1$.

Note: In general it is impossible to find F in a closed form from the given function elements f_1, f_2, etc.

8.3.3

Exercises

1: Show that the functions defined by
$$1 + az + a^2 z^2 + \ldots$$
and
$$\frac{1}{1-z} - \frac{(1-a)z}{(1-z)^2} + \frac{(1-a)^2 z^2}{(1-z)^3} - \ldots$$
are analytic continuations of each other.

2: If $F(\alpha, \beta, \gamma, z)$ denotes the series

$$1 + \frac{\alpha\beta}{1 \cdot \gamma}z + \frac{\alpha(\alpha+1)\beta(\beta+1)}{1 \cdot 2 \cdot \gamma(\gamma+1)}z^2 + \ldots ,$$

show that the series

$$f(z) = F(a, 1, c, z)$$

and

$$g(z) = \frac{1}{1-z} F\left(c - a, 1, c, \frac{z}{z-1}\right)$$

have a common domain of convergence, that in this domain they both satisfy the equation

$$z(1-z)\frac{d^2u}{dz^2} + [c - (a+2)z]\frac{du}{dz} - au = 0,$$

and that $f(0) = g(0)$, $f'(0) = g'(0)$, and hence that the two functions are analytic continuations of each other.

3: Show that the series

$$\sum_{1}^{\infty} \frac{z^n}{n} \quad \text{and} \quad i\pi + \sum_{1}^{\infty}(-1)^n \frac{(z-2)^n}{n}$$

have no common domain of convergence, yet are analytic continuations of each other.

4: Prove that if a function f is entire, real-valued on the real axis, and assumes pure imaginary values on the imaginary axis, then f is an odd function for all $z \in \mathcal{C}$.

5: Find a series representation about $z = -1$ for the function

$$f(z) = \int_0^\infty t^2 e^{-zt} dt, \quad 0 < t < \infty,$$

analytic in $R(z) > 0$. What is its analytic continuation to the whole plane?

8.4 The principle of Reflection

Some functions have the property that if $w = f(z)$, then $\bar{w} = f(\bar{z})$; for example, the functions z, $z^2 + 1$, e^z, $\sin z$. On the other hand, the functions iz, e^{iz}, $z^2 + i$ do not have this property. The following theorems show that a function has the above property if and only if the function is real for real z.

8.4

Theorem *Let $f(z)$ be a function analytic in a domain D intersected by the real axis and let $f(z)$ be real on the real axis. Then $f(z)$ takes conjugate values for conjugate values of z, that can be reached by conjugate paths from the real axis.*

Analytic Continuation

PROOF. Let $z_0 \in D$ and be on the real axis. Then we can expand

$$f(z) = \sum_{n=0}^{\infty} a_n(z - z_0)^n$$

for sufficiently small $|z - z_0|$. All coefficients a_n are real since

$$a_0 = f(z_0) \quad \text{and} \quad a_1 = f'(z_0).$$

Thus a_0 is real,

$$a_1 = \lim_{z \to z_0} \frac{f(z) - f(z_0)}{z - z_0}$$

through real values, and we see that a_1 is real. Similarly a_n is real for all n. Consequently the result is true inside the circle of convergence. The general result follows by continuation since the power series about conjugate points will always have conjugate coefficients. □

A sort of converse of this result is given by the so-called Schwarz principle of reflection.

8.4.2

Theorem *Let $f(z)$ be analytic in a domain D and let $f_1(z)$ be analytic in a domain D_1; also, let $D \cap D_1 = \emptyset$, while $\overline{D} \cap \overline{D}_1 \equiv C$ a rectifiable Jordan arc such that $f(z)$ and $f_1(z)$ are continuous on C with $f(z) = f_1(z)$ along C. Then the two functions are analytic continuations of each other.*

PROOF. We approach the proof through the following lemma: Suppose that a domain D has a segment l of a straight line as part of its boundary. Suppose that $w = f(z)$ is analytic in D, continuous on l and such that w describes a straight line λ in the w-plane as z describes l. Let $z \in D$, z_1 be the reflection of z in l, and w_1 be the reflection of w in λ. Then $w_1 = w_1(z_1)$ is an analytic continuation of w.

We see that w_1 is an analytic function of z_1 since if w' corresponds to z' and w'_1, z'_1 are their reflections, then

$$|z'_1 - z_1| = |z' - z|,$$
$$|w'_1 - w_1| = |w' - w|,$$
$$\arg(z'_1 - z_1) = 2\alpha - \arg(z' - z),$$
$$\arg(w'_1 - w_1) = 2\beta - \arg(w' - w),$$

where α, β are the angles between l, λ and the real axis, respectively. Now,

$$\lim_{z' \to z} \frac{w' - w}{z' - z}$$

exists, since $w = f(z)$ is analytic in D; consequently

$$\lim_{z' \to z} \left| \frac{w' - w}{z' - z} \right| \quad \text{and} \quad \lim_{z' \to z} \{\arg(w' - w) - \arg(z' - z)\}$$

exist. Thus

$$\lim_{z_1' \to z_1} \left| \frac{w_1' - w_1}{z_1' - z_1} \right| \quad \text{and} \quad \lim_{z_1' \to z_1} \{\arg(w_1' - w) - \arg(z_1' - z_1)\}$$

exist and consequently

$$\lim_{z_1' \to z_1} \frac{w_1' - w_1}{z_1' - z_1}$$

exists, that is, w_1 is an analytic function of z_1. Further, $w_1 = w$ on λ.

We now show that the two functions are analytic continuations of each other. Take any point $\xi \in l \cap D$ and describe a circle Γ about ξ such that $E \subset \overline{D} \cup \overline{D}_1$, where E consists of Γ together with its interior points. Let $\gamma_1 = \partial(E \cap \overline{D})$ and $\gamma_2 = \partial(E \cap \overline{D}_1)$.

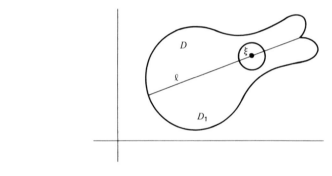

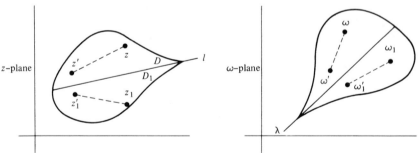

Let $\varphi(z) = w$ in int$\{\overline{D} \cap E\}$ and $\varphi(z) = w_1$ in int$\{\overline{D}_1 \cap E\}$.

Thus $\varphi(z)$ is continuous, and we are required to show that $\varphi(z)$ is an analytic function of z in E.

Let $z_0 \in \text{int}\{E \cap \overline{D}\}$, then

$$\varphi(z_0) = \frac{1}{2\pi i} \int_{\gamma_1} \frac{\varphi(z)}{z - z_0} dz,$$

and since $\varphi(z)/(z - z_0)$ is analytic in D_1, we have

$$0 = \frac{1}{2\pi i} \int_{\gamma_2} \frac{\varphi(z)}{z - z_0} dz.$$

Analytic Continuation

By adding these two equations we get

$$\varphi(z_0) = \frac{1}{2\pi i} \int_E \frac{\varphi(z)}{z - z_0} \, dz. \tag{8.1}$$

We obtain the same formula if $z_0 \in l \cap \text{int } E$.

However, the right-hand side of (8.1) is an analytic function of z_0 inside E, which proves the lemma. The theorem now follows. □

8.5

We are now in a position to discuss more carefully one or two topics mentioned earlier.

8.5.1 Zeros of Analytic Functions

In 6.4 we defined the concept of a *zero of an analytic function*. We now point out that zeros of an analytic function $f \not\equiv 0$ cannot have a limit point belonging to the domain D in which f is analytic, since otherwise f would be identically zero by the identity theorem (see 8.2). In particular, if a is a zero of $f(z)$, it is isolated, that is, there exists $\delta > 0$ such that $|f(z)| > 0$ for $0 < |z - a| < \delta$.

8.5.2 Harmonic Functions

It is convenient now to prove the following theorem on the existence of a conjugate harmonic function.

8.5.3

Theorem *If a harmonic function $u = u(x, y)$ is defined in a simply connected domain D, then there exists a conjugate harmonic function $v = v(x, y)$ in D.*

PROOF. Consider the simply connected domain D. Let u be harmonic in D. We are required to find a function v such that $u_x = v_y$ and $u_y = -v_x$.

Let $P(x_0, y_0)$ be a fixed point of D and $Q(x, y)$ a point in a neighborhood $N(z_0) \in D$; it is joined to P by a line segment from (x_0, y_0) to (x_0, y) to (x, y), the line segments lying in D.

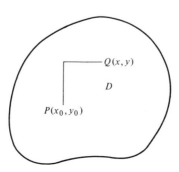

Consider $v_x = -u_y$. Integrating with respect to x, we get

$$v(x, y) = -\int_{x_0}^{x} u_y(x, y)\, dx + \varphi(y) \tag{8.2}$$

and for all $\varphi(y)$ we have $v_x = -u_y$. Since we also require that $u_x = v_y$, we must find $\partial/\partial y$ of (8.2), that is,

$$-\frac{\partial}{\partial y}\int_{x_0}^{x} u_y(x, y)\, dx + \varphi'(y) = u_x(x, y). \tag{8.3}$$

Since f is assumed to be analytic, derivatives of all orders exist. In particular, u_{yy} is continuous and u is harmonic, thus,

$$\int_{x_0}^{x} -u_{yy}(x, y)\, dx = \int_{x_0}^{x} u_{xx}(x, y)\, dx$$

$$= u_x(x, y) - u_x(x_0, y)$$

Substituting in (8.3), we get

$$\varphi'(y) = u_x(x_0, y).$$

Therefore,

$$v(x, y) = -\int_{x_0}^{x} u_y(x, y)\, dx + \int_{y_0}^{y} u_x(x_0, y)\, dy + k,$$

where $v(x_0, y_0) = k$. This v clearly satisfies the Cauchy–Riemann equations in $N(z_0)$. By the analytic continuation it is clear that the result is true for the simply connected domain D. □

8.6 Natural Boundaries

Functions that have a Taylor expansion throughout a circle of radius $R(>0)$, such that there exists no finite arc of the circle without singularities of the function, cannot be analytically continued in any direction. Functions with this property are said to have a *natural boundary*. Thus, if the singularities of the function are dense enough on the circle of convergence, the circle is called the *natural boundary of the function*. It can be shown (see exercises) that there exists at least one singularity on the circle of convergence, i.e., the one nearest the point about which the function is expanded. Certain power series $\sum_n a_n z^n$ that have *gaps* (in the sense that sequences of coefficients are zero) may have singularities dense enough to make the circle of convergence a natural boundary. Consider the following example:

$$f(z) = \sum_{n=1}^{\infty} z^{2^n} = z^2 + z^4 + z^8 + \cdots$$

Then $f(z) = z^2 + f(z^2)$, and clearly f, has singularities at $z^2 = 1$ on its circle of convergence $|z| = 1$. Similarly,

$$f(z^2) = z^4 + f(z^4) + \cdots + f(z^{2^{n-1}}) = z^{2^n} + f(z^{2^n}).$$

Then $z^n = 1 = e^{2k\pi i} = e^{i 2^n \theta}$ implies that $\theta = 2k\pi/2^n$.

Analytic Continuation

Thus, given a fixed value θ_1 of θ on $|z| = 1$, we can choose n and k so that the singularity is arbitrarily close to the point on $|z| = 1$ corresponding to θ_1. Thus $z^{2^n} = 1$ exhibits singularities that are dense enough on $|z| = 1$, so that there is no finite arc, across which analytic continuation is possible.

It can be shown that power series with sufficiently large gaps will always have the circle of convergence as a natural boundary.

The following theorem was established by J. Hadamard; we shall not prove it here.*

8.6.1

Theorem *If in the power series*

$$f(z) = \sum_{n=0}^{\infty} a_n z^n \quad \text{and}$$

$a_n = 0$ *(except when n belongs to a sequence n_K such that $n_{K+1} > (1 + \theta)$ $\cdot n_K$, $\theta > 0$), then the circle of convergence of the series is a natural boundary of the function.*

8.6.2

Exercises

1: Show that $|z| = 1$ is a natural boundary for $f(z) = \sum_{n=1}^{\infty} z^{n!}$.

2: Show that there always exists a singularity on the circle of convergence of the power series $\sum_{n=0}^{\infty} a_n z^n$.

3: Show that the function $\sum_{n=1}^{\infty} z^{2^n}/n^2$ is continuous in and on the unit circle, but every point of the circle is a singularity.

4: If

$$\left| \frac{a_n}{a_{n+1}} \right| = \left\{ 1 + \frac{c}{n} + o\left(\frac{1}{n}\right) \right\} R,$$

where $c > 1$, then $\sum_n a_n z^n$ converges absolutely everywhere on its circle of convergence (of radius R).

With the development of the concept of analytic continuation we are now in a position to study functions that are no longer single-valued in a domain.

8.7 Logarithmic Function

We have studied e^z for complex z and concluded that $e^z = e^x \operatorname{cis} y$, where e^x is the real exponential and $\operatorname{cis} y = \cos y + i \sin y$. Since e^z is analytic everywhere, we can find a Taylor expansion about $z = 0$ and

---

* See E. C. Titchmarsh, *Theory of Functions*. Oxford: Oxford University Press, 1960, p. 223.

hence $e^z = \sum_{n=0}^{\infty} z^n/n!$. Since $e^z = e^{z+2ki\pi}$, $k = 0, \pm 1, \ldots$, let us write $\exp z = e^{z+2k\pi i}$ for $k = 0$. Thus we shall agree to call $e^z \equiv \exp z$ the principal value of the exponential function.

Now suppose $z = \exp w$. Set $|z| = r$ and $\theta = \arg z$, where $-\pi < \theta \leq \pi$. Write $w = u + iv$. Then if $r = |z| > 0$,

$$z = e^u \operatorname{cis} v = \exp(u)\operatorname{cis} v,$$

and

$$r = \exp(u), \qquad \theta = v + 2n\pi \quad (n \text{ integer}).$$

Thus $z = e^w$ defines w as a many-valued function of z given by

$$w = \operatorname{Log} z = u + iv = \log r + i(\theta + 2l\pi), \quad (l \text{ integer}).$$

The *principal value* of $\operatorname{Log} z$, written $\log z$, is given by $\log z = \log r + i\theta$, $-\pi < \theta \leq \pi$, that is, corresponding to $l = 0$.

8.7.1

Definition of $\log z$ as an integral

The function $1/z \in A$ in a simply connected domain D whose boundary consists of the half-line $x \leq 0$, $y = 0$. Hence, the function

$$F(z) = \int_1^z d\xi/\xi$$

is analytic in D, the path of the integral lies in D, and

$$F'(z) = 1/z, \quad \forall z \in D.$$

Let $|z| = r(>0)$, $\arg z = \theta$, $-\pi < \theta < \pi$, and take the integral along the path shown in the figure. Then

$$F(z) = \int_1^r \frac{dt}{t} + \int_0^\theta \frac{1}{re^{i\varphi}} ire^{i\varphi} d\varphi = \log r + i\theta.$$

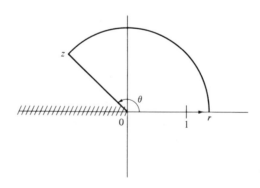

Note that $e^{i\varphi}$ is to be taken as an abbreviation for $\exp(i\theta)$ since $e^{i\varphi}$ is many-valued. In general,

$$\begin{aligned}e^{i\varphi} &= \exp\{i\varphi \operatorname{Log} e\} \\ &= \exp\{i\varphi(1 + i2k\pi)\} \\ &= \exp(i\varphi)\exp(-2k\varphi\pi), \quad k = 0, \pm 1, \ldots\end{aligned}$$

Analytic Continuation

Thus $F(z) = \log z$ and $\log z \in A$ in D, with

$$\frac{d}{dz} \log z = \frac{1}{z}.$$

If z is a negative real number, we define $\log z$ as

$$\log z = \int_1^z d\xi/\xi, \text{ where}$$

the integral is taken along a rectifiable Jordan curve of the form shown in the diagram (i.e., it meets the negative real axis at the point z only).

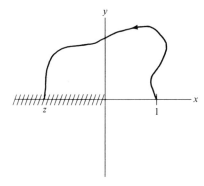

For a contour C (rectifiable Jordan curve that joins I to z and does not pass through the origin) it can be proved that

$$\int_C \frac{d\xi}{\xi} = \log z + 2p\pi i$$

where p is the number of times C crosses the negative real axis from above to below less the number of times C crosses the negative real axis from below to above.* In the case shown,

$$\int_C \frac{d\xi}{\xi} = \log z + 2\pi(5-3)i = \log z + 4\pi i.$$

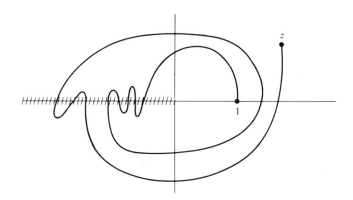

---

* Hardy, G. H., *A Course of Pure Mathematics*, 9th Ed., Cambridge Univ. Press, (1948) 449–454.

We shall return to a further discussion of the logarithmic function after we have studied many-valued functions.

8.7.2 Logarithmic Series

Since $\log(1 + z) \in A$ in a domain bounded by the half-line $y = 0$, $x < -1$, that is, it is analytic in $|z| < 1$, we have, by Taylor's theorem,

$$\log(1 + z) = \sum_{n=1}^{\infty} (-1)^{n+1} \frac{z^n}{n} \quad (|z| < 1).$$

8.8 General Power

$$a^z \quad (a \neq 0)$$

8.8.1
Definition

$$a^z = \exp\{z \operatorname{Log} a\}$$
$$= \exp\{z \log|a| + 2p\pi i z\} \quad p \text{ integer.}$$

The principal value $P.V.$ of a^z is $\exp z \log|a|$. Thus

$$e^z = \exp\{z \operatorname{Log} e\} = \exp\{z(1 + 2p\pi i)\} \cdot P.V. \text{ of } e^z = \exp z.$$

8.9 The Binomial Theorem

Since the $P.V.$ of $(1 + z)^m$ is

$$\exp\{m \log(1 + z)\} \in A \text{ in } |z| < 1,$$

we have, by Taylor's theorem,

$$P.V.(1 + z)^m \quad (m, \text{real}) = \sum_{r=0}^{\infty} \binom{m}{r} z^r \quad \text{for } |z| < 1.$$

8.10 Multiply Valued Functions, Branches, and Branch Points

If we take a function $f(z)$ such that more than one value of $w = f(z)$ corresponds to some values of z in the domain of definition, then we say that $w = f(z)$ is a multiply valued, or many-valued, function. An important class of many-valued functions arises when we consider the inverse of a single-valued analytic function.

8.10.1

Consider the equation $w^2 = z$. Clearly z is a single-valued analytic function of w; however, if we take the inverse $w = \sqrt{z}$, we see that two values of w correspond to each value of $z \neq 0$.

Let us write $z = re^{i\theta}$, then

$$w = \sqrt{r} \, e^{i(\theta + 2k\pi)/2}, \quad k = 0, 1.$$

If for example we restrict θ to the range $0 < \theta < 2\pi$, we obtain two single-valued functions of w, namely,

$$w_1 = \sqrt{r}\, e^{i\theta/2} \quad \text{and} \quad w_2 = \sqrt{r}\, e^{i[(\theta/2)+\pi]}, \quad 0 < \theta < 2\pi.$$

Thus $w_1 = -w_2$. We note that although w_1 is defined and single-valued for all values of z, it is not continuous at any point z_0 on the positive real axis, since $\theta \to 0$ and $w_1 \to \sqrt{r}$ as $z \to z_0$ from above; while $\theta \to 2\pi$ and $w_1 \to -\sqrt{r}$ as $z \to z_0$ from below. In order to make w_1 continuous, we restrict ourselves to the domain D_0: $0 < \theta < 2\pi$, $r > 0$, that is, the z-plane with the nonnegative half of the real axis excluded. Thus the values of w are given in D_0 by two single-valued functions: w_1 and $w_2 = -w_1$.

The nonnegative half of the real axis is called a *branch cut* for w_1 and w_2.

We show that w_1 is an analytic function of z in D_0. Since $z = w_1^2$, we get $dz/dw_1 = 2w_1$. Since w_1 is a continuous function of z in D_0, we have

$$\lim_{\Delta z \to 0} \Delta w_1 = 0 \quad \text{and} \quad \lim_{\Delta w_1 \to 0} \Delta z = 0.$$

Thus

$$\lim_{\Delta z \to 0} \frac{\Delta w_1}{\Delta z} = \lim_{\Delta w_1 \to 0} \frac{\Delta w_1}{\Delta z} = \lim_{\Delta w_1 \to 0} \left(1 / \frac{\Delta z}{\Delta w_1}\right)$$

$$= 1 / \lim_{\Delta w_1 \to 0} \frac{\Delta z}{\Delta w_1} = \frac{1}{dz/dw_1} = \frac{1}{2w_1}.$$

For $z \in D_0$, $w_1 \neq 0$; thus dw_1/dz exists (and is finite). Consequently, w_1 is a single-valued analytic function of $z \in D_0$. Also, since $w_2 = -w_1$, w_2 is also a single-valued analytic function of $z \in D_0$.

Each of the functions w_1 and w_2 is called a *branch* in D_0 of the many-valued function w.

This now prompts us to make the following definition.

8.10.2

Definition Given a many-valued function $w = f(z)$ defined in a domain D of the complex plane, let $D_0 \subseteq D$. Then, by a *branch* of w in D_0 we mean a single-valued analytic function $w_1 = g(z)$ defined in D_0, such that $g(z)$ is one of the values of $f(z)$ for any $z \in D_0$.

Different authors go to considerable lengths to clarify the concept of a many-valued function and its associated mappings; however, a clear and concise formulation of the concept cannot be adequately handled without the further concept of a *Riemann surface*. We shall not develop the concept of a Riemann surface here, but point out that several elegant texts exist on the subject.*

---

* Springer, G., Introduction to Riemann Surfaces, Addison-Wesley Publishing Company, Inc., Reading, Mass., (1957).

8.10.3 The Function $w^2 = z$

We return to the example $w^2 = z$ and instead of taking $0 < \theta < 2\pi$, we restrict θ to $-\pi < \theta < \pi$. Setting $z = re^{i\theta}$ and denoting by D_π the domain consisting of the z-plane with the nonpositive half of the real axis excluded, we again find the values of w in D_π by the branches:

$$w_1 = \sqrt{r}\, e^{i\theta/2} \quad \text{and} \quad w_2 = -w_1; \quad r > 0, \quad -\pi < \theta < \pi.$$

Actually, any ray $\theta = \alpha$ determines a domain D_α: $\alpha < \theta < \alpha + 2\pi$, $r > 0$, in which $w = \sqrt{z}$ has two branches, namely, $w_1 = \sqrt{r}\, e^{i(\theta - \alpha)}$ and $w_2 = \sqrt{r}\, e^{i(\theta - \alpha + \pi)}$. If we examine the behavior of $w = \sqrt{z}$ as z moves around a circle with center at the origin and radius r_0, starting from a fixed point $z_0 = r_0 e^{i\theta_0}$ and increasing θ continuously, then the point z traverses the circle counterclockwise. Let us write $W = \sqrt{r_0}\, e^{i\theta/2}$; then W is a single-valued continuous function of θ, varying continuously as z moves around the circle and $W^2 = z$. As θ increases by 2π, the function W becomes multiplied by -1 since

$$\sqrt{r_0}\, e^{i(\theta + 2\pi)/2} = -\sqrt{r_0}\, e^{i\theta/2}.$$

Thus, if the value of W corresponding to the point z_0 is originally given by one of the branches of w in D_α, where $\alpha = 0, \pi$, then we observe that, after encircling the origin once and returning to the point z_0, the value of w assigned to it is now given by another branch. If we traverse a circle or any closed Jordan curve (on which z_0 is a point and which does not intersect the branch cut) by considering the geometry of the situation, it is clear that a changeover to a different branch does *not* occur.

This property that distinguishes the origin from all other points is one of the main characteristics of what we call a *branch point*.

8.10.4 The Function $w = \operatorname{Log} z$

The origin $z = 0$ is a branch point for the function $w = \operatorname{Log} z$ which has an infinite number of branches in D_π:

$$w_n = \log r + i \arg z + i 2n\pi; \quad n = 0, \pm 1, \ldots$$

There is no unique way of dividing up a function into branches; however, whenever we fix a branch cut, then every point z_0 on the cut is a singularity of every branch that determines the function (not necessarily an *infinity* of the function), since the derivative does not exist at the point z_0.

We now show that if $f \in A$ and is nonzero in some suitable domain, then we can always find a single-valued branch of $\log f(z)$ in that domain.

8.10.5

Lemma *If $f(z)$ is analytic and nonzero on a simply-connected domain Ω, then there exists a single-valued function $F(z)$ defined on Ω such that $f(z) = e^{F(z)}$ on Ω and $F(z_0) = \log f(z_0)$, where z_0 is some point of Ω. We say* that $F(z)$ is a branch of $\log f(z)$ in Ω that assumes the value $\log f(z_0)$ at z_0.

PROOF. Since f'/f is analytic in Ω, we have

$$\int_\gamma f'/f \, dz = 0$$

for any closed rectifiable Jordan curve γ in Ω. Hence

$$\int_{z_0}^\xi \frac{f'(z)}{f(z)} \, dz$$

does not depend on the path in Ω joining z_0 to ξ. Put

$$F(\xi) = \log f(z_0) + \int_{z_0}^\xi \frac{f'(z)}{f(z)} \, dz, \quad \xi \in \Omega.$$

Then

$$\frac{dF}{dz} = \frac{f'(z)}{f(z)} \ (z \in \Omega), \quad \text{and} \quad F(z_0) = \log f(z_0).$$

Hence, $g(z) = e^{-F(z)}f(z)$ has a zero derivative in Ω and is, therefore, constant. Since

$$g(z_0) = e^{-\log f(z_0)}f(z_0) = 1,$$

it follows that $e^{F(z)} = f(z)$ in Ω. □

A more recondite way of defining branches is to define an analytic function in a more general sense. Thus we could define an analytic function as the set of all function elements that can be obtained by a continuation from any element of the function, i.e., in the sense we have been discussing analytic continuation in the previous section. Let us define a further concept.

8.10.6

Definition Two analytic functions f and g with circular domains of analyticity D_1 and D_2, respectively, are called *immediate continuations* of each other if and only if
  i. $D_1 \cap D_2 \neq \emptyset$ and
  ii. $g(z) = f(z), \forall z \in D_1 \cap D_2$.

Thus it follows from our previous analysis that for any analytic function f with circular domain of analyticity D_0 and any circular domain D there exists at most one immediate continuation of f from D_0 into D. Analytic continuation is then immediate continuation of immediate continuation in a finite number of steps.

For example,
$$f_1(z) = \sum_{n=0}^{\infty} z^n \quad \text{in} \quad D_0 = \{z \mid |z| < 1\}$$
can be immediately continued to
$$f_2(z) = \frac{1}{1-i} \sum_{n=0}^{\infty} \left(\frac{z-i}{1-i}\right)^n \quad \text{in} \quad D_1 = \{z \mid |z-i| < \sqrt{2}\}$$
and $f_1(z) = f_2(z)$ in the intersection of these domains. We repeat that in general we cannot expect to find the immediately continuation of an analytic function (assuming it has such a continuation) in a closed form.

Thus, given an analytic function f defined in a domain D, let g be the immediate continuation of f into a domain S. Thus, $D \cap S \neq \emptyset$. Let h be another immediate continuation of f into a domain T. Thus, $D \cap T \neq \emptyset$. In general, it may happen that $g \neq h$ in $S \cap T$; then we say that f is many-(two in this case) valued.

8.11

As a further and slightly more complicated example, we study the nature of the branches and branch points of the many-valued function
$$w = \sqrt{z^2 - 1}$$
given by
$$w^2 = z^2 - 1.$$
Let
$$z - 1 = \rho_1 e^{i\theta_1}$$
and
$$z + 1 = \rho_2 e^{i\theta_2}.$$
Then $w^2 = z^2 - 1$ corresponds to $w^2 = \rho_1 \rho_2 e^{i(\theta_1 + \theta_2)}$. Consequently
$$w = \sqrt{\rho_1 \rho_2} \exp\left\{i\left(\frac{\theta_1 + \theta_2}{2} + k\pi\right)\right\}, \quad k = 0, 1.$$

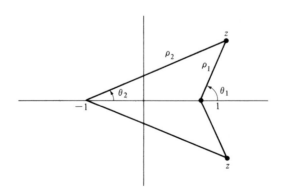

If we make the restrictions
$$-\pi < \theta_1 \leq \pi \quad \text{and} \quad 0 \leq \theta_2 < 2\pi,$$
we obtain two single-valued functions of z:
$$w_1 = \sqrt{\rho_1 \rho_2}\, e^{i(\theta_1 + \theta_2)/2} \quad \text{and} \quad w_2 = -w_1.$$
If we examine $\lim(\theta_1 + \theta_2)$ as z approaches any point on the real axis to the right of $z = 1$, then $\theta_1 + \theta_2 \to 0$ if approached from above, and 2π if approached from below. Consequently, the two limiting values of $(\theta_1 + \theta_2)/2$ differ by π. Thus w_1 (and also w_2) is discontinuous at such a point. Similarly w_1 (and w_2) is discontinuous at points on the real axis to the left of $z = -1$. Hence, in order to make w_1 and w_2 continuous, we restrict ourselves to the domain D_1 consisting of all points of the z-plane except the points $z = x + 0i$, $|x| \geq 1$, that is, the complex plane slit or cut along $y = 0$, $|x| \geq 1$. As in 8.10.1, we can show that $w_1 \in A$ in D_1 with $dw_1/dz = z/w_1$. Furthermore, for each $z \in D_1$, the value of w_1 is one of the values of the multi-valued function $w = \sqrt{z^2 - 1}$. Hence w_1 (and w_2) is a branch in D_1 of the multi-valued function $w = \sqrt{z^2 - 1}$. The half-lines $x \geq 1$, $y = 0$ and $x \leq -1$, $y = 0$ are branch cuts for the branches w_1 and w_2.

We can make the following restrictions on θ_1 and θ_2:
$$-\pi < \theta_1 \leq \pi \quad \text{and} \quad -\pi < \theta_2 \leq \pi.$$
Then we again have two single-valued functions of z:
$$W_1 = \sqrt{\rho_1 \rho_2}\, e^{i[(\theta_1 + \theta_2)/2]} \quad \text{and} \quad W_2 = -W_1.$$
If z_0 is any point on the real axis such that $|z_0| > 1$, then W_1 approaches the same value regardless of the manner in which $z \to z_0$. However, if z_0 is real and $|z_0| < 1$, then W_1 approaches different values depending on whether $z \to z_0$ from above or below. Thus in order to make W_1 continuous, we restrict ourselves to the domain D_2 consisting of all points in the z-plane except the points $z = x + 0i$, $|x| \leq 1$. As above, W_1 and W_2 are branches in D_2 of the multi-valued function $w = \sqrt{z^2 - 1}$. The branch cut is now the segment $-1 \leq x \leq 1$, $y = 0$.

Note that each point $z \neq \pm 1$ is in some domain (say D_1 or D_2) in which there are two branches of the function w given by $w^2 = z^2 - 1$. However, this is not true for the points $z = 1$ and $z = -1$. Applying an argument similar to that used in 8.10.3, we see that $z = 1$ and $z = -1$ are branch points of the multi-valued function $w = \sqrt{z^2 - 1}$.

8.12

Since we have the concept of a general power of a complex number, we terminate this chapter with an application of an infinite series to convergence.

EXAMPLE. Show that the series $\sum_{n=1}^{\infty} 1/n^{1+i\alpha}$ diverges for all real α.

SOLUTION 1. Let $\mu = \alpha + i\beta$, where α, β real and $\alpha > 0$. Let

$$I_n = \frac{1}{n^\mu} - \int_n^{n+1} \frac{dx}{x^\mu} = \int_n^{n+1} dx \int_n^x \frac{\mu\, dy}{y^{\mu+1}} \, ;$$

then

$$|I_n| \leq |\mu| \int_n^{n+1} dx \int_n^x \frac{dy}{y^{\alpha+1}} = \frac{|\mu|}{\alpha} \int_n^{n+1} \left(\frac{1}{n^\alpha} - \frac{1}{x^\alpha} \right) dx$$

$$\leq \frac{|\mu|}{\alpha} \left(\frac{1}{n^\alpha} - \frac{1}{(n+1)^\alpha} \right).$$

Thus

$$|I_1| + \cdots + |I_n| \leq \frac{|\mu|}{\alpha}\left(1 - \frac{1}{(n+1)^\alpha}\right) \to \frac{|\mu|}{\alpha} \quad \text{as} \quad n \to \infty$$

and $\sum_n I_n$ is convergent (absolutely). Consequently,

$$I_1 + I_2 + \cdots + I_n = \frac{1}{1^\mu} + \frac{1}{2^\mu} + \cdots + \frac{1}{n^\mu} - \int_1^{n+1} \frac{dx}{x^\mu} \to l \quad \text{(finite)}$$

as $n \to \infty$. Now, if $\alpha = 1$, then

$$\int_1^{n+1} \frac{dx}{x^\mu} = \int_1^{n+1} \frac{dx}{x^{1+i\beta}} = \frac{1}{i\beta}\left[1 - \frac{1}{(n+1)^{i\beta}} \right]$$

$$= \frac{1}{i\beta}\left[1 - \cos(\beta \log \overline{n+1}) + i \sin(\beta \log \overline{n+1}) \right]$$

$$\nrightarrow \text{limit as } n \to \infty,$$

since it oscillates finitely. Thus,

$$\frac{1}{1^\mu} + \cdots + \frac{1}{n^\mu} \nrightarrow \text{limit}$$

and $\sum_n 1/n^\mu$ diverges.

SOLUTION 2. Consider the sequence $\{n^{-i\alpha}\}$, where α is real. The sequence clearly diverges.

Thus $\sum_{n=2}^{k} \{ n^{-i\alpha} - (n-1)^{-i\alpha} \} = k^{-i\alpha} - 1$ diverges since it does not tend to a finite limit as $k \to \infty$.

Now

$$n^{-i\alpha} - (n-1)^{-i\alpha} = n^{-i\alpha}\left[1 - \left(1 - \frac{1}{n}\right)^{-i\alpha} \right]$$

$$= n^{-i\alpha}\left\{ -\frac{i\alpha}{n} + O\!\left(\frac{1}{n^2}\right)\right\},$$

since
$$\left| \frac{1}{n^r} \binom{-i\alpha}{r} \right| = O(1) \quad \text{as } r \to \infty;$$
thus,
$$a_n = n^{-i\alpha} - (n-1)^{-i\alpha} = -\frac{i\alpha}{n^{1+i\alpha}} + O\left\{ \frac{1}{n^{2+i\alpha}} \right\}$$
and $\sum_n 1/n^{1+i\alpha}$ diverges, since $\sum_n a_n$ diverges.

9

Laurent Series, Singularities

9.1

We now study the behavior of a function that may have points in the plane at which it ceases to be analytic. We first develop a power-series expansion (in positive and negative powers) of a function that is analytic in some annulus of the plane, but ceases to be analytic at the center of the annulus.

9.1.1

Theorem (Laurent) *If $f(z) \in A$ in $r < |z - a| < R$, where $0 \leqslant r < R$ (R may be finite or infinite), then*

$$f(z) = \sum_{n=0}^{\infty} a_n(z-a)^n + \sum_{n=1}^{\infty} a_{-n}(z-a)^{-n} \quad \text{for} \quad r < |z-a| < R,$$

where

$$a_n = \frac{1}{2\pi i} \int_C \frac{f(z)}{(z-a)^{n+1}} dz, (n = 0, \pm 1, \ldots),$$

and C is the circle $|z - a| = \rho$, $0 \leqslant r < \rho < R$.

Note: By Cauchy's theorem, a_n is independent of the choice of ρ, provided $r < \rho < R$.

PROOF. Take h such that $r < |h| < R$. Then, choose ρ_1, ρ_2 such that $r < \rho_1 < |h| < \rho_2 < R$. Let C_1, C_2 be the circles $|z - a| = \rho_1$, $|z - a| = \rho_2$, respectively. By Cauchy's integral formula,

$$f(a+h) = \frac{1}{2\pi i} \left[\int_{C_2} \frac{f(z)\,dz}{(z-a-h)} - \int_{C_1} \frac{f(z)}{(z-a-h)} dz \right].$$

Laurent Series, Singularities

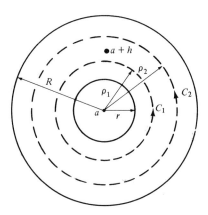

Now

$$\left|\frac{z-a}{h}\right| = \frac{\rho_1}{|h|} < 1 \quad \text{on} \quad C_1$$

and f is bounded on C_1, thus

$$\sum_{n=0}^{\infty}\left[\frac{z-a}{h}\right]^n f(z)$$

is a uniformly convergent series of continuous functions on C_1. Consequently,

$$-\int_{C_1}\frac{f(z)}{(z-a-h)}\,dz = \frac{1}{h}\int_{C_1} f(z)\left[1 - \frac{z-a}{h}\right]^{-1} dz$$

$$= \frac{1}{h}\int_{C_1} f(z) \sum_{n=0}^{\infty}\left[\frac{z-a}{h}\right]^n dz$$

$$= \sum_{n=0}^{\infty}\left[\frac{1}{h^{n+1}}\int_{C_1} f(z)(z-a)^n\,dz\right]$$

$$= 2\pi i \sum_{n=1}^{\infty} a_{-n} h^{-n}.$$

Similarly,

$$\int_{C_2}\frac{f(z)}{(z-a-h)}\,dz = \sum_{n=0}^{\infty} 2\pi i a_n h^n$$

and we have

$$f(a+h) = \sum_{n=0}^{\infty} a_n h^n + \sum_{n=1}^{\infty} a_{-n} h^{-n}.$$

Since both series are absolutely convergent for $r < |h| < R$, we can write without ambiguity that

$$f(a+h) = \sum_{n=-\infty}^{\infty} a_n h^n.$$

Set $z = a + h$, then

$$f(z) = \sum_{n=-\infty}^{\infty} a_n (z-a)^n, \quad \text{for} \quad r < |z-a| < R. \qquad \square$$

9.1.2 Extension of Cauchy's Inequalities

Theorem *Under the conditions for Laurent's theorem, that is, for $r < \rho < R$ and $n = 0, \pm 1, \ldots$,*

$$|a_n| \leq \frac{M(\rho)}{\rho^n}, \quad \text{where} \quad M(\rho) = \max_{|z-a|=\rho} |f(z)|.$$

PROOF. For $r < \rho < R$ and $n = 0 \pm 1, \ldots$, we have

$$a_n = \frac{1}{2\pi i} \int_C \frac{f(z)}{(z-a)^{n+1}} dz,$$

where C is the circle $|z - a| = \rho$. On C:

$$\left| \frac{f(z)}{(z-a)^{n+1}} \right| \leq \frac{M(\rho)}{\rho^{n+1}}, \quad \text{thus} \quad |a_n| \leq \frac{1}{2\pi} \frac{M(\rho)}{\rho^{n+1}} 2\pi\rho = \frac{M(\rho)}{\rho^n}. \qquad \square$$

We now illustrate Laurent's theorem by examples.

9.1.3

EXAMPLE 1. Given $f(z) = e^{(u/2)[z-(1/z)]}$, where $|z| > 0$, show that f can be expressed by the series

$$\sum_{n=-\infty}^{\infty} J_n(u) z^n, \quad \text{where} \quad J_n(u) = \frac{1}{2\pi} \int_0^{2\pi} \cos(u \sin\theta - n\theta) \, d\theta.$$

Show also that $J_n(u) = (-1)^n J_{-n}(u)$. We call $J_n(u)$ the *Bessel function* of order n.

SOLUTION. $f(z)$ is clearly analytic in $|z| > 0$, thus we can expand by Laurent's theorem and get

$$f(z) = \sum_{n=-\infty}^{\infty} a_n z^n,$$

where

$$a_n = \frac{1}{2\pi i} \int_C \frac{e^{(u/2)[z-(1/z)]}}{z^{n+1}} dz \quad (\text{when } a = 0).$$

Laurent Series, Singularities

Choose C: $|z| = 1$, then

$$a_n = \frac{1}{2\pi i} \int_0^{2\pi} \frac{e^{(u/2)}(e^{i\theta} - e^{-i\theta})}{e^{ni\theta + i\theta}} i e^{i\theta} \, d\theta$$

$$= \frac{1}{2\pi} \int_0^{2\pi} e^{(u/2) 2i \sin \theta} [\cos n\theta - i \sin n\theta] \, d\theta$$

$$= \frac{1}{2\pi} \int_0^{2\pi} [\cos(u \sin \theta) + i \sin(u \sin \theta)][\cos n\theta - i \sin n\theta] \, d\theta$$

$$= \frac{1}{2\pi} \int_0^{2\pi} [\cos(u \sin \theta - n\theta) + i \sin(u \sin \theta - n\theta)] \, d\theta.$$

Thus, by transforming the integral ($\theta = 2\pi - \varphi$) we obtain

$$I\{a_n\} = \frac{1}{2\pi} \int_0^{2\pi} \sin(u \sin \theta - n\theta) \, d\theta = 0$$

Consequently,

$$a_n \equiv J_n(u) = \frac{1}{2\pi} \int_0^{2\pi} \cos(u \sin \theta - n\theta) \, d\theta.$$

Further, since

$$e^{(u/2)[z - (1/z)]} = \sum_{n=-\infty}^{\infty} J_n(u) z^n,$$

changing the index of summation n to $-n$, we have

$$e^{(u/2)[z - (1/z)]} = \sum_{n=\infty}^{-\infty} J_{-n}(u) z^{-n}.$$

Now replace z by $-1/z$; then

$$e^{(u/2)[z - (1/z)]} = \sum_{n=-\infty}^{\infty} J_{-n}(u)(-1)^n z^n$$

and $J_n(u) = (-1)^n J_{-n}(u)$.

We can go further and show, for example, that

$$J_n(u) = \frac{u^n}{2^n} \sum_{k=0}^{\infty} \frac{(-1)^k u^{2k}}{2^{2k}(n+k)! k!},$$

where n is a nonnegative integer. Since

$$e^{uz/2} e^{-u/2z} = \sum_{n=-\infty}^{\infty} J_n(u) z^n$$

and both Taylor series for the exponentials converge absolutely for all z, then

$$\sum_{n=0}^{\infty} \frac{u^n z^n}{2^n n!} \sum_{m=0}^{\infty} \frac{(-1)^m u^m}{2^m z^m m!} = \sum_{n=-\infty}^{\infty} C_n(u) z^n,$$

where C_n is given as follows.
We choose the coefficient of z^n from this formal product:

$$z^n: \quad \frac{u^n}{2^n n!} \cdot 1 + \frac{u^{n+1}}{2^{n+1}(n+1)!}(-1)\frac{u}{2 \cdot 1!} + \cdots$$

$$+ \frac{u^{n+k}}{2^{n+k}(n+k)!} \frac{(-1)^k u^k}{2^k k!} + \cdots.$$

Thus

$$C_n = \frac{u^n}{2^n} \sum_{k=0}^{\infty} \frac{u^{2k}(-1)^k}{2^{2k}(n+k)!k!} \equiv J_n(u).$$

This can all be generalized for arbitrary real or complex n.*

It is not always necessary to use the integral representation to determine the coefficients a_n in the Laurent expansion of a function analytic in a given annulus.

EXAMPLE 2. Find the Laurent series expansion of

$$f(z) = \frac{1}{(z-1)(z-2)} \in A$$

in a) $|z| < 1$, b) $1 < |z| < 2$, and c) $|z| > 2$. Consider

$$f(z) = \frac{1}{z-2} - \frac{1}{z-1}.$$

a)

$$\frac{1}{z-2} = -\frac{1}{2}\frac{1}{1-(z/2)} = -\frac{1}{2}(1-z/2)^{-1}$$

$$= -\frac{1}{2}\left(1 + \frac{z}{2} + \frac{z^2}{4} + \cdots + \frac{z^n}{2^n} + \cdots\right)$$

and

$$\frac{1}{z-1} = -(1-z)^{-1} = -(1 + z + z^2 + \cdots + z^n + \cdots);$$

---

* See, for example, F. Bowman, *Introduction to Bessel functions*. New York: Dover Publications, Inc., 1958.

hence
$$f(z) = \frac{1}{2} + \frac{3z}{4} + \frac{7z}{8} + \cdots$$

b)
$$\frac{1}{z-2} = -\frac{1}{2}\left(1 + \frac{z}{2} + \frac{z^2}{4} + \cdots\right) \quad \text{if} \quad |z| < 2$$

and
$$\frac{1}{z-1} = \frac{1}{z}\left(1 - \frac{1}{z}\right)^{-1}$$
$$= \frac{1}{z}\left(1 + \frac{1}{z} + \frac{1}{z^2} + \cdots + \frac{1}{z^n} + \cdots\right) \quad \text{if} \quad |z| > 1;$$

consequently,
$$f(z) = -\frac{1}{2}\left(1 + \frac{z}{2} + \frac{z^2}{4} + \cdots\right) - \left(\frac{1}{z} + \frac{1}{z^2} + \cdots\right).$$

c)
$$\frac{1}{z-1} = \frac{1}{z}\left(1 + \frac{1}{z} + \frac{1}{z^2} + \cdots\right)$$

and
$$\frac{1}{z-2} = \frac{1}{z}\left(1 - \frac{2}{z}\right)^{-1} = \frac{1}{z}\left(1 + \frac{2}{z} + \frac{4}{z^2} + \cdots\right), \quad |z| > 2;$$

thus
$$f(z) = \frac{1}{z^2} + \frac{3}{z^3} + \cdots.$$

9.1.4
Exercises

Expand the following functions in a Laurent series in the given domain:

1: $\dfrac{z^2 - 2z + 5}{(z-2)(z^2+1)}$ in $|z| > 2$ and $1 < |z| < 2$.

2: $z^2 e^{1/z}$ in $|z| > 0$.

3: $\dfrac{1}{(z^2+1)^2}$ in the neighborhood of $z = i$.

4: $\sin z \sin \dfrac{1}{z}$ in $0 < |z| < \infty$.

5: $\dfrac{z-1}{(z+1)(2z-3)}$ in $\begin{cases} |z| < 1 \\ 1 < |z| < 3/2 \\ |z| > 3/2 \end{cases}$.

6: $\log \dfrac{z-a}{z-b}$ in the neighborhood of infinity.

7: Show that the Laurent expansion of $\operatorname{cosec} z$ about $z = 0$ is given by

$$\operatorname{cosec} z = \frac{1}{z} + \frac{z}{6} + \frac{7z^3}{360} + \cdots$$

What is the radius of convergence of this series?

8: Show that

$$[\log(1+z)]^2 = 2 \sum_{n=1}^{\infty} \frac{(-1)^{n-1}}{n+1} H_n z^{n+1},$$

$|z| < 1$ and $H_n = 1 + \dfrac{1}{2} + \cdots + \dfrac{1}{n}$.

9: Show that the Laurent series expansion of $f(z) = 1/(e^z - 1)$ about the origin is

$$\sum_{n=0}^{\infty} \frac{B_n}{n!} z^{n-1}.$$

The numbers B_n are called Bernoulli numbers. Find the first six Bernoulli numbers. Deduce that

$$z \coth z = \sum_{n=0}^{\infty} \frac{B_{2n}}{(2n)!} (2z)^{2n}.$$

10: Show that the Laurent series expansion of $f(z) = \operatorname{sech} z$ about the origin is

$$\sum_{n=0}^{\infty} \frac{E_n}{n!} z^n.$$

The numbers E_n are called Euler numbers. Find the first six Euler numbers. Show that $E_{2k-1} = 0$; $k = 0, 1, 2, \ldots$. Deduce that

$$\sec z = \sum_{n=0}^{\infty} (-1)^n \frac{E_{2n}}{(2n)!} z^{2n}.$$

11: Show that

$$\tan z = \sum_{n=1}^{\infty} (-1)^{n-1} \frac{B_{2n}}{(2n)!} (2^{2n} - 1) 2^{2n} z^{2n-1}.$$

Laurent Series, Singularities

As in the case of the Taylor development of a function, we have the equivalent of Liouville's theorem as follows.

9.1.5

Theorem *If $f(z) \in A$ in $|z| > r$ and $f(z) = O(|z|^k)$ as $|k| \to \infty$, where k is an integer (positive, negative, or zero), then*

$$f(z) = \sum_{n=-\infty}^{k} a_n z^n \quad \text{for} \quad |z| > r,$$

where

$$a_n = \frac{1}{2\pi i} \int_C \frac{f(z)}{z^{n+1}} dz \ldots, \tag{9.1}$$

C being the circle $|z| = \rho > r$.

PROOF. By Laurent's theorem,

$$f(z) = \sum_{-\infty}^{\infty} a_n z^n \quad \text{for} \quad |z| > r,$$

where a_n is given by (9.1). By Cauchy's inequalities,

$$|a_n| \leq \frac{M(\rho)}{\rho^n} \quad \text{for} \quad \rho > r \quad \text{and all} \quad n.$$

From the second condition it follows that there exist K, r_0 ($K > 0, r_0 > r$) such that

$$|f(z)| \leq K|z|^k \quad \text{for} \quad |z| \geq r_0.$$

Thus,

$$\frac{M(\rho)}{\rho^k} \leq K \quad \text{and} \quad |a_n| \leq \frac{K\rho^k}{\rho^n} \quad \text{when} \quad \rho \geq r_0.$$

When $\rho \to \infty$, it follows that $a_n = 0$ for $n > k$. □

9.1.6

Theorem *If $f(z) \in A$ in $0 < |z| < R$ and $f(z) = O(|z|^k)$ as $|z| \to 0$, where k is an integer (positive, negative, or zero), then*

$$f(z) = \sum_{n=k}^{\infty} a_n z^n, \quad \text{for} \quad 0 < |z| < R,$$

where

$$a_n = \frac{1}{2\pi i} \int_C \frac{f(z)}{z^{n+1}} dz \quad \text{(by (9.1))},$$

C being the circle $|z| < \rho$ ($0 < \rho < R$).

PROOF. As above,

$$f(z) = \sum_{-\infty}^{\infty} a_n z^n \quad \text{for} \quad 0 < |z| < R,$$

where a_n is given by (9.1). By Cauchy's inequalities,

$$|a_n| \leq \frac{M(\rho)}{\rho^n} \quad \text{for} \quad 0 < \rho < R \text{ and all } n.$$

By the second condition, there exists $K, r_0 (K > 0, 0 < r_0 < R)$ such that

$$|f(z)| \leq K|z|^k \quad \text{for} \quad 0 < |z| \leq r_0.$$

Thus

$$\frac{M(\rho)}{\rho^k} \leq K \quad \text{and} \quad |a_n| \leq \frac{K\rho^k}{\rho^n} \quad \text{when} \quad 0 < \rho \leq r_0.$$

When $\rho \to +0$, it follows that $a_n = 0$ for $n < k$. □

With these results we can now prove a remarkable theorem of Weierstrass concerning the values taken by an entire function.

9.1.7

Theorem *Suppose $f(z)$ is entire and not a polynomial and c is an arbitrary complex number. Then given ϵ, $R > 0$, there exists z such that $|z| > R$ and $|f(z) - c| < \epsilon$.*

PROOF. The result is immediate if $f(z)$ assumes the value c outside every circle $|z| = r$ $(r > 0)$.

Now suppose that there exists $r > 0$ such that $|f(z) - c| \geq \delta > 0$ for $|z| > r$. Then

$$\frac{1}{f(z) - c} \in A \quad \text{in} \quad |z| > r.$$

Assume, if possible, that

$$\left| \frac{1}{f(z) - c} \right| = O(1) \quad \text{as} \quad |z| \to \infty;$$

then, by the extensions to Liouville's theorem,

$$\frac{1}{f(z) - c}$$

has a Laurent expansion of the form

$$\sum_{n=-\infty}^{0} a_n z^n = \sum_{n=0}^{\infty} \frac{b_n}{z^n} \quad \text{for} \quad |z| > r.$$

Laurent Series, Singularities   151

Let b_m be the first nonvanishing coefficient, then for $|z| > r$

$$\frac{z^m}{f(z) - c} = b_m + b_{m+1}\frac{1}{z} + b_{m+2}\frac{1}{z^2} + \cdots \to b_m(\neq 0) \quad \text{as} \quad |z| \to \infty$$

(by the continuity of the power series $b_m + b_{m+1}\xi + b_{m+2}\xi^2 + \cdots$, where $\xi = 1/z$ at the point $\xi = 0$).

Thus $f(z) - c = O(|z|^m)$ as $|z| \to \infty$, whence $f(z)$ is a polynomial of degree $\leq m$. The contradiction shows that

$$\left|\frac{1}{f(z) - c}\right|$$

is not bounded above in $|z| > r$. That is, given $R, \epsilon > 0$, there exists z such that $|z| > R$ and

$$\frac{1}{|f(z) - c|} > \frac{1}{\epsilon} \quad \text{or} \quad |f(z) - c| < \epsilon. \qquad \square$$

REMARK. A much more difficult theorem of E. Picard says the following: "A nonconstant entire function assumes every finite complex value with one possible exception." For example, $e^z \neq 0$ in the finite plane.

An alternative statement of Theorem 9.1.7 was suggested by Weierstrass:

If R is an arbitrary positive number with $f(z)$ and c as before and S is the set of values of $f(z)$ for $|z| > R$, then the closure $\bar{S}$ is the whole complex plane.

Thus Weierstrass shows that the set of values of f is dense everywhere, whereas Picard shows that we can omit at most one value.

9.2 Convergence of Laurent series of the form $\sum_{-\infty}^{\infty} a_n z^n$.

By definition, $\sum_{-\infty}^{\infty}$ is convergent if and only if $\sum_{-\infty}^{-1}$ and $\sum_{0}^{\infty}$ are both convergent. Now, $\sum_{n=0}^{\infty} a_n z^n$ and $\sum_{n=1}^{\infty} a_{-n} z^{-n}$ are power series in z and $\xi = 1/z$ respectively. Suppose these two series have finite nonzero radii of convergence $|z| = R$, $|\xi| = 1/r$. Then $\sum_{n=0}^{\infty} a_n z^n$ is both uniformly and absolutely convergent in the domain $|z| \leq R'$ ($0 \leq R' < R$) and hence may be differentiated term by term for $|z| < R$.

$\sum_{n=1}^{\infty} a_{-n} z^{-n}$ is uniformly and absolutely convergent in $|1/z| \leq 1/r'$, that is, $|z| \geq r'$, where $1/r' < 1/r$ or $r' > r$. Thus its sum is an analytic function in $|z| > r$ and may be differentiated term by term in $|z| > r$.

If $r < R$, then $\sum_{n=-\infty}^{\infty} a_n z^n$ is absolutely convergent in the annulus $r < |z| < R$ and is uniformly convergent in $r' \leq |z| \leq R'$, where

$$r < r' \leq R' < R.$$

The sum is analytic in $r < |z| < R$ and the series may be differentiated term by term in $r < |z| < R$.

Once we have the Laurent expansion of a function f in some annulus, we would expect the expansion to be unique by virtue of our previous study of analytic continuation. This expectation is confirmed as follows.

9.2.1

Theorem *If* $\sum_{-\infty}^{\infty} a_n(z-a)^n = \sum_{-\infty}^{\infty} b_n(z-a)^n$, *where* $0 \leq r < |z-a| < R$ *and both series are convergent in this domain, then* $a_n = b_n$ *for* $n = 0, \pm 1, \ldots$.

PROOF. Take ρ such that $r < \rho < R$. Let C be the circle $|z-a| = \rho$. Then

$$\sum_{1}^{\infty} a_{-n}(z-a)^{-n}, \quad \sum_{0}^{\infty} a_n(z-a)^n, \quad \sum_{1}^{\infty} b_{-n}(z-a)^{-n}, \quad \sum_{0}^{\infty} b_n(z-a)^n$$

are all uniformly convergent series of continuous functions on C. Since $|1/(z-a)^{n+1}|$ is bounded on C and $1/(z-a)^{n+1}$ is continuous on C (for all n), we may multiply the original equation throughout by $1/(z-a)^{n+1}$ and integrate term by term along C, obtaining $a_n = b_n$, $n = 0, \pm 1, \ldots$. □

We now turn our attention to the study of points at which a function ceases to be analytic.

9.3 Singularities

9.3.1

Definition The point $z = a$ is called an *isolated singularity* of $f(z)$ if there exists $\delta > 0$, such that $f(z) \in A$ in $0 < |z-a| < \delta$, but not in $|z-a| < \delta$.

In this case, for $0 < |z-a| < \delta$ we have

$$f(z) = \sum_{-\infty}^{\infty} a_n(z-a)^n$$

where

$$a_n = \frac{1}{2\pi i} \int_C \frac{f(z)}{(z-a)^{n+1}} dz$$

and C is the circle $|z-a| = \rho$ $(0 < \rho < \delta)$.

9.3.2

Definition If $f(z) = \sum_{-\infty}^{\infty} a_n(z-a)^n$ for $0 < |z-a| < \delta$, then

$$\sum_{-\infty}^{-1} a_n(z-a)^n = \sum_{1}^{\infty} \frac{a_{-n}}{(z-a)^n}$$

is called the *principal part* of $f(z)$ at $z = a$.

Laurent Series, Singularities

Note that the principal part of $f(z)$ at $z = a$ is a power series in $\xi = 1/(z - a)$ convergent for $|\xi| > 1/\delta$, so that the radius of convergence is infinite.

There are three cases to consider:

1. $a_n = 0$, $\forall n < 0$ (the principal part is identically zero);
2. There exists a positive integer m such that $a_{-m} \neq 0$ but $a_{-n} = 0$ for $n > m$ (the principal part is a polynomial in ξ of degree ≥ 1);
3. $\sum_{n=1}^{\infty} a_{-n}/(z - a)^n$ contains an infinite number of terms for $0 < |z - a| < \delta$.

Case 1. For $0 < |z - a| < \delta$ we have

$$f(z) = \sum_{n=0}^{\infty} a_n(z - a)^n \to a_0 \quad \text{as} \quad z \to a$$

by continuity of the power series at a point interior to the circle of convergence. Write

$$g(z) = \begin{cases} f(z) & \text{for } 0 < |z - a| < \delta, \\ a_0 & \text{for } z = a. \end{cases}$$

Then $g(z) \in A$ in $|z - a| < \delta$. We call $z = a$ a *removable* singularity of $f(z)$.
For example, consider the function $\sin z/z$. In $|z| > 0$,

$$\frac{\sin z}{z} = \sum_{n=1}^{\infty} (-1)^{n-1} \frac{z^{2n-2}}{(2n-1)!}.$$

If we define $(\sin z)/z = 1$ for $z = 0$, then we can write

$$g(z) = \begin{cases} \dfrac{\sin z}{z} & \text{for } z \neq 0, \\ 1 & \text{for } z = 0, \end{cases}$$

and $g(z) \in A$ in $|z| \geq 0$.

Case 2. For $0 < |z - a| < \delta$, $\delta > 0$, $m > 0$ we have

$$f(z) = \sum_{n=-m}^{\infty} a_n(z - a)^n$$

$$= \frac{b_m}{(z - a)^m} + \frac{b_{m-1}}{(z - a)^{m-1}} + \cdots + \frac{b_1}{z - a} + \sum_{n=0}^{\infty} a_n(z - a)^n,$$

where $b_m \neq 0$, $a_{-n} = b_n$ for $n = 1, 2, \ldots, m$.

We then say that $f(z)$ has a *pole* of order m at $z = a$. In this case, for $0 < |z - a| < \delta$ we have

$$f(z) = \frac{1}{(z - a)^m} [b_m + b_{m-1}(z - a) + \cdots] = \frac{\varphi(z)}{(z - a)^m},$$

where $\varphi(z) \in A$ for $|z - a| < \delta$ and $\varphi(a) = b_m \neq 0$.

Thus $(z - a)^m f(z)$ has a removable singularity at $z = a$ and
$$|(z - a)^m f(z)| \to |b_m| \neq 0 \quad \text{as} \quad z \to a.$$
Consequently we have
$$|f(z)| = O(|z - a|^{-m}) \quad \text{as} \quad z \to a.$$
Since
$$|f(z)| = \frac{|\varphi(z)|}{|z - a|^m} \quad \text{and} \quad |\varphi(z)| \to |b_m| \neq 0 \quad \text{as} \quad z \to a,$$
it follows that $|f(z)| \to \infty$ as $z \to a$.

For example,
$$f(z) = \frac{1}{(z - 2)^3 (z - 4)^4}$$
has a pole of order three at $z = 2$ and a pole of order four at $z = 4$.

Case 3. Here $f(z) = \sum_{-\infty}^{\infty} a_n (z - a)^n$, where an infinite subset of the coefficients $\{a_{-1}, a_{-2}, \ldots, a_{-n}, \ldots\}$ is nonzero. Then $f(z)$ is said to have an *essential* singularity at $z = a$.

Essential singularities can be isolated or not. Thus
$$e^{1/z} = 1 + \frac{1}{z} + \frac{1}{2! z^2} + \cdots$$
has an isolated essential singularity at $z = 0$, whereas $\operatorname{cosec}(1/z)$ has an essential singularity at $z = 0$; it is not isolated since $\sin(1/z) = 0$ for $z = 1/(n\pi)$, $n = 0, \pm 1, \ldots$.

Thus $z = 0$ is a point of accumulation of poles of this function. All other poles are isolated. We note that the point $z = \infty$ is also a pole (i.e., an isolated singularity). The behavior of a function at an essential singularity is quite complicated, as we shall presently see.

9.3.3

Definition (Case 2 or 3). If $f(z)$ has a pole or isolated essential singularity at $z = a$, the coefficient $b_1 = a_{-1}$ of $1/(z - a)$ in the Laurent expansion of $f(z)$ about $z = a$ is called the *residue* of $f(z)$ at $z = a$.

Thus
$$b_1 = \frac{1}{2\pi i} \int_C f(z) \, dz,$$
where $n = -1$ in the expression for
$$a_n = \frac{1}{2\pi i} \int_C \frac{f(z)}{(z - a)^{n+1}} \, dz$$
and C is the circle $|z - a| = \rho$, $0 < \rho < \delta$.

Laurent Series, Singularities

Note: In the case of a pole of order one (called a *simple* pole) we have $(z - a)f(z) \to b_1$ as $z \to a$. Thus the residue of $f(z)$ at a simple pole $z = a$ is $\lim_{z \to a}(z - a)f(z)$.

9.3.4 Conditions for a Pole or Isolated Essential Singularity

Suppose $f(z) \in A$ for $0 < |z - a| < \delta$ ($\delta > 0$).

1. *Pole.* If $f(z) = O\{(z - a)^k\}$ (k an integer) as $z \to a$, then by the extension of Liouville's theorem,

$$f(z) = \sum_{n=k}^{\infty} a_n(z - a)^n, \quad 0 < |z - a| < \delta.$$

Hence if $f(z) = O\{(z - a)^{-m}\}$ (for m a positive integer) as $z \to a$ and $|f(z)| \to \infty$ as $z \to a$, then $f(z)$ has a pole at $z = a$ of order not exceeding m.

If the poles were of order $m - 1$, then $f(z) = O\{(z - a)^{-m+1}\}$ as $z \to a$. Thus if

$$f(z) = O\{(z - a)^{-m}\} \quad \text{but} \quad f(z) \neq O\{(z - a)^{-m+1}\} \quad \text{as } z \to a,$$

then $f(z)$ has a pole of order m at $z = a$.

2. *Isolated Essential Singularity* (i.e.s). At an isolated essential singularity, $f(z) \neq O\{(z - a)^{-m}\}$ for any positive integer m, that is, $|(z - a)^m f(z)|$ is not bounded in any neighborhood of $z = a$. Thus, given any positive integer m and $K, \eta > 0$, there exists z such that

$$0 < |z - a| < \eta \quad \text{and} \quad |f(z)| > \frac{K}{|z - a|^m}.$$

9.3.5 Point at Infinity

We assume that there exists one point at infinity in the complex plane. Suppose $f(z)$ is defined for $0 < R < |z| < \infty$. Then $f(1/\xi)$ is defined for $0 < |\xi| < 1/R$. We say that $f(z)$ has a property at ∞, if $f(1/\xi)$ has the corresponding property at $\xi = 0$, assuming that $f(1/\xi)$ is definable by continuity at $\xi = 0$.

Suppose $f(z) \in A$ in $|z| > R$ with the Laurent expansion

$$f(z) = \sum_{n=-\infty}^{\infty} a_n z^n \quad |z| > R.$$

Then

$$f\left(\frac{1}{\xi}\right) = \sum_{n=-\infty}^{\infty} a_n \left(\frac{1}{\xi^n}\right), \quad 0 < |\xi| < \frac{1}{R}.$$

Thus, if $f(1/\xi)$ has an isolated singularity at $\xi = 0$, we say that $f(z)$ has an isolated singularity at $z = \infty$, and $z = \infty$ is an i.e.s. if $\sum_{n=1}^{\infty} a_n z^n$ contains an infinite number of nonzero terms. If the series contains only a finite number of (nonzero) terms, then $z = \infty$ is a pole. Thus a polynomial of degree n has a pole of order n $(n \geq 1)$ at $z = \infty$. An entire function (not a polynomial) has an i.e.s. at $z = \infty$.

We study the behavior of a function in the neighborhood of an i.e.s. with the following theorem.

9.3.6

Theorem (Weierstrass). *If $z = a$ is an i.e.s. of $f(z)$, then given any complex number c and arbitrary positive numbers ϵ, η, there exists a point z such that*

$$|f(z) - c| < \epsilon \quad \text{and} \quad 0 < |z - a| < \eta.$$

PROOF. Suppose $f(z) \in A$ for $0 < |z - a| < \delta$. The result is immediate if $f(z)$ assumes the value c inside every domain $0 < |z - a| < \rho$, $\rho > 0$.

Now suppose there exists $\rho > 0$ and $\rho < \delta$ such that $|f(z) - c| \geq \delta > 0$ and $0 < |z - a| < \rho$. Then

$$\frac{1}{f(z) - c} \in A \quad \text{in} \quad 0 < |z - a| < \rho.$$

Suppose if possible that

$$\frac{1}{f(z) - c} = O(1) \quad \text{as} \quad z \to a.$$

Then by the extension of Liouville's theorem, it has a Laurent expansion of the form

$$\sum_{n=0}^{\infty} a_n(z - a)^n, \quad 0 < |z - a| < \rho.$$

Let a_m ($m \geq 0$) be the first nonzero coefficient in this series. Then, by continuity of the power series at any point within the open circle of convergence, we have

$$\frac{(z - a)^{-m}}{f(z) - c} = a_m + a_{m+1}(z - a) + \cdots \to a_m \neq 0 \quad \text{as} \quad z \to a.$$

Hence $f(z) - c = O(|z - a|^{-m})$ as $z \to a$ and thus

$$f(z) = O(|z - a|^{-m}) \quad \text{as} \quad z \to a,$$

and

$$f(z) = \sum_{n=-m}^{\infty} a_n(z - a)^n \quad \text{and} \quad 0 < |z - a| < \delta.$$

Consequently $f(z)$ does not have an essential singularity at $z = a$. The contradiction proves the theorem. □

Note 1: As a consequence, at an isolated essential singularity $f(z)$ is at the same time unbounded and arbitrarily close to zero.

Laurent Series, Singularities

Note 2: Theorem 9.1.6 is clearly a special case of Theorem 9.3.6, since an entire function (not a polynomial) has an i.e.s. at $z = \infty$.

We sum up our observations as follows:

1. If $|f(z)|$ approaches a finite limit as $z \to a$ (isolated singularity), then the singularity is removable.
2. If $|f(z)| \to \infty$ as $z \to a$ (isolated singularity), then $z = a$ is a pole.
3. If $|f(z)|$ oscillates as $z \to a$ (isolated singularity), then $z = a$ is an essential singularity.

As before with the Weierstrass theorem for an entire function, Picard's theorem says something much stronger. We only state the theorem as follows.*

9.3.7

Theorem *In every neighborhood of an i.e.s., an analytic function takes on every value (with one possible exception) an infinite number of times.*

EXAMPLE. $f(z) = e^{1/z}$ has an i.e.s. at $z = 0$. We show that in every neighborhood of $z = 0$ there exists an infinite number of values of z that satisfy $e^{1/z} = i$.

SOLUTION
$$e^{1/z} = i = e^{(i\pi/2) + 2k\pi i}, \quad k = 0, \pm 1, \ldots$$
thus
$$\frac{1}{z} = \frac{i\pi}{2} + 2k\pi i, \quad k = 0, \pm 1, \ldots .$$

Consequently $e^{1/z} = i$ for an infinite sequence $\{z_k\}$ approaching $z = 0$, namely,
$$\{z_k\} = \left\{ \frac{2}{(1 + 4k)\pi i} \right\}.$$

The exceptional value in this case is $z = 0$ itself. □

9.3.8 Limit Points of Zeros and Poles

If $f(z) \in A$ and $f(z) \not\equiv 0$ in a domain D, then the zeros of $f(z)$ have no limit point belonging to D. Another way of saying this is that a function $f(z) \not\equiv 0$ cannot be analytic at any point a which is a limit point of zeros.

EXAMPLE
$$\sin\frac{1}{z} = \frac{1}{z} - \frac{1}{3! z^3} + \cdots, \quad |z| > 0$$

---

*For a proof, see A.S.B. Holland, *Introduction to the Theory of Entire Functions*. New York: Academic Press, 1973.

has an i.e.s. at $z = 0$:

$$\sin \frac{1}{z} = 0 \quad \text{at} \quad z = \frac{1}{n\pi}; \quad n = \pm 1, \pm 2, \ldots,$$

and these zeros have a limit point at $z = 0$.

A limit point of zeros of a function $f(z)$ cannot be a pole, since if $\{z_n\}$ is a sequence of distinct points such that $a = \lim_{n \to \infty} z_n$ and $f(z_n) = 0$ for $n = 1, 2, \ldots$, then $|f(z)| \not\to \infty$ as $z_n \to a$.

Also, a limit point $z = a$ of zeros of $f(z)$, not identically zero, cannot be a removable singularity, for otherwise it is readily seen that $f(z)$ would be identically zero in some domain $0 < |z - a| < \eta$.

If $z = a$ is a limit point of poles of $f(z)$, then there exist no positive δ such that $f(z) \in A$ in $0 < |z - a| < \delta$, that is, a is not an isolated singularity. Such a point is called an *essential singularity*.

9.3.9
Exercises

Find the singular points of the following functions, explain their nature, and discuss the behavior at infinity:

1: $\dfrac{1}{z - 2z^2}$ 2: $\dfrac{1}{z(z^2 + 9)^2}$ 3: $\dfrac{z^6}{1 + z^6}$

4: $\dfrac{2z^5}{(1 - z)^3}$ 5: $\dfrac{z}{e^z}$ 6: $\cot z - \dfrac{2}{z}$

7: $\dfrac{1}{\sin z - \sin a}$ 8: $\csc \dfrac{1}{z - 1}$ 9: $\sin \dfrac{1}{z} + \dfrac{1}{z^2}$

10: $e^{-1/z}$ 11: $e^{\tan(1/z)}$ 12: $\sin \dfrac{1}{\sin(1/z)}$

13: $\dfrac{e^{1/z}}{e^z - 1}$ 14: $\tan^2 z$

9.4 Meromorphic Functions, Theory of Residues, and Evaluation of Integrals

9.4.1

Definition If $f(z) \in A$ in a domain D (which may be the whole finite plane), except at singularities which are poles, then we say that $f(z)$ is *meromorphic* in D.

Let us write $f \in M$ in D to mean "f is meromorphic in a domain D." Thus every entire function is meromorphic but the converse is not necessarily true.

Since a limit point of poles is an essential singularity, the poles of f cannot have a limit point $z_0 \in D$. Hence the number of poles in any bounded closed region $\mathcal{R} \subseteq D$ is finite.

9.4.2 Rational Functions

Consider the expression $f(z) = P(z)/Q(z)$, where

$$P(z) = a_0 + a_1 z + \cdots + a_p z^p \quad a_p \neq 0,$$
$$Q(z) = b_0 + b_1 z + \cdots + b_q z^q, \quad b_q \neq 0.$$

Then $f(z) \in A$ except at the zeros of $Q(z)$. Suppose $z = a$ is a zero of order n of $Q(z)$. Then $Q(z) = (z-a)^n \theta(z)$, where $\theta(a) \neq 0$, and hence $|\theta(z)| > 0$ for $|z - a| < \delta$ (positive). Also, $P(z) = (z-a)^m \varphi(z)$, where m is zero or a positive integer and $\varphi(a) \neq 0$. Then

$$\frac{P(z)}{Q(z)} = \frac{1}{(z-a)^{n-m}} \frac{\varphi(z)}{\theta(z)} = \frac{1}{(z-a)^{n-m}} \psi(z),$$

where $\psi(z) \in A$ for $|z-a| < \delta$ and $\psi(a) \neq 0$. Thus $P(z)/Q(z)$ has a pole at $z = a$, unless $m \geq n$ when there exists a removable singularity at $z = a$. Hence $P(z)/Q(z)$ is a meromorphic function with at most q poles.

EXAMPLE. $f(z) = \sin(\pi/z)$ has simple zeros at $z = 1/n$, $n = \pm 1, \pm 2, \ldots$, because

$$f'(z) = -\frac{\pi}{z^2} \cos \frac{\pi}{z} \neq 0 \quad \text{at} \quad z = \frac{1}{n},$$

thus $\operatorname{cosec}(\pi/z)$ has simple poles at $z = 1/n$, $n = \pm 1, \pm 2, \ldots$.

9.4.3 Decomposition of a Rational Function into Partial Fractions

Consider $f(z) = P(z)/Q(z)$, where P and Q are as before. Let P and Q have no common factors and $q \geq 1$. Then $f(z)$ has a finite number of poles $\gamma_1, \ldots, \gamma_v$, $v \leq q$, and no other singularities for finite z.

Let $h_r(z)$ be the principal part of $f(z)$ at γ_r, $r = 1, 2, \ldots, v$. Then

$$g(z) = f(z) - \sum_{r=1}^{v} h_r(z)$$

is analytic except perhaps at points γ_r. Clearly $f(z) - h_r(z)$ has a removable singularity at γ_r, while $h_i(z) \in A$ at γ_r if $i \neq r$, $i, r = 1, \ldots, v$. Thus $g(z)$ has a removable singularity at $\gamma_1, \ldots, \gamma_v$. Defining $g(z)$ by continuity at these points γ_r, we see that the function $g(z)$ so obtained is an entire function.

Now,
$$\left.\begin{array}{l} f(z) = O(|z|^{p-q}) \\ h_r(z) = O\left(\dfrac{1}{|z|}\right) \end{array}\right\} \text{ as } |z| \to \infty.$$

Hence, if $p > q$, then $g(z) = O(|z|^{p-q})$ and, by the extensions of Liouville's theorem, $g(z)$ is a polynomial of degree not exceeding $p - q$. If $p \leqslant q$, then $g(z) = O(1)$ as $|z| \to \infty$ and $g(z) = \text{const}\,(0$ if $p < q)$. Thus,

$$f(z) = g(z) + \sum_{r=1}^{v} h_r(z),$$

where $g(z)$ is a polynomial, and

$$h_r(z) = \frac{c_1}{z - \gamma_r} + \cdots + \frac{c_{m_r}}{(z - \gamma_r)^{m_r}}, \quad r = 1, \ldots, v,$$

where m_r is the order of the zero of Q at $z = \gamma_r$.

9.4.4

Theorem (Residue). *Let D be a bounded domain, whose boundary C consists of a finite number of closed rectifiable Jordan curves. Suppose $f(z) \in A$ in D except for a finite number of isolated singularities at $z_1, z_2, \ldots, z_n$. Let $f(z)$ be continuous on $\overline{D} - \{z_1, z_2, \ldots, z_n\}$. Then*

$$\int_C f(z)\,dz = 2\pi i \sum_{r=1}^{n} b_r,$$

where b_r is the residue of $f(z)$ at z_r, each portion of C being traversed in the positive sense.

PROOF. Let γ_r be the circle $|z - z_r| = \rho_r$, $r = 1, \ldots, n$, where the positive numbers $\rho_1, \ldots, \rho_n$ are chosen such that all the sets $\{z \mid |z - z_r| \leqslant \rho_r\} \subseteq D$ and are disjoint. Then, by Cauchy's theorem,

$$\int_C f(z)\,dz = \sum_{r=1}^{n} \int_{\gamma_r} f(z)\,dz,$$

since $f(z) \in A$ in the domain D bounded by $C, \gamma_1, \ldots, \gamma_n$, and f is continuous on $\overline{D}$.

Since $\int_{\gamma_r} f(z)\,dz = 2\pi i b_r$, $r = 1, \ldots, n$, then

$$\int_C f(z)\,dz = 2\pi i \sum_{r=1}^{n} b_r. \qquad \square$$

9.4.5 Residue at a Pole of Order $m (\geqslant 1)$

Let $f(z)$ have a pole of order m at $z = z_0$; then

$$f(z) = \frac{b_m}{(z - z_0)^m} + \frac{b_{m-1}}{(z - z_0)^{m-1}} + \cdots + \frac{b_1}{(z - z_0)} + \sum_{n=0}^{\infty} a_n(z - z_0)^n,$$

where $b_m \neq 0$ and $0 < |z - z_0| < r$.

Let $\varphi(z) = (z - z_0)^m f(z)$; then $\varphi(z) = b_m + b_{m-1}(z - z_0) + \cdots$ has a removable singularity at z_0 since we define $\varphi(z_0) = b_m$. Thus $\varphi(z) \in A$ in $|z - z_0| < r$. The residue of f at $z = z_0$ is the coefficient b_1, that is, the coefficient of $(z - z_0)^{m-1}$ in the Taylor expansion of $\varphi(z)$ about z_0. Consequently

$$b_1 = \frac{\varphi^{[m-1]}(z_0)}{(m-1)!} = \frac{1}{(m-1)!} \frac{d^{m-1}}{dz^{m-1}} \left[(z-z_0)^m f(z)\right]_{z=z_0}.$$

For a simple pole we get $m = 1$ and

$$b_1 = \lim_{z \to z_0} \left[(z - z_0)f(z)\right].$$

Let us write Res $f(z_0)$ to mean the residue of the function $f(z)$ at the pole $z = z_0$.

EXAMPLE 1

$$f(z) = \frac{z+2}{z^3 + z^2} = \frac{z+2}{z^2(z+1)}$$

has a pole of order 1 at $z = -1$;

$$\text{Res } f(-1) = \lim_{z \to -1} \frac{z+2}{z^2} = 1$$

and a pole of order 2 at $z = 0$;

$$\text{Res } f(0) = \lim_{z \to 0} \frac{d}{dz} \frac{z+2}{z+1} = -1.$$

EXAMPLE 2

$$f(z) = \frac{9}{(z-7)^5} + \frac{1}{(z-7)^2} + z + (z-7)^3$$

has a pole of order 5 at $z = 7$; Res $f(7) = 0$ since f is in the form of a Laurent expansion about $z = 7$ and the coefficient of $1/(z - 7)$ is zero.

Suppose $f(z) = g(z)/h(z)$, where $g, h \in A$ in $|z - a| < \delta (> 0)$ with $g(z) \neq 0$, and $z = a$ is a simple zero of h. Then the residue of $f(z)$ at $z = a$ is given by

$$\lim_{z \to a} (z - a)f(z) = \lim_{z \to a} \frac{g(z)}{\dfrac{h(z) - h(a)}{z - a}} = \frac{g(a)}{h'(a)}.$$

Note that $h(a) = 0$.

EXAMPLE 3

$$f(z) = \frac{1}{z^2 \sin z};$$

we require the residue of $f(z)$ at $z = \pi$. We identify $g(z) = 1/z^2$,

$h(z) = \sin z$, and then
$$\text{Res } f(\pi) = \frac{1}{\pi^2}\left[\frac{1}{\cos z}\right]_{z=\pi} = -\frac{1}{\pi^2}.$$
In fact, the residue at the simple zeros of $\sin z$ (namely, at $z = n\pi$, where $n = \pm 1, \pm 2, \ldots$) is given by
$$\text{Res } f(n\pi) = \frac{(-1)^n}{n^2\pi^2}.$$
At $z = 0$, the function f has a pole of order 3. We can either evaluate
$$\frac{1}{2!}\frac{d^2}{dz^2}\left(\frac{z}{\sin z}\right)$$
or find the Laurent expansion of f about zero:
$$\sin z = z - \frac{z^3}{3!} + \cdots,$$
thus
$$\operatorname{cosec} z = \frac{1}{z}\left[1 - \frac{z^2}{6} + \frac{z^4}{24} \cdots\right]^{-1}$$
$$= \frac{1}{z}\left[1 + \left(\frac{z^2}{6} + \cdots\right) + \left(\frac{z^2}{6} + \cdots\right)^2 + \cdots\right],$$
consequently,
$$f(z) = \frac{\operatorname{cosec} z}{z^2} = \frac{1}{z^3}\left[1 + \frac{z^2}{6} + \cdots\right] = \frac{1}{z^3} + \frac{1}{6z} + \cdots$$
and the coefficient of $1/z$ is clearly $\frac{1}{6}$.

There is no quick or easy way to obtain the residue of a function at a pole of high order. We can look at the problem in a slightly different way.

Let $f(z) = P(z)/Q(z)$, with $P(z_0) \ne 0$, $Q(z_0) = 0$, $Q'(z_0) \ne 0$. Thus f has a simple pole at $z = z_0$. If $P, Q \in A$ in $|z - z_0| < r$, we can write a Taylor series for P and Q as follows:
$$f(z) = \frac{P(z)}{Q(z)} = \frac{P(z_0) + P'(z_0)(z - z_0) + \cdots}{Q'(z_0)(z - z_0) + \cdots};$$
thus,
$$(z - z_0)f(z) = \frac{P(z_0) + P'(z_0)(z - z_0) + \cdots}{Q'(z_0) + \frac{1}{2}Q''(z_0)(z - z_0) + \cdots}.$$
Since the numerator and denominator are power series in $(z - z_0)$ and $P(z_0) \ne 0$, we have
$$\lim_{z \to z_0}(z - z_0)f(z) = \frac{P(z_0)}{Q'(z_0)},$$
which confirms our previous result. However, if $P(z_0) \ne 0$, $Q(z_0) = 0$,

Laurent Series, Singularities

$Q'(z_0) = 0$, and $Q''(z_0) \neq 0$, then clearly f has a pole of order 2. By the same method as above, we can calculate the residue of f at the pole $z = z_0$ of order 2. Prove that

$$\text{Res } f(z_0) = \frac{2P'(z_0)}{Q''(z_0)} - \frac{2P(z_0)Q'''(z_0)}{3[Q''(z_0)]^2}.$$

EXAMPLE 4. Consider

$$f(z) = \frac{1}{z(e^z - 1)}.$$

This has a pole of order 2 (why?) at $z = 0$.

a. The value of

$$\frac{d}{dz}[z^2 f(z)] = \frac{e^z - 1 - ze^z}{(e^z - 1)^2}$$

is indeterminate at $z = 0$. Using L'Hôpital's rule (twice), we get the value at $z = 0$ as $-\frac{1}{2}$.

b. $P(z) = 1$, $Q(z) = z(e^z - 1)$, $P'(0) = 0$, $Q'(0) = 0$, $Q''(0) = 2$, $Q'''(0) = 3$. Thus by the above formula,

$$\text{Res } f(0) = -\frac{2}{3} \cdot \frac{1 \cdot 3}{(2)^2} = -\frac{1}{2}.$$

c. Using a Laurent series as far as the term in $1/z$, we obtain

$$f(z) = \frac{1}{z^2\left(1 + \dfrac{z}{2} + \cdots\right)}.$$

Thus

$$f(z) = \frac{1}{z^2}\left(1 - \frac{z}{2} + \cdots\right) = \frac{1}{z^2} - \frac{1}{2z} + \cdots$$

and the coefficient of $1/z$ is $-\frac{1}{2}$.

All three methods in the last example are of approximately the same length. If power series (about the appropriate point) are readily available, method (c) can be used to obtain the coefficient of $1/(z - z_0)$ without too much difficulty; otherwise it is necessary to use the general method given at the beginning of the section. For high-order poles, method (b) becomes far too cumbersome.

9.4.6 Application of Residues to Partial Fractions

Residues can be used to obtain the coefficients in partial-fraction expansion of functions as follows. Suppose we are given $1/[z^2(z + 1)]$. By the general theory of partial fractions, we know that

$$f(z) = \frac{1}{z^2(z + 1)} = \frac{A}{z^2} + \frac{B}{z} + \frac{C}{z + 1}.$$

Thus the right-hand side gives the principal parts of f at each singularity, namely, $z = 0, 1$. Consequently C is the residue of f at the simple pole $z = 1$, that is,

$$\lim_{z \to 1} (z + 1)f(z) = \lim_{z \to 1} \frac{1}{z^2} = 1 = C;$$

B is the residue of f at the pole $z = 0$ of order 2, that is,

$$\lim_{z \to 0} \frac{d}{dz}\left[\frac{1}{z+1}\right] = -1 = B;$$

A is the residue of $zf(z)$ at $z = 0$ since

$$zf(z) = \frac{1}{z(z+1)} = \frac{A}{z} + B + \frac{Cz}{z+1} = \frac{A}{z} + B + C\left[1 - \frac{1}{z+1}\right].$$

Thus

$$A = \lim_{z \to 0} \frac{1}{z+1} = 1 \quad \text{and} \quad f(z) = \frac{1}{z^2} - \frac{1}{z} + \frac{1}{z+1}.$$

9.4.7
Exercises

Use residues to find the partial fraction decomposition of

1: $\dfrac{1}{(z-1)^2(z+1)^2}$

2: $\dfrac{2}{(z-3)(3z+4)}$

3: $\dfrac{z^2 - 1}{z^3 - 1}$

4: $\dfrac{2z^2 + 3z + 1}{(z-2)(z^2+4)}$

9.4.8 Residue at Infinity

Let $f(z)$ have an isolated singular point at $z = \infty$. The residue at infinity is defined by the expression

$$\text{Res } f(\infty) = -\frac{1}{2\pi i} \int_C f(z)\, dz,$$

where C is the circle $|z| = R$, so large that the only singularity of f in $|z| \geqslant R$ is the point $z = \infty$. The minus sign is used because the point at infinity, at which the residue is being computed, is exterior to the circle and the circle is negatively oriented with respect to its exterior.

The residue at infinity can also be defined as $-a_{-1}$, where a_{-1} is the

Laurent Series, Singularities

coefficient of $1/z$ in the Laurent expansion of $f(z)$ at infinity. Hence, $-\operatorname{Res} f(\infty)$ is the coefficient of ξ in the expansion of $f(1/\xi)$ in the neighborhood of the origin. It is the coefficient of $1/\xi$ in the expansion of

$$F(\xi) = \frac{1}{\xi^2} f\left(\frac{1}{\xi}\right)$$

in the neighborhood of $z = 0$.

Using the change of variable $z = 1/\xi$, we get

$$\int_{|z|=R} f(z)\,dz = \int_{|\xi|=1/R} f\left(\frac{1}{\xi}\right) \frac{d\xi}{\xi^2} ;$$

thus the definition of $\operatorname{Res} f(\infty)$ above, which is also $\operatorname{Res} F(0)$, is consistent. The value of

$$-\frac{1}{2\pi i} \int_C f(z)\,dz$$

clearly does not depend on R.

EXAMPLE 5. Find the residue at infinity ($z = \infty$) of

$$f(z) = \sqrt{(z-a)(z-b)} , \quad \text{if} \quad \sqrt{} \equiv +\sqrt{} ,$$

SOLUTION

$$f\left(\frac{1}{\xi}\right) = \left[(1 - a\xi)^{1/2}(1 - b\xi)^{1/2}\right]\frac{1}{\xi}$$

$$= \frac{1}{\xi}\left[1 - \frac{1}{2}a\xi + \frac{\frac{1}{2}(\frac{1}{2}-1)}{2}a^2\xi^2 + \cdots\right]$$

$$\times \left[1 - \frac{1}{2}b\xi + \frac{\frac{1}{2}(\frac{1}{2}-1)}{2}b^2\xi^2 + \cdots\right].$$

The coefficient of $1/\xi$ in $(1/\xi^2)f(1/\xi)$ is given by

$$\frac{1}{\xi^3}\left[-\frac{1}{8}a^2 - \frac{1}{8}b^2 + \frac{1}{4}ab\right]\xi^2 = \frac{1}{\xi}\left(-\frac{1}{8}\right)(a^2 - 2ab + b^2).$$

Thus $\operatorname{Res} f(\infty) = \frac{1}{8}(a-b)^2$.

EXAMPLE 6. Find the residue of

$$f(z) = \frac{z^\alpha}{1 - \sqrt{z}} \quad \text{at} \quad z = 1 \quad \text{and} \quad 0 < \alpha < 1.$$

SOLUTION. We assume that $z^\alpha = \exp(\alpha \log z)$ and we take the square root to be positive. Also, we take a branch of the log function that excludes the negative real axis.

The order of the pole at $z = 1$ is 1 since

$$\frac{z^\alpha}{1-\sqrt{z}} = O\{(z-1)^{-1}\} \quad \text{as} \quad z \to 1.$$

Thus

$$\frac{(z-1)z^\alpha}{1-\sqrt{z}} = -(\sqrt{z}+1)z^\alpha \to -2 \exp(\alpha \log 1) \quad \text{as} \quad z \to 1$$

and Res $f(1) = -2e^{\alpha 2\pi k i}$; $k = 0, \pm 1, \ldots$. If we take the principal value of z^α, then the residue is -2.

EXAMPLE 7. Find the residue of $\sin[z/(1+z)]$ at $z = -1$.

SOLUTION

$$\sin\frac{z}{1+z} = \sin\left[1 - \frac{1}{1+z}\right] = \left[1 - \frac{1}{1+z}\right] - \frac{1}{3!}\left[1 - \frac{1}{1+z}\right]^3 + \cdots.$$

The coefficient of $1/(1+z)$ can be picked out from each bracket as follows:

$$-1 + \frac{3}{3!} - \frac{5}{5!} \cdots = -\left(1 - \frac{1}{2!} + \frac{1}{4!} + \cdots\right) = -\cos 1 = \operatorname{Res} f(-1).$$

Exercises

Find the residues of the following functions at the isolated singularities.

1: $\sin z \sin \dfrac{1}{z}$ 2: $\dfrac{\sqrt{z}}{1-z}$ 3: $e^{z+1/z}$

4: $\sec \dfrac{1}{z}$ 5: $\log z \sin \dfrac{1}{1-z}$

9.5 Evaluation of Integrals

We now study an important application of the theory of residues, viz., the evaluation of so-called nonelementary integrals. Some of the following integrals can be evaluated by techniques of real-variable theory, however, the introduction of suitable contours in a complex integral will often shorten the calculations as well as give results of great generality.

9.5.1 Improper Integrals

The Cauchy principal value (C.V.) of $\int_{-\infty}^{\infty} f(x)\,dx$ is given by

$$\text{C.V.}\int_{-\infty}^{\infty} f(x)\,dx = \lim_{R \to \infty}\left[\int_{-R}^{0} f(x)\,dx + \int_{0}^{R} f(x)\,dx\right],$$

provided the limit exists, that is, $\int_{-\infty}^{\infty} f(x)\,dx$ converges in the Cauchy sense when we evaluate the right-hand side.

Laurent Series, Singularities

If we wish to say that $\int_{-\infty}^{\infty} f(x)\,dx$ is just "convergent," then we define

$$\int_{-\infty}^{\infty} f(x)\,dx = \lim_{R_1 \to \infty} \int_{-R_1}^{0} f(x)\,dx + \lim_{R_2 \to \infty} \int_{0}^{R_2} f(x)\,dx. \quad (9.2)$$

Convergence of $\int_{-\infty}^{\infty}$ implies convergence in the Cauchy sense, however, the converse is not necessarily true; for example,

$$\int_{-R}^{R} 2x\,dx = x^2\Big]_{-R}^{R} = 0,$$

thus C.V. $\int_{-\infty}^{\infty} 2x\,dx$ exists and equals zero, but neither limit in (9.2) exists. The C.V. for odd functions is zero.

Case when the integrand approaches infinity at $c \in [a,b]$, where a,b are finite. Consider

$$\int_{0}^{c-\epsilon} f(x)\,dx + \int_{c+\eta}^{b} f(x)\,dx, \quad \text{where} \quad \epsilon > 0, \quad \eta > 0.$$

If the first term approaches a limit as $\epsilon \to 0$, and the second term approaches a limit as $\eta \to 0$, then we define

$$\int_{b}^{a} f(x)\,dx = \lim_{\epsilon \to 0^+} \int_{a}^{c-\epsilon} f(x)\,dx + \lim_{\eta \to 0^+} \int_{c+\eta}^{b} f(x)\,dx.$$

If the limits fail to exist, then

$$\lim_{\epsilon \to 0^+} \left[\int_{a}^{c-\epsilon} f(x)\,dx + \int_{c+\epsilon}^{b} f(x)\,dx \right]$$

may exist and is called the Cauchy principal value, that is, C.V. $\int_{a}^{b} f(x)\,dx$. For example, $\int_{-1}^{1} dx/x$ does not exist, but C.V. $\int_{-1}^{1} dx/x = 0$.

Note: $\int_{\eta}^{1} dx/x = \log(1/\eta) \to \infty$ as $\eta \to 0^+$, however

$$\int_{-1}^{-\epsilon} \frac{1}{x}\,dx + \int_{\epsilon}^{1} \frac{1}{x}\,dx = \log(-\epsilon) - \log(-1) + \log 1 - \log \epsilon$$

$$= \log(-1) + \log \epsilon - \log(-1) + \log \frac{1}{\epsilon}$$

$$= \log \epsilon + \log \frac{1}{\epsilon} = 0.$$

9.5.2 Integrals Involving Rational Functions

Theorem *Let $P(x)$, $Q(x)$ be polynomials of degree p, q respectively. Suppose $Q(x)$ has no real zeros and P, Q have real coefficients. Then*

a) $\int_{-\infty}^{\infty} [P(x)/Q(x)]\,dx$ *exists when* $q \geq p + 2$;

b) $\int_{-\infty}^{\infty} [P(x)/Q(x)e^{mix}]\,dx$ *exists when* $q \geq p + 1$, $m \neq 0$, *and m is real.*

We require the following form of the second mean-value theorem for

integrals, viz., Bonnet's form:*
1. If $\varphi(x) \geq 0$ and decreases or $\varphi \leq 0$ and increases, and if $\psi(x) \in C[a,b]$, then there exists $\xi \in [a,b]$ such that

$$\int_a^b \varphi(x)\psi(x)\,dx = \varphi(a)\int_a^\xi \psi(x)\,dx.$$

2. If $\varphi(x) \geq 0$ and increases or $\varphi(x) \leq 0$ and decreases, and if $\psi(x) \in C[a,b]$, then there exists $\xi \in [a,b]$ such that

$$\int_a^b \varphi(x)\psi(x)\,dx = \varphi(b)\int_\xi^b \psi(x)\,dx.$$

PROOF

a) There exists $k, r_0 > 0$ such that

$$\left|\frac{P(x)}{Q(x)}\right| \leq \frac{k}{|x|^2} \quad \text{for} \quad |x| \geq r_0.$$

Now, $\int_{r_0}^\infty dx/x^2$ exists, hence, by the comparison test, $\int_{r_0}^\infty |P(x)/Q(x)|\,dx$ exists, whence $\int_{r_0}^\infty [P(x)/Q(x)]\,dx$ exists.

Similarly, $\int_{-\infty}^{-r_0} [P(x)/Q(x)]\,dx$ exists. Since $P/Q \in C[-r_0, r_0]$, then the integral over $[-r_0, r_0]$ exists. Thus $\int_0^\infty (P/Q)\,dx$ and $\int_{-\infty}^0 (P/Q)\,dx$ exist, so that $\int_{-\infty}^\infty (P/Q)\,dx$ exists.

b) Assume that the coefficients of P, Q are real. Otherwise we must consider $\Re\{P/Q\}$ and $\Im\{P/Q\}$ separately. Let $f(x) = P(x)/Q(x)$; then $f'(x) = \theta(x)/[Q(x)]^2$, where θ is a polynomial of degree not exceeding $p + q - 1$. Thus $f'(x)$ is of constant sign for $x > a$, where a is the greatest zero (in absolute value) of $\theta(x)$, so that $f(x)$ is monotonic for $x \geq |a|$.

Now, $F(x) = \int_0^x f(t)\,dt$ approaches a finite limit as $x \to \infty$ if and only if for any $\epsilon > 0$ there exists $x_0 > 0$ such that $|F(x_2) - F(x_1)| < \epsilon$ whenever $x_1, x_2 > x_0$.

Take x_1, x_2 such that $x_2 > x_1 > |a|$. Since $q \geq p + 1$, the function $f(x) \to 0$ as $x \to \infty$; thus either f is positive and decreases to $+0$ as $x \to \infty$ for $x > |a|$ or f is negative and increases to -0 as $x \to \infty$ for $x > |a|$.

By Bonnet's form of the second mean-value theorem, there exists ξ such that $x_1 < \xi < x_2$ and $\int_{x_1}^{x_2} f(x) \cos mx\,dx = f(x_1)\int_{x_1}^\xi \cos mt\,dt$. Thus,

$$\left|\int_{x_1}^{x_2} f(x) \cos mx\,dx\right| \leq 2\frac{|f(x_1)|}{|m|}.$$

However, from our assumption we know that, given $\epsilon > 0$, there exists $x_0(\epsilon) > 0$ such that $|f(x)| < \epsilon|m|/2$ for $x > x_0$; thus,

$$\left|\int_{x_1}^{x_2} f(x) \cos mx\,dx\right| < \epsilon \quad \text{when} \quad x_2 > x_1 > \max(x_0, |a|),$$

therefore $\int_0^\infty f(x)\cos mx\,dx$ exists.

---

* See D. V. Widder, *Advanced Calculus*. 2nd Ed., Prentice-Hall, p. 163, 1961.

Similarly, $\int_0^\infty f(x)\sin mx\, dx$ exists, therefore $\int_0^\infty f(x)e^{imx}\, dx$ exists. Similarly, $\int_{-\infty}^0 f(x)e^{imx}\, dx$ exists and thus $\int_{-\infty}^\infty f(x)e^{imx}\, dx$ exists. □

Two theorems now arise corresponding to a general class of integrals.

9.5.3

Theorem *If $P(x), Q(x)$ are polynomials of degree p, q respectively, with $q \geqslant p + 2$, and $Q(x)$ has no real zeros, then*

$$\int_{-\infty}^\infty \frac{P(x)}{Q(x)}\, dx = 2\pi i \sigma,$$

where σ is the sum of the residues of $P(z)/Q(z)$ at the zeros of $Q(z)$ in the domain $\Im\{z\} > 0$.

PROOF. Since Q has only a finite number of zeros, there exists $\rho > 0$ such that Q has no zeros in $|z| > \rho$. Take $R > \rho$ and let C_R be the semicircle $|z| = R, 0 \leqslant \arg z \leqslant \pi$.

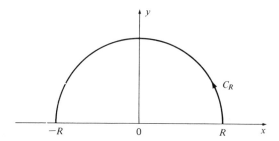

By the theorem of residues, we have

$$\int_{-R}^R \frac{P(x)}{Q(x)}\, dx + \int_{C_R} \frac{P(z)}{Q(z)}\, dz = 2\pi i \sigma. \tag{9.3}$$

Since $q \geqslant p + 2$, there exists $K, r_0 > 0$ such that $|P/Q| \leqslant K/|z|^2$ when $|z| > r_0$. Thus, if $R > \max(r_0, \rho)$, then

$$\left|\int_{C_R} \frac{P(z)}{Q(z)}\, dz\right| \leqslant \frac{K}{R^2}\pi R = \frac{K\pi}{R} \quad \text{and} \quad \int_{C_R} \to 0 \quad \text{as} \quad R \to \infty.$$

Letting $R \to \infty$ in (9.3), we get

$$\text{C.V.} \int_{-\infty}^\infty \frac{P(x)}{Q(x)}\, dx = 2\pi i \sigma.$$

Since C.V. exists, it follows that

$$\int_{-\infty}^\infty \frac{P(x)}{Q(x)}\, dx = 2\pi i \sigma. \qquad \square$$

9.5.4

Jordan's Lemma Let C_R be the semicircle $|z| = R$, where $R > 0$, $0 \leqslant \arg z \leqslant \pi$. If

1. $f(z)$ is continuous on C_R for $R \geqslant R_0$,
2. $|f(z)| \leqslant M_R$ on C_R, where $M_R \to 0$ as $R \to \infty$,
3. $m > 0$,

then

$$\int_{C_R} f(z) e^{imz} \, dz \to 0 \quad \text{as} \quad R \to \infty.$$

PROOF.

$$\int_{C_R} f(z) e^{imz} \, dz = \int_0^\pi f(Re^{i\theta}) e^{imR \cos \theta} Rie^{i\theta} \, d\theta$$

$$= iR \int_0^\pi f(Re^{i\theta}) e^{-mR \sin \theta} e^{i(mR \cos \theta + \theta)} \, d\theta.$$

Thus

$$\left| \int_{C_R} f(z) e^{imz} \, dz \right| \leqslant R \int_0^\pi M_R e^{-mR \sin \theta} \cdot 1 \, d\theta$$

$$= 2R \int_0^{\pi/2} M_R e^{-mR \sin \theta} \, d\theta,$$

since $\int_{\pi/2}^\pi = \int_0^{\pi/2}$.

Now $(\sin \theta)/\theta$ decreases (strictly) from 1 to $2/\pi$ in $(0, \pi/2]$, so that $-\sin \theta \leqslant -2\theta/\pi$ in $[0, \pi/2]$; thus

$$\left| \int_{C_R} f(z) e^{imz} \, dz \right| \leqslant 2R \int_0^{\pi/2} M_R e^{-2mR\theta/\pi} \, d\theta$$

$$= 2RM_R \frac{\pi}{2mR} \left[-e^{-2mR\theta/\pi} \right]_0^{\pi/2}$$

$$= \frac{\pi M_R}{m} (1 - e^{-mR}) \to 0 \quad \text{as} \quad R \to \infty. \qquad \square$$

9.5.5

Theorem If P, Q are polynomials (with real coefficients) of degree p, q respectively, with $q \geqslant p + 1$, and if Q has no real zeros, then

$$\int_{-\infty}^\infty \frac{P(x)}{Q(x)} e^{imx} \, dx = 2\pi i \sigma, \quad m > 0,$$

where σ is the sum of the residues of $[P(z)/Q(z)] e^{imz}$ at the zeros of $Q(z)$ in the domain $\Im(z) > 0$.

Laurent Series, Singularities

PROOF. The proof follows immediately from Jordan's lemma on setting $f(z) = P(z)/Q(z)$. □

EXAMPLE 1. (Function with a nonreal singularity.) Consider

$$I = \int_{-\infty}^{\infty} \frac{\cos 2x}{1+x^2}\, dx.$$

If we choose $f(z) = e^{i2z}/(z^2+1)$, then using Jordan's lemma we have

$$I_1 = \int_{-\infty}^{\infty} \frac{e^{i2x}}{1+x^2}\, dx = 2\pi i\sigma,$$

where $\sigma = \operatorname{Res} f(i)$ since $z = i$ is the only pole of f in $I(z) > 0$. Now, $\sigma = \lim_{z\to i} e^{2iz}/z+i = e^{-2}/2i$, thus $I_1 = 2\pi i e^{-2}/2i = \pi/e^2$. Taking real and imaginary parts of I_1, we obtain

$$I = \int_{-\infty}^{\infty} \frac{\cos 2x}{1+x^2}\, dx = \frac{\pi}{e^2}$$

and

$$\int_{-\infty}^{\infty} \frac{\sin 2x}{1+x^2}\, dx = 0.$$

EXAMPLE 2. (Function with a real singularity.) Consider

$$I = \int_0^{\infty} \frac{\sin x}{x}\, dx.$$

We show that

$$\int_0^{\infty} \frac{\sin x}{x}\, dx = \frac{\pi}{2},$$

since by continuity $(\sin x)/x = 1$ at $x = 0$. Let $f(z) = e^{iz}/z$ (expression $(\sin z)/z$ would not do; why?) With obvious notation, using Cauchy's theorem, we obtain

$$\left[\int_{-R}^{-\epsilon} + \int_{\epsilon}^{R} + \int_{C_R} - \int_{C_\epsilon}\right] f(z)\, dz = 0 \qquad (9.4)$$

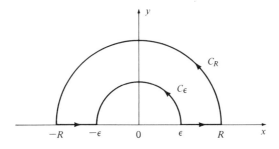

By Jordan's lemma, $\int_{C_R} f(z)dz \to 0$ as $R \to \infty$. Now, for $|z| > 0$,

$$f(z) = \frac{1}{z} + \varphi(z),$$

where $\varphi(z)$, defined by continuity at $z = 0$, is analytic for all z. Thus there exists $K > 0$ such that $|\varphi(z)| < K$ for $|z| \leq 1$. Thus,

$$\left|\int_{C_\epsilon} \varphi(z)\,dz\right| \leq K\pi\epsilon \quad \text{for} \quad 0 < \epsilon < 1.$$

Consequently,

$$\lim_{\epsilon \to +0} \int_{C_\epsilon} \varphi(z)\,dz = 0.$$

Now

$$\int_{C_\epsilon} \frac{1}{z}\,dz = \int_0^\pi \frac{1}{\epsilon e^{i\theta}} \cdot \epsilon i e^{i\theta}\,d\theta = \pi i$$

and

$$\int_{-R}^{-\epsilon} f\,dz + \int_{\epsilon}^{R} f\,dz = \int_{\epsilon}^{R} \frac{e^{ix} - e^{-ix}}{x}\,dx = 2i\int_{\epsilon}^{R} \frac{\sin x}{x}\,dx;$$

thus, by (9.4),

$$\lim_{\substack{\epsilon \to +0 \\ R \to \infty}} 2i \int_{\epsilon}^{R} \frac{\sin x}{x}\,dx = \pi i,$$

and consequently

$$\int_0^\infty \frac{\sin x}{x}\,dx = \frac{\pi}{2}.$$

9.5.6
Exercises

Verify the following integrals:

1: $\int_0^\infty \frac{\sin^2 x}{x^2}\,dx = \frac{\pi}{2}.$

2: $\int_0^\infty \frac{\sin^3 x}{x^3}\,dx = \frac{3\pi}{8}.$

3: Verify that $\int_0^\infty \frac{\sin mx}{x}\,dx$, $m > 0$, is independent of m.

We now demonstrate two theorems that are useful in evaluating integrals.

9.5.7

Theorem Let $f(z)$ be continuous in the domain
$$E = [z \mid 0 < |z| < b, \ b > 0, \ \alpha \leqslant \arg z \leqslant \beta].$$
Let C_R be the arc $|z| = R$, $0 < R < \infty$, $\alpha \leqslant \arg z \leqslant \beta$. Then, if $zf(z)$ approaches a finite constant k uniformly as $|z| \to +0$, we have
$$\lim_{R \to +0} \int_{C_R} f(z)\,dz = ki(\beta - \alpha).$$

PROOF. For $z \in C_R$, we have
$$f(z) = \frac{k}{z} + \frac{\mu}{z}, \quad \text{where} \quad |\mu| \leqslant \epsilon (\epsilon > 0),$$
when $z \in C_R$, $0 < R < R_0$, and $0 < R_0 < b$. Hence, for $0 < R < R_0$,
$$\left| \int_{C_R} f(z)\,dz - ki(\beta - \alpha) \right| = \left| \int_{C_R} \left[\frac{k}{z} + \frac{\mu}{z} \right] dz - ki(\beta - \alpha) \right|$$
$$= \left| \int_{C_R} \frac{\mu}{z} \, dz + \int_\alpha^\beta \frac{kRie^{i\theta}d\theta}{Re^{i\theta}} - ki(\beta - \alpha) \right|$$
$$= \left| \int_{C_R} \frac{\mu}{z} \, dz \right| \leqslant \epsilon(\beta - \alpha). \qquad \square$$

9.5.8

Theorem Let $f(z)$ be continuous on the domain
$$E = \{z \mid a < |z| < \infty, \ \alpha \leqslant \arg z \leqslant \beta\}.$$
Let C_R be the arc $|z| = R$, $0 < R < \infty$, $\alpha \leqslant \arg z \leqslant \beta$. Then if $zf(z)$ approaches a finite constant k uniformly as $|z| \to \infty$, we have
$$\lim_{R \to \infty} \int_{C_R} f(z)\,dz = ki(\beta - \alpha).$$

PROOF. Similar to the previous theorem. $\qquad \square$

EXAMPLE 3. Find the C.V. $\int_{-\infty}^\infty [5x^3/(1 + x + x^2 + x^3 + x^4)]\,dx$.

SOLUTION. Choose
$$f(z) = \frac{5z^3(z-1)}{z^5 - 1},$$
which has a removable singularity at $z = 1$. The poles of f occur at $z^5 = e^{2n\pi i}$, where $n = 0, 1, 2, 3, 4$, that is, at
$$z = 1, e^{2\pi i/5}, e^{4\pi i/5}, e^{6\pi i/5}, e^{8\pi i/5}.$$
In the upper half-plane we take, say, $z = e^{2\pi i/5} = w_1$ and $z = e^{4\pi i/5} = w_2$.

Then

$$\text{Res } f(w_1) = \frac{5w_1^3(w_1 - 1)}{5w_1^4} = 1 - \frac{1}{w_1} \quad \text{and} \quad \text{Res } f(w_2) = 1 - \frac{1}{w_2}.$$

With a semicircle of radius R in the upper half-plane, we get

$$\lim_{R \to \infty} \left[\int_{-R}^{R} + \int_{C_R} \right] f(z)\, dz = 2\pi i \left[2 - \frac{1}{w_1} - \frac{1}{w_2} \right] = 2\pi i \left[2 - e^{-2\pi i/5} - e^{-4\pi i/5} \right].$$

We examine $\int_{C_R}$ and see, by the previous theorem, that $zf(z) \to 5$ as $z \to \infty$; thus

$$\int_{C_R} f(z)\, dz = 5\pi i.$$

Consequently,

$$\int_{-\infty}^{\infty} f(z)\, dz = 2\pi i \left[2 - \cos\frac{2\pi}{5} + i\sin\frac{2\pi}{5} - \cos\frac{4\pi}{5} + i\sin\frac{4\pi}{5} \right] - 5\pi i.$$

The real part of the right-hand side becomes

$$-2\pi\left[\sin\frac{2\pi}{5} + \sin\frac{4\pi}{5} \right] = \int_{-\infty}^{\infty} \frac{5x^3}{1 + x + x^2 + x^3 + x^4}\, dx;$$

the imaginary part becomes

$$-\pi - 2\pi\cos\frac{2\pi}{5} - 2\pi\cos\frac{4\pi}{5} = 0.$$

9.5.9 Integrals Involving Rational Functions

1. Consider for example

$$I = \int_{0}^{\infty} \frac{x^2}{x^4 + x^2 + 1}\, dx.$$

Examine first,

$$\int_{C} \frac{z^2}{z^4 + z^2 + 1}\, dz,$$

where C is to be determined. The function

$$f(z) = \frac{z^2}{z^4 + z^2 + 1}$$

is meromorphic with poles at $z = \alpha, \beta, \bar{\alpha}, \bar{\beta}$, where

$$\alpha = \frac{1 + i\sqrt{3}}{2}; \quad \beta = \frac{-1 + i\sqrt{3}}{2}.$$

We take a semicircle C with radius $R > 1$ and a straight line $[-R, R]$, enclosing the poles of f at α and β. No matter how large R is, the contour never includes any other singular points of f.

Laurent Series, Singularities

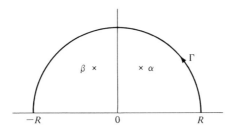

Thus $\int_C f(z)dz = 2\pi i \sigma$ where σ is the sum of residues of f in C, that is, at $z = \alpha, \beta$. We show that

$$\lim_{R \to \infty} \int_\Gamma f(z)\,dz = 0 \quad \text{and} \quad \lim_{R \to \infty} \int_C f(z)\,dz = \lim_{R \to \infty} \int_{-R}^R f(x)\,dx = 2\pi i \sigma.$$

We have:

$$\operatorname{Res} f(\alpha) = \lim_{z \to \alpha} (z - \alpha)f(z) = \frac{\alpha}{4\alpha^2 + 2} = \frac{\alpha}{2i\sqrt{3}};$$

$$\operatorname{Res} f(\beta) = \lim_{z \to \beta} (z - \beta)f(z) = \frac{\beta}{4\beta^2 + 2} = -\frac{\beta}{2i\sqrt{3}}.$$

Thus,

$$\sigma = \frac{1}{2i\sqrt{3}} (\alpha - \beta) = \frac{1}{2i\sqrt{3}}.$$

Now,

$$\int_C = \int_\Gamma + \int_{-R}^R \quad \text{and} \quad |z^4 + z^2 + 1| \geq |z^4| - |z^2| - 1;$$

for z with a sufficiently large modulus, this is positive. Consequently,

$$\left|\int_\Gamma f(z)dz\right| \leq \int_\Gamma \frac{|z|^2}{|z|^4 - |z|^2 - 1} |dz|.$$

Let $z = Re^{i\theta}$; then for sufficiently large R we get

$$\left|\int_\Gamma f(z)\,dz\right| \leq \int_\Gamma \frac{R^2}{R^4 - R^2 - 1} R\,d\theta = \frac{\pi R^3}{R^4 - R^2 - 1} \to 0 \quad \text{as} \quad R \to \infty.$$

Next,

$$2\pi i \sigma = \lim_{R \to \infty} \int_{-R}^R f(x)\,dx$$

(or use Theorem 9.5.3). On $[-R, R]$, $z = x$ and thus

$$2\pi i \frac{1}{2i\sqrt{3}} = \text{C.V.} \int_{-\infty}^\infty \frac{x^2}{x^4 + x^2 + 1}\,dx$$

and

$$\frac{\pi}{2\sqrt{3}} = \int_0^\infty \frac{x^2}{x^4 + x^2 + 1} \, dx.$$

2. Consider

$$I = \int_{-\infty}^\infty \frac{dx}{(x^2 + 1)^n}, \quad n \geq 1.$$

As before, we examine

$$f(z) = \frac{1}{(z^2 + 1)^n},$$

and choose a semicircle C_R of radius $R > 1$ in the upper half-plane. Since $2n \geq 2$, we can apply Theorem 9.5.2. The poles in the upper half-plane are not simple. The only root of $(z^2 + 1)^n = 0$ in the upper half-plane is at $z = i$ and is of multiplicity n.

We require the $(n-1)$th derivative of

$$\varphi(z) = \frac{(z - i)^n}{(z^2 + 1)^n} \quad \text{at} \quad z = i.$$

Thus

$$\sigma = \frac{1}{(n-1)!} \frac{d^{n-1}}{dz^{n-1}} \varphi(i).$$

Now, $\varphi(z) = (z + i)^{-n}$, thus

$$\varphi^{[n-1]}(z) = (-1)^{n-1} n(n+1) \cdots (n+n-2)(z+i)^{-n-(n-1)}$$

and

$$\sigma = \frac{1}{(n-1)!} \frac{(2n-2)!}{(n-1)!} \frac{1}{2^{2n-1}} i^{-2n+1+2n-2}$$

$$= -\frac{(2n-2)!}{2^{2n-1}[(n-1)!]^2} i.$$

Consequently,

$$I = \int_{-\infty}^\infty \frac{dx}{(x^2 + 1)^n} = 2\pi i \sigma = \frac{\pi}{2^{2n-2}} \frac{(2n-2)!}{[(n-1)!]^2}.$$

Note: It is possible to evaluate this integral by other methods. For example, transforming $x = \tan \theta$ and using a beta-function (to be discussed later).

9.5.10
Exercises

Verify the following integrals:

1: $\int_0^\infty \dfrac{x^2+1}{x^4+1}\,dx = \dfrac{\pi}{\sqrt{2}}.$

2: $\int_{-\infty}^\infty \dfrac{dx}{(x^2+a^2)(x^2+b^2)} = \dfrac{\pi}{ab(a+b)},\ a>0,\ b>0.$

3: $\int_{-\infty}^\infty \dfrac{x\sin x}{x^2-2x+10}\,dx = \dfrac{\pi}{3e^3}(3\cos 1 + \sin 1).$

4: $\int_{-\infty}^\infty \dfrac{x\sin 3x}{x^2+9}\,dx = \dfrac{\pi}{2}e^{-9}.$

5: $\int_{-\infty}^\infty \dfrac{\cos ax}{1+x^3}\,dx = \dfrac{\pi}{3}[\sin a + e^{-a\sqrt{3}/2}(\sin a/2 + \sqrt{3}\cos a/2)],$
$a \geq 0.$

6: $\int_0^\infty \dfrac{\sin ax}{x(x^2+b^2)^2}\,dx = \dfrac{\pi}{4b^4}[2-(2+|ab|)e^{-|ab|}]\sin a.$

Evaluate the following integrals:

7: $\int_0^\infty \dfrac{x^6}{(x^4+1)^2}\,dx.$

8: $\int_0^\infty \dfrac{x(x^2+1)\sin x}{x^4+x^2+1}\,dx.$

9: $\int_0^\infty \dfrac{x^4}{x^6+2}\,dx.$

10: Show that

$$\int_0^\infty \dfrac{x^4+3x^2+1}{(x^4+x^2+1)^2}\cos x\,dx = \dfrac{\pi}{3}\left[1+\dfrac{2}{\sqrt{3}}\right]e^{-\sqrt{3}/2}\cos\dfrac{1}{2}.$$

11: Show that if a and m are positive, then

$$\int_0^\infty \frac{\sin^2 mx}{x^2(a^2+x^2)^2} dx = \frac{\pi}{8a^5}\left[e^{-2ma}(2ma+3)+4ma-3\right].$$

9.5.11 Integrals Involving Trigonometric Functions

We study integrals of the form

$$I = \int_{|z|=1} f(\sin\theta, \cos\theta)\, d\theta.$$

Integrals involving other trigonometric functions can normally be reduced to combinations of $\sin\theta$ and $\cos\theta$.

We first observe that if $z = e^{i\theta}$, then

$$\sin\theta = \frac{1}{2i}\left[z - \frac{1}{z}\right] \quad \text{and} \quad \cos\theta = \frac{1}{2}\left[z + \frac{1}{z}\right].$$

Thus, in many cases involving integrals of various combinations of $\cos\theta$ and $\sin\theta$, we can reduce the problem to integration of rational functions of z.

1. Consider

$$I = \int_0^{2\pi} \frac{d\theta}{a + \cos\theta}, \quad a > 1.$$

Let $z = e^{i\theta}$, then

$$I = -2i\int_C \frac{dz}{z^2 + 2az + 1},$$

where C is the circle $|z| = 1$.

By the residue theorem, $I = -2i \cdot 2\pi\sigma = 4\pi\sigma$, where σ is the sum of the residues of $f(z) = (z^2 + 2az + 1)^{-1}$ at the poles in C. Poles occur at $z = -a \pm \sqrt{a^2 - 1}$. Since $a > 1$,

$$z_1 = -a - \sqrt{a^2 - 1} \quad \text{and} \quad z_2 = -a + \sqrt{a^2 - 1}$$

are respectively exterior and interior to C (why?). Hence,

$$\operatorname{Res} f(z_2) = \lim_{z\to z_2} \frac{1}{2z + 2a} = \frac{1}{2\sqrt{a^2 - 1}}.$$

Thus

$$\int_0^{2\pi} \frac{d\theta}{a + \cos\theta} = \frac{2\pi}{\sqrt{a^2 - 1}}, \quad a > 1.$$

What can you say about I for other values of a?

2. Consider

$$I = \int_0^{2\pi} \frac{\cos nx}{\cosh a + \cos x}\, dx, \quad a > 0.$$

Laurent Series, Singularities

Let $z = e^{ix}$, then $dz = iz\,dx$ and $\cos nx = (z^n + z^{-n})/2$. Then,

$$I = -\frac{i}{2}\int_C \frac{1}{z}[z^n + z^{-n}]\left[\frac{e^a + e^{-a} + z + z^{-1}}{2}\right]^{-1} dz$$

$$= -i\int_C \frac{z^{2n} + 1}{z^n}[z(e^a + e^{-a}) + z^2 + 1]^{-1} dz, \quad C: |z| = 1.$$

Now poles occur at $z = 0$ (multiplicity n) and at $z^2 + 2z\cosh a + 1 = 0$, that is,

$$z = \left[-2\cosh a \pm \sqrt{4(\cosh^2 a - 1)}\right]\frac{1}{2}$$

$$= -\cosh a \pm \sinh a.$$

Let

$$P_1 = -\cosh a - \sinh a = (-e^a - e^{-a} - e^a + e^{-a})\frac{1}{2} = -e^a$$

and

$$P_2 = -\cosh a + \sinh a = (-e^{-a} - e^a + e^a - e^{-a})\frac{1}{2} = -e^{-a}.$$

For $a > 0$, we have $e^a > 1$; then $-e^a < -1$ and P_1 is outside $|z| = 1$. Also, $0 < e^{-a} < 1$; then $-1 < -e^{-a} < 0$ and P_2 is inside $|z| = 1$.

Taking

$$f(z) = \frac{z^{2n} + 1}{z^n(z^2 + 2z\cosh a + 1)} = \frac{z^{2n} + 1}{z^n(z - P_1)(z - P_2)},$$

we require residues at $z = 0$, P_2:

$$\operatorname{Res} f\{P_2\} = \sigma_2 = \frac{P_2^{2n} + 1}{P_2^n(P_2 - P_1)};$$

$$\operatorname{Res} f\{0\} = \sigma_1 = \frac{1}{(n-1)!}\frac{d^{n-1}}{dz^{n-1}}\left[(z^{2n} + 1)(z^2 + 2z\cosh a + 1)^{-1}\right]_{z=0},$$

which is not a satisfactory expression to evaluate. We rewrite $f(z)$ to obtain the coefficient of $1/z$:

$$f(z) = [z^n + z^{-n}]\frac{1}{P_1 - P_2}\left[\frac{1}{P_2 - z} - \frac{1}{P_1 - z}\right]$$

$$= [z^n + z^{-n}]\frac{1}{P_1 - P_2}\left[\frac{1}{P_2}\sum_0^\infty z^k P_2^{-k} - \frac{1}{P_1}\sum_0^\infty z^k P_1^{-k}\right].$$

Note: Expansions are valid in a neighborhood of the origin. The coefficient of $1/z$ is

$$\frac{1}{P_1 - P_2}\left[\frac{1}{P_2^n} - \frac{1}{P_1^n}\right] = \sigma_1.$$

Thus,

$$I = -i \cdot 2\pi i(\sigma_1 + \sigma_2) = 2\pi \frac{1}{P_1 - P_2}\left[P_2^{-n} - P_1^{-n} - P_1^n - P_2^{-n}\right]$$

$$= \frac{2\pi}{-e^a + e^{-a}}\left[(-1)(P_1^{-n} + P_1^n)\right] = \frac{2\pi}{2\sinh a}\left[(-1)^n e^{-na} + (-1)^n e^{-na}\right]$$

$$= (-1)^n \frac{2\pi e^{-na}}{\sinh a}.$$

9.5.12

Exercises

Verify the following integrals:

1: $\int_0^{2\pi} \frac{d\theta}{(a + b\cos\theta)^2} = \frac{2\pi a}{(a^2 - b^2)^{3/2}}$, where $a > b > 0$.

2: $\int_0^{2\pi} \frac{d\theta}{1 - 2a\cos\theta + a^2} = \begin{cases} 2\pi/(1 - a^2), & |a| < 1 \\ 2\pi/(a^2 - 1), & |a| > 1 \end{cases}$ and a complex.

What is the principal value if $|a| = 1$?

3: $\int_0^\pi \frac{\sin^2\theta}{a + b\cos\theta}\, d\theta = \frac{\pi}{b^2}(a - \sqrt{a^2 - b^2})$, $a > |b| > 0$.

4: $\int_0^{2\pi} (\cos\theta)^{2m}\, d\theta = \frac{\pi}{2^{2m-1}}\binom{2m}{m}$, $m = 1, 2, \ldots$.

5: $\int_0^{2\pi} \frac{\cos^3\theta}{1 - 2a\cos\theta + a^2}\, d\theta = \frac{\pi}{2} \frac{(a^3 + 3a)}{(1 - a^2)}$, $0 < a < 1$.

9.5.13

We now study several special integrals and contours that indeed aid us in evaluating the integrals; however, the choice of a suitable contour depends, to a great extent, on experience. We will observe that sometimes a suitably indented circle, semicircle, rectangle, sector of a circle, or some other planar figure will aid us in our evaluation.

9.5.14

Consider $I = \int_0^\infty e^{-x^2}\, dx$. There are several ways to evaluate this integral and we first choose the method involving double integrals.

Let $C_R, C_{\sqrt{2}R}, \square_R$ be, respectively, the first quadrant of a circle of radius

Laurent Series, Singularities

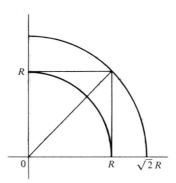

R, the first quadrant of a circle of radius $\sqrt{2}\,R$, and a square with vertices at $(0,0),(R,0),(R,R),(0,R)$.

Let $I_R = \int_0^R e^{-x^2}\,dx$, then

$$I_R^2 = \int_0^R e^{-x^2}\,dx \int_0^R e^{-y^2}\,dy$$

$$= \int_0^R \int_0^R e^{-(x^2+y^2)}\,dx\,dy = \iint_{\square_R} e^{-(x^2+y^2)}\,dx\,dy.$$

Transform $x = r\cos\theta$, $y = r\sin\theta$; then the Jacobian $|J| = r$. Now,

$$\iint_{C_R} e^{-(x^2+y^2)}\,dx\,dy \leqslant \iint_{\square_R} e^{-(x^2+y^2)}\,dx\,dy \leqslant \iint_{C_{\sqrt{2}R}} e^{-(x^2+y^2)}\,dx\,dy$$

and

$$\iint_{C_R} e^{-(x^2+y^2)}\,dx\,dy \leqslant I_R^2 \leqslant \iint_{C_{\sqrt{2}R}} e^{-(x^2+y^2)}\,dx\,dy.$$

Thus,

$$\int_0^{\pi/2} \int_0^R e^{-r^2} r\,dr\,d\theta \leqslant I_R^2 \leqslant \int_0^{\pi/2} \int_0^{\sqrt{2}R} e^{-r^2} r\,dr\,d\theta,$$

that is,

$$\int_0^{\pi/2} \int_0^R \left[\frac{-e^{-r^2}}{2}\right] d\theta \leqslant I_R^2 \leqslant \int_0^{\pi/2} \left[\frac{-e^{-r^2}}{2}\right]_0^{\sqrt{2}R} d\theta$$

and

$$\frac{\pi}{2} \cdot \left[\frac{1}{2} - \frac{e^{-R^2}}{2}\right] \leqslant I_R^2 \leqslant \frac{\pi}{2} \cdot \left[\frac{1}{2} - \frac{e^{-2R^2}}{2}\right].$$

As $R \to \infty$,

$$I_R^2 \to \left[\int_0^\infty e^{-x^2}\,dx\right]^2;$$

consequently,

$$\frac{\pi}{4} \leq \left[\int_0^\infty e^{-x^2}dx\right]^2 \leq \frac{\pi}{4} \quad \text{or} \quad I \equiv \int_0^\infty e^{-x^2}dx = \frac{\sqrt{\pi}}{2}.$$

It is possible, in a round about way, to evaluate $\int_0^\infty e^{-x^2}dx$ using contour integration as follows.

Choose a rectangle $HKPQ$ and a sector of a circle of radius R. Consider

$$\int_C \frac{e^{iz^2}}{\sin z\sqrt{\pi}} dz,$$

where C is the contour of rectangle $HKPQ$; we have

$$\int_{PQ} + \int_{HK} = 2\int_{-R}^{R} e^{i(\frac{1}{4}\pi - y^2)} i\, dy.$$

On QH,

$$|\sin z\sqrt{\pi}| > \sinh R\sqrt{\pi} \quad \text{and} \quad |e^{iz^2}| = e^{-2Rx},$$

thus

$$\left|\int_{QH} \frac{e^{iz^2}}{\sin z\sqrt{\pi}} dz\right| \leq \int_{-\sqrt{\pi}/2}^{\sqrt{\pi}/2} \frac{e^{-2Rx}}{\sinh R\sqrt{\pi}} dx = \frac{1}{R}$$

(similarly for KP).

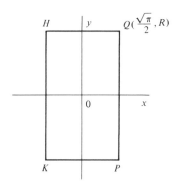

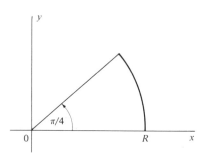

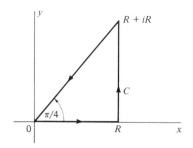

Laurent Series, Singularities

The residue of $e^{iz^2}/\sin\sqrt{\pi}\, z$ at $z = 0$ is $1/\sqrt{\pi}$. Using the residue theorem and letting $R \to \infty$, we have

$$\int_0^\infty e^{i[(\pi/4) - y^2]} dy = \frac{\sqrt{\pi}}{2},$$

which implies that

$$\int_0^\infty \cos\left[\frac{\pi}{4} - y^2\right] dy = \frac{\sqrt{\pi}}{2} \quad \text{and} \quad \int_0^\infty \sin\left[\frac{\pi}{4} - y^2\right] dy = 0.$$

Solve:

$$\int_0^\infty \cos y^2\, dy = \int_0^\infty \sin y^2\, dy = \frac{\sqrt{\pi}}{2\sqrt{2}}.$$

Now using the contour of the sector we can show that

$$\int_0^\infty \cos x^2\, dx = \int_0^\infty \sin x^2\, dx = \frac{1}{\sqrt{2}} \int_0^\infty e^{-x^2} dx.$$

and thus

$$\int_0^\infty e^{-x^2} dx = \frac{\sqrt{\pi}}{2}.$$

9.5.15 The Fresnel Integrals

$$I_1 = \int_0^\infty \cos x^2\, dx, \qquad I_2 = \int_0^\infty \sin x^2\, dx.$$

The integrals I_1 and I_2 are called the Fresnel integrals.* We evaluated them above by using a rectangular contour whose dimensions were not particularly "motivated." Indeed, the function $f(z) = e^{iz^2}/\sin z\sqrt{\pi}$ was not particularly "motivated" either.

Let us evaluate the Fresnel integrals in a more direct way. Consider $\int_C e^{-z^2} dz$, where C is the boundary of the triangle in the figure below.

$$\int_C e^{-z^2} dz = \int_0^R e^{-x^2} dx + i \int_0^R e^{-(R+iy)^2} dy + \int_R^0 e^{-(1+i)^2 x^2}(1 + i)\, dx,$$

since $z = (1 + i)x$ on the line at $\pi/4$ to the x-axis. The function e^{-z^2} has no poles in the finite complex plane; thus, by Cauchy's theorem,

$$\int_0^R e^{-x^2} dx + (1 + i) \int_R^0 e^{-2ix^2} dx + i \int_0^R e^{-(R+iy)^2} dy = 0.$$

Also,

$$\left| i \int_0^R e^{-(R^2 - y^2)} e^{-2iRy} dy \right| \leq \int_0^R e^{-(R^2 - y^2)} dy$$

---

*The biographies of Fresnel and other mathematicians can be studied in books like *Men of Mathematics*, by E. T. Bell. Penguin Books, 1953.

and $R^2 - y^2 \geq R(R - y)$ for $R \geq y \geq 0$. Thus
$$\int_0^R e^{-(R+iy)^2} dy \to 0 \quad \text{as} \quad R \to \infty$$
(transform $R(R - y) = u$ etc.) and
$$\int_0^\infty e^{-x^2} dx = (1 + i) \int_0^\infty e^{-2ix^2} dx = \frac{\sqrt{\pi}}{2} \quad \text{(by 9.5.14)}.$$
Hence, $(1 + i) \int_0^\infty (\cos 2x^2 - i \sin 2x^2) dx = \sqrt{\pi}/2$. Equating the real and imaginary parts, we get
$$\int_0^\infty \cos x^2 \, dx = \frac{\sqrt{\pi}}{2\sqrt{2}} = \int_0^\infty \sin x^2 \, dx.$$

Note: We could have used C as the boundary of the sector of a circle of radius R and central angle $\pi/4$.

An integral related to $\int_0^\infty e^{-x^2} dx$ is the following.

9.5.16
Consider
$$\int_{-\infty}^\infty e^{-x^2} \cos 2\lambda x \, dx.$$

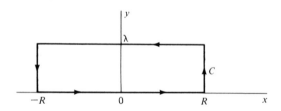

We integrate e^{-z^2} along the rectangle in the figure:
$$\int_C e^{-z^2} dz = \int_{-R}^R e^{-x^2} dx + i \int_0^\lambda e^{-(R+iy)^2} dy$$
$$+ \int_R^{-R} e^{-(x+i\lambda)^2} dx + i \int_\lambda^0 e^{-(-R+iy)^2} dy = 0.$$
Since e^{-z^2} is analytic in and on C, then by Cauchy's theorem we get
$$\int_{-R}^R e^{-x^2} dx + i \int_0^\lambda e^{-(R+iy)^2} dy - e^{\lambda^2} \int_{-R}^R e^{-x^2} e^{2i\lambda x} dx + i \int_\lambda^0 e^{-(-R+iy)^2} dy = 0.$$
Let us write this as $I_1 + I_2 + I_3 + I_4 = 0$. Consider
$$|I_2| = \left| i \int_0^\lambda e^{-(R+iy)^2} dy \right| \leq \int_0^\lambda e^{-R^2+y^2} dy \leq M e^{-R^2},$$

Laurent Series, Singularities

where $M = \int_0^\lambda e^{y^2} dy$ is finite. Thus $I_2 \to 0$ as $R \to \infty$. Similarly, $I_4 \to 0$ as $R \to \infty$. Thus,

$$\int_{-\infty}^{\infty} e^{-x^2} dx = e^{\lambda^2} \int_{-\infty}^{\infty} e^{-x^2}(\cos 2\lambda x + i \sin 2\lambda x) dx.$$

Consequently,

$$\int_{-\infty}^{\infty} e^{-x^2} \sin 2\lambda x \, dx = 0 \quad \text{and} \quad \int_{-\infty}^{\infty} e^{-x^2} dx = e^{\lambda^2} \int_{-\infty}^{\infty} e^{-x^2} \cos 2\lambda x \, dx = \sqrt{\pi} \,;$$

hence,

$$\int_{-\infty}^{\infty} e^{-x^2} \cos 2\lambda x \, dx = e^{-\lambda^2} \sqrt{\pi}.$$

9.5.17
Exercises

Verify the following integrals:

1: $\quad \displaystyle\int_0^\infty \frac{\cos x}{\sqrt{x}} dx = \int_0^\infty \frac{\sin x}{\sqrt{x}} dx = \sqrt{\frac{\pi}{2}}.$

2: $\quad \displaystyle\int_0^\infty \frac{\sin x^2}{x} dx = \frac{\pi}{4}.$

We proceed with a short lemma that can help us to evaluate the contribution to an integral around the arc of a circle of arbitrarily small radius.

9.5.18

Lemma *Let $f(z)$ have a simple pole at $z = \alpha$ with Res $f(\alpha) = a_{-1}$, and let $k = k(\epsilon, \varphi)$ be an arc of the circle $|z - \alpha| = \epsilon$ subtending an angle φ at the center. Then*

$$\lim_{\epsilon \to 0^+} \int_k f(z) dz = i\varphi a_{-1}.$$

This says that, provided the pole is simple and the radius $\epsilon \to 0$, integration over a fraction of a circle enclosing the pole gives a corresponding fraction of $2\pi i \cdot a_{-1}$.

PROOF. Let

$$f(z) = g(z) + \frac{a_{-1}}{z - \alpha},$$

where $g(z) \in A$ in $|z - \alpha| \leq \delta (> 0)$. If $0 < \epsilon < \delta$, then

$$\int_k f(z) dz = \int_k g(z) dz + a_{-1} \int_k \frac{dz}{z - \alpha}.$$

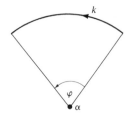

Now, $|\int_k g(z)\,dz| \leq M\epsilon\varphi$, where M is an upper bound for $|g|$ in $|z - \alpha| \leq \delta$. Since on k we may write $z - \alpha = \epsilon e^{i\theta}$, then

$$\int_k \frac{dz}{z - \alpha} = \int_{\theta_0}^{\theta_0+\varphi} \frac{\epsilon i e^{i\theta}}{\epsilon e^{i\theta}}\,d\theta = i\varphi.$$

Thus,

$$\lim_{\epsilon \to 0} \int_k f(z)\,dz = ia_{-1}\varphi. \qquad \square$$

The use of this lemma is illustrated by the following class of examples.

9.5.19

Consider the following function with *real zeros*:

$$\text{C.V.} \int_{-\infty}^{\infty} \frac{\cos x}{a^2 - x^2}\,dx, \quad a > 0.$$

In the upper half-plane, indent the semicircle (of radius R with center at the origin) by little circles about the poles at $x = \pm a$. Call this indented domain D and its boundary C.

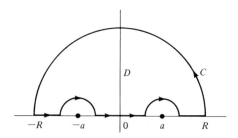

Let

$$I \equiv \int_C \frac{e^{iz}}{a^2 - z^2}\,dz = 0,$$

since $e^{iz}/(a^2 - z^2)$ has no poles in D. Let the radii of the circles about $-a, a$ be η, ϵ, respectively. Then by Cauchy's theorem we have

$$\left[\int_{-R}^{-a-\eta} + \int_{-a+\eta}^{a-\epsilon} + \int_{a+\epsilon}^{R}\right] \frac{e^{ix}}{a^2 - x^2}\,dx + J_1 + J_2 + J_3 = 0,$$

Laurent Series, Singularities

where J_1, J_2, J_3 are integrals over the semicircles with radii η, ϵ, and R, respectively.

Now
$$\lim_{R \to \infty} J_3 = 0, \quad \text{since } |e^{iz}| = e^{-y} \leq 1,$$

as y is assumed to be nonnegative, and

$$\left| \frac{e^{iz}}{a^2 - z^2} \right| \leq \frac{1}{a^2 - R^2}$$

(or use Jordan's lemma).

Residues at $-a$ and a are $e^{-ia}/2a$ and $e^{ia}/(-2a)$ respectively. Thus by the lemma,

$$\lim_{\eta \to 0^+} J_1 = \frac{-\pi i e^{-ia}}{2a} \quad \text{and} \quad \lim_{\epsilon \to 0^+} J_2 = \frac{\pi i e^{ia}}{2a}, \quad \text{where } \varphi = -\pi \text{ in } J_1, J_2,$$

since they are traversed clockwise. Hence, after letting $\epsilon \to 0^+$, $\eta \to 0^+$, $R \to \infty$, we get

$$\int_{-\infty}^{\infty} \frac{e^{ix}}{a^2 - x^2} dx + \frac{\pi i}{2a}(e^{ia} - e^{-ia}) = 0.$$

Thus,

$$\int_{-\infty}^{\infty} \frac{\cos x + i \sin x}{a^2 - x^2} dx = \frac{\pi \sin a}{a} \quad \text{and} \quad \int_{-\infty}^{\infty} \frac{\cos x}{a^2 - x^2} dx = \frac{\pi \sin a}{a}.$$

Note 1: We could have evaluated $\int_{-\infty}^{\infty} \cos x \, dx / (a^2 - x^2)$ by using partial fractions and invoking $\int_{-\infty}^{\infty} \sin x \, dx / x = \pi$.

Note 2: We choose $f(z) = e^{iz}/(a^2 - z^2)$ rather than
$$f(z) = \cos z / (a^2 - z^2)$$
since $\cos z$ does not admit a suitable dominant on the circle of radius R.

We now study a further class of integrals, where the integrand is not single-valued.

9.5.20

Let us show that

$$\int_0^\infty \frac{x^{a-1}}{1+x} dx = \frac{\pi}{\sin a\pi}, \quad 0 < a < 1.$$

Consider $f(z) = e^{az}/(1 + e^z)$ (transform of the above with $x = e^z$). Let C be the boundary of the rectangle in the figure. The only pole within C is at $z = \pi i$. Thus,

$$\text{Res } f(\pi i) = -e^{a\pi i}, \quad \text{that is,} \quad \lim_{z \to \pi i} \left[\frac{z - \pi i}{e^z + 1} \cdot e^{az} \right]$$

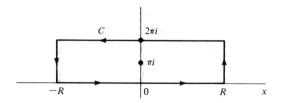

Along the line $x = R$ we have $z = R + iy$, thus,

$$|f(z)| = \left|\frac{e^{az}}{1 + e^z}\right| \leq \frac{e^{aR}}{1 - e^R}$$

and $\lim_{R \to \infty} f(z) = 0$, since $0 < a < 1$.

Along the line $x = -R$ we have $z = -R + iy$, thus,

$$|f(z)| \leq \frac{e^{-aR}}{1 - e^{-R}} \to 0 \text{ as } R \to \infty.$$

Consequently, integrals along $x = R, -R$ tend to zero for $R \to \infty$. By the residue theorem,

$$\frac{1}{2\pi i} \int_{-\infty}^{\infty} \frac{e^{ax}}{1 + e^x} dx - \frac{1}{2\pi i} \int_{-\infty}^{\infty} \frac{e^{a(x + 2\pi i)}}{1 + e^x} dx = -e^{a\pi i},$$

that is,

$$\frac{1 - e^{2a\pi i}}{2\pi i} \int_{-\infty}^{\infty} \frac{e^{ax}}{1 + e^x} dx = -e^{a\pi i},$$

therefore,

$$\int_{-\infty}^{\infty} \frac{e^{ax}}{1 + e^x} dx = \frac{\pi}{\sin a\pi}.$$

Transforming $y = e^x$, we get $y = 0$ when $x = -\infty$, $y = \infty$ when $x = \infty$, and

$$\int_0^{\infty} \frac{y^{a-1}}{1 + y} dy = \frac{\pi}{\sin a\pi}, \quad 0 < a < 1.$$

9.5.21
Exercises

1: Evaluate $\int_{-\infty}^{\infty} \frac{\cosh ax}{\cosh x} dx$. [Answer: $\pi \sec \frac{\pi a}{2}$.]

2: Evaluate $\int_0^{\infty} \frac{y^{a-1}}{1+y} dy$ for $a = p + iq$, $0 < p < 1$.

[Answer: $\int_0^{\infty} \frac{y^{p-1}}{1+y} \cos(q \log y) dy = \frac{\pi \sin \pi p \cosh \pi q}{\sin^2 \pi p + \sinh^2 \pi q}$.]

Laurent Series, Singularities

[Hint: It is instructive to evaluate $\int_0^\infty \frac{y^{a-1}}{1+y} dy$ directly by using

$$f(z) = \frac{z^\lambda}{1+z}, \quad -1 < \lambda < 0.$$

3: Show that $\displaystyle\int_0^\infty \frac{x \cos ax}{\sinh x} dx = \frac{\pi^2 e^{\pi a}}{(e^{\pi a}+1)^2}.$

4: Show that $f(x) = \operatorname{sech} x \sqrt{\pi/2}$ satisfies the equation

$$f(t) = \sqrt{\frac{2}{\pi}} \int_0^\infty f(x) \cos xt \, dx.$$

5: Show that $\displaystyle\int_0^\infty \frac{\sinh ax}{\sinh \pi x} dx = \frac{1}{2} \tan \frac{a}{2}, \quad -\pi < a < \pi.$

6: Prove that $\displaystyle\int_0^\infty \frac{\sinh^2 mx}{x \sinh x} dx = \frac{1}{2} \log \sec m\pi$, where $-\frac{1}{2} < m < \frac{1}{2}$.

7: Prove that $\displaystyle\int_0^\infty \frac{\cosh bx}{\cosh \pi x} \cos ax \, dx = \frac{\cosh(a/2)\cos(b/2)}{\cosh a + \cos b}$, where a, b are real and $-\pi < b < \pi$.

8: Prove that, if $-b < a < b$, then

$$\int_0^\infty \frac{\cos ax}{\cos bx} \frac{dx}{1+x^2} = \frac{\pi}{2} \frac{\cosh a}{\cosh b}$$

and

$$\int_0^\infty \frac{\cos ax}{\sin bx} \frac{x \, dx}{1+x^2} = \frac{\pi}{2} \frac{\cosh a}{\sinh b}.$$

9: Prove that $\displaystyle\int_0^\infty \frac{\cosh cx}{\cosh^2 \pi x} dx = \frac{c}{2\pi \sin(c/2)}$, where $-2\pi < c < 2\pi$.

10: Prove that P.V. $\displaystyle\int_0^\infty \frac{x^{a-1}}{1-x} dx = \pi \cot a\pi.$

11: Prove that P.V. $\displaystyle\int_0^\infty \frac{x \sin ax}{x^2 - r^2} dx = \frac{\pi}{2} \cos ar.$

9.5.22
Consider

$$\int_0^\infty \frac{x^\lambda}{1+x} \, dx, \quad -1 < \lambda < 0.$$

We study $\int_C z^\lambda \, dz/(1+z)$ around some appropriate contour C. The function z^λ has a branch-point at the origin which we exclude.

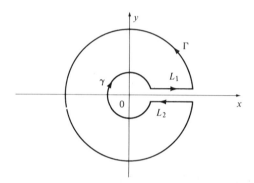

Choose a contour C comprised of a circle Γ of radius R and center $(0,0)$, a circle γ of radius $r < R$ and center $(0,0)$, and lines L_1 and L_2 such that they coincide with the real axis (when $r \to 0$) (see the figure). Lines L_1, L_2 coincide in the z-plane but not on the (many-valued) Riemann surface of z^λ. Thus we require that $z^\lambda = x^\lambda$ when $x > 0$ on L_1 and $z^\lambda = (ze^{2\pi i})^\lambda$ when $x > 0$ on L_2.

Since $z^\lambda = \exp \lambda(\log z + 2k\pi i)$, the case of $k = 1$ corresponds to the first Riemann surface. Basically we imagine the z-plane slit along the positive real axis. The line L_1 is on the upper half of the branch cut and L_2 is on the lower half. Now z^λ is single-valued and analytic on (and in) C, where C is considered to be drawn on the Riemann surface of z^λ. Thus,

$$\int_C \frac{z^\lambda}{1+z} \, dz = 2\pi i \sigma,$$

where σ is the sum of the residues at the poles, that is, at $z = -1$.
Thus $\sigma = (-1)^\lambda = e^{\lambda \pi i}$ and

$$\int_C \frac{z^\lambda}{1+z} \, dz = \int_r^R \frac{x^\lambda}{1+x} \, dx + \int_\Gamma \frac{z^\lambda}{1+z} \, dz + \int_R^r \frac{x^\lambda e^{2\pi i \lambda}}{1+x} \, dx + \int_\gamma \frac{z^\lambda}{1+z} \, dz.$$

For $z = Re^{i\theta}$:

$$\left| \int_\Gamma \frac{z^\lambda}{1+z} \, dz \right| \leq \int_0^{2\pi} \frac{R^\lambda}{R-1} R \, d\theta = \frac{2\pi R^{\lambda+1}}{R-1} \to 0 \quad \text{as } R \to \infty \quad \text{since } \lambda < 0.$$

Laurent Series, Singularities

For $z = re^{i\theta}$:

$$\left| \int_\gamma \frac{z^\lambda}{1+z} dz \right| \leq \int_0^{2\pi} \frac{r^\lambda}{r-1} r\, d\theta = \frac{2\pi r^{\lambda+1}}{r-1} \to 0 \quad \text{as } r \to 0 \quad \text{since } \lambda > -1.$$

Thus,

$$2\pi i e^{\lambda \pi i} = \int_0^\infty \frac{x^\lambda}{1+x} dx - e^{2\lambda \pi i} \int_0^\infty \frac{x^\lambda}{1+x} dx$$

and

$$\int_0^\infty \frac{x^\lambda}{1+x} dx = -\frac{\pi}{\sin \lambda \pi}; \quad -1 < \lambda < 0.$$

Write $\lambda = a - 1$; then we obtain

$$\int_0^\infty \frac{y^{a-1}}{1+y} dy = \frac{\pi}{\sin \pi a}.$$

Further, transforming $x = u^{2n}$, we get

$$2n \int_0^\infty \frac{u^{2n\lambda + 2n - 1}}{1 + u^{2n}} du = -\frac{\pi}{\sin \lambda \pi},$$

and letting $2n\lambda + 2n - 1 = 2m$, we get

$$\lambda = \frac{2m - 2n + 1}{2n} = \frac{2m+1}{2n} - 1.$$

Thus,

$$\int_0^\infty \frac{u^{2m}}{1 + u^{2n}} du = -\frac{1}{2n} \pi \operatorname{cosec}\left[\frac{2m+1}{2n} \pi - \pi \right]$$
$$= \frac{\pi \operatorname{cosec}[(2m+1)\pi/2n]}{2n}$$

Note: $\int_0^\infty x^{a-1} dx/(1-x)$ can be found by indenting the semicircle at $x = 1$ and $x = 0$ or by considering a branch cut along the negative real axis and the residue at $z = 1$.

We shall now study another class of integrals that is, integrals of multiple-valued functions.

9.5.23 Integrals of Functions Involving Logarithms

Consider

$$I = \int_0^{\infty^+} \frac{\log x}{(x+a)^2 + b^2} dx, \quad a, b \text{ real}, b > 0.$$

Take

$$f(z) = \frac{(\log z)^2}{(z+a)^2 + b^2},$$

where $\log z = \log|z| + i\theta$; $z = |z|e^{i\theta}$, $0 < \theta < 2\pi$, that is, we take the branch of $\log z$ such that $z \neq 0$.

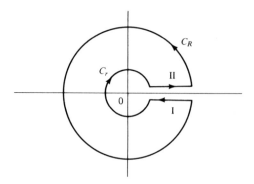

We make a slit along the positive x-axis and exclude the origin. On the upper and lower edges of the cut (II, I), where $x = z$, the function $(\log z)^2$, takes value $(\log x)^2$ and $(\log x + 2\pi i)^2$ respectively. Integrals involving $(\log x)^2$ cancel along I and II. The domain contains two singularities, namely, $z = -a \pm ib$, with residues

$$\frac{(\log z)^2}{d[(z+a)^2 + b^2]/dz} = \frac{[\log(-a+ib)]^2}{2ib} \quad \text{and} \quad \frac{[\log(-a-ib)]^2}{-2ib}.$$

We choose R such that the circle encloses these two singularities. Let $\rho = \sqrt{a^2 + b^2}$, $\varphi = \tan^{-1}(b/a)$, $0 < \varphi < \pi$, and σ be the sum of the residues. Then

$$2\pi i\sigma = 2\pi i \left\{ \frac{[\log \rho + i(\pi - \varphi)]^2}{2ib} + \frac{[\log \rho + i(\pi + \varphi)]^2}{-2ib} \right\}$$

$$= \frac{\pi}{b} 4\varphi(\pi - i\log \rho) = \left[\int_I + \int_{C_r} + \int_{II} + \int_{C_R} \right] f(z)\, dz.$$

For $\int_{C_R}$ let $z = Re^{i\theta}$, $0 < \theta < 2\pi$; then

$$|f(z)| = O\left\{ \frac{(\log z)^2}{z^2} \right\}, \quad |z| \to \infty.$$

Thus

$$\int_{C_R} |f||dz| = O\left\{ \frac{(\log R)^2}{R^2} \right\} \cdot 2\pi R \to 0 \quad \text{as} \quad R \to \infty.$$

OR use inequalities:

$$|\log z|^2 = (\log R)^2 + \theta^2 \leq (\log R)^2 + 64,$$

$$|(z+a)^2 + b^2| \geq |z + a|^2 - b^2.$$

Laurent Series, Singularities

For $\int_{C_r}$, let $z = re^{i\theta}$, $0 < \theta < 2\pi$, $0 < r < 1$; then

$$\left|\int_{C_R} f(z)\,dz\right| \leq \int_{C_r} \frac{(\log r)^2 + 64}{|z^2 + a^2 + 2az + b^2|}\,|dz|$$

$$\leq \int_{C_r} \frac{(\log r)^2 + 64}{|z^2 + 2az| - (a^2 + b^2)}\,|dz| \leq \frac{[(\log r)^2 + 64]2\pi r}{r^2 - 2|a|r - (a^2 + b^2)}$$

and $|\int_{C_r} f(z)\,dz| \to 0$ as $r \to 0$. Consequently,

$$\int_0^\infty \frac{(\log x)^2}{(x+a)^2 + b^2}\,dx + \int_0^\infty \frac{(\log x + 2\pi i)^2}{(x+a)^2 + b^2}\,dx$$

$$= -4\pi i \int_0^\infty \frac{\log x}{(x+a)^2 + b^2}\,dx + 4\pi^2 \int_0^\infty \frac{dx}{(x+a)^2 + b^2}$$

$$= \frac{-4\pi\varphi}{b}(i\log\rho - \pi).$$

Equating the imaginary parts, we get

$$\int_0^\infty \frac{\log x}{(x+a)^2 + b^2}\,dx = \varphi\,\frac{\log\rho}{b} = \frac{1}{2b}\log(a^2 + b^2)\tan^{-1}\frac{b}{a}.$$

Let us now consider an integral that occurs in later analysis; we shall obtain an interesting by-product of this result.

9.5.24

Consider $I = \int_0^{2\pi} \log|1 - e^{i\theta}|\,d\theta$. Choose $f(z) = [\log(1-z)]/z$ that has a removable singularity at $z = 0$. Choose C to be the circle $|z| = 1$ indented at $z = 1$ as in the diagram. Let the indenting circle C_ϵ be of radius ρ and subtend an angle 2ϵ at the origin.

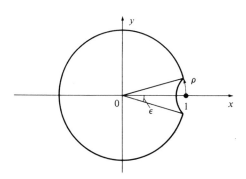

The function $f(z)$ is single-valued in the plane cut from 1 to ∞ and is real for $0 < z < 1$. Thus,

$$\int_C \frac{\log(1-z)}{z} \, dz = 0.$$

Consequently,

$$i \int_0^{2\pi - \epsilon} \log(1 - e^{i\theta}) \, d\theta + \int_{C_\epsilon} \log(1-z) \frac{dz}{z} = 0.$$

Now C_ϵ is centered at $z = 1$ and passes through the two points at which $|z| = 1$, $\arg z = -\epsilon$ and $|z| = 1$, $\arg z = \epsilon$. The radius of C_ϵ is given by $\rho = 2 \sin(\epsilon/2)$ and therefore $\rho \to 0$ as $\epsilon \to 0^+$.

We write $C_\epsilon : 1 - z = \rho e^{i\varphi}$, where $|\varphi| \leq \pi$; then

$$|\log(1-z)| = |\log \rho + i\varphi| \leq [(\log \rho)^2 + \pi^2]^{1/2} \leq \sqrt{2} \, |\log \rho|$$

if $|\log \rho| \geq \pi$ (which is possible since $\rho \to 0$). Thus,

$$\left| \int_{C_\epsilon} \log(1-z) \frac{dz}{z} \right| \leq \frac{\pi \rho \sqrt{2} \, |\log \rho|}{1 - \rho} \to 0 \quad \text{since} \quad \epsilon \to 0$$

and

$$\int_0^{2\pi} \log|1 - e^{i\theta}| \, d\theta = 0.$$

Evaluating this last integral, we see that $\int_0^\pi \log \sin \theta \, d\theta = -\pi \log 2$.

EXAMPLE. Show that

$$\int_0^\infty \frac{\log(1 + x^2)}{1 + x^2} \, dx = \pi \log 2$$

and deduce that $\int_0^\pi \log \sin x \, dx = -\pi \log 2$.

SOLUTION. Consider

$$f(z) = \frac{\log(i + z)}{z^2 + 1}$$

taken around a semicircle C of radius R with its center at the origin and in the upper half-plane.

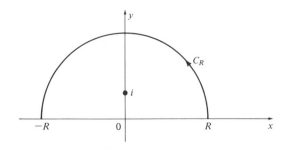

The function $f(z)$ has a pole of order 1 at $z = i$ (inside C). The residue of $f(z)$ at $z = i$ is

$$\left.\frac{\log(i+z)}{2z}\right|_{z=i} = \frac{\log 2i}{2i}$$

Thus

$$\int_C f(z)\,dz = 2\pi i\,\frac{\log 2i}{2i} = \pi\left[\log 2 + \frac{i\pi}{2}\right]$$

$$= \int_{-R}^{R} \frac{\log(x+i)}{x^2+1}\,dx + \int_{C_R} f(z)\,dz.$$

Now,

$$\int_{-R}^{0} \frac{\log(x+i)}{x^2+1}\,dx + \int_{0}^{R} \frac{\log(x+i)}{x^2+1}\,dx$$

$$= \int_{0}^{R} \frac{\log(x+i)}{x^2+1}\,dx + \int_{0}^{R} \frac{\log(i-x)}{x^2+1}\,dx$$

$$= \int_{0}^{R} \frac{\log[-(1+x^2)]}{x^2+1}\,dx$$

$$= \int_{0}^{R} \frac{\log(1+x^2)}{1+x^2}\,dx + \int_{0}^{R} \frac{i\pi}{1+x^2}\,dx.$$

Also, by Theorem 9.5.8 we obtain that $zf(z) \to 0$ as $z \to \infty$, hence

$$\lim_{R \to \infty} \int_{C_R} f(z)\,dz = 0.$$

Thus,

$$\int_{0}^{\infty} \frac{\log(1+x^2)}{1+x^2}\,dx + i\pi \tan^{-1} x \Big|_{0}^{\infty} = \pi \log 2 + \frac{i\pi^2}{2},$$

that is,

$$\int_{0}^{\infty} \frac{\log(1+x^2)}{1+x^2}\,dx + \frac{i\pi^2}{2} = \pi \log 2 + \frac{i\pi^2}{2},$$

and the result follows.

Now transform

$$x = \tan\theta \quad \text{and} \quad \pi \log 2 = -2\int_{0}^{\pi/2} \log\cos\theta\,d\theta.$$

Also transform

$$\theta = \frac{\pi}{2} - \varphi \quad \text{and} \quad \pi \log 2 = -2\int_{0}^{\pi/2} \log\sin\theta\,d\theta,$$

from which we can deduce that $\int_{0}^{\pi} \log\sin\theta\,d\theta = -\pi\log 2$. (Why?)

9.5.25
Exercises

Verify the following integrals:

1: $\displaystyle\int_0^\infty \frac{\log x}{x^2 + a^2}\, dx = \frac{\pi}{2|a|} \log|a|$, where $a \neq 0$ and real.

2: $\displaystyle\int_1^\infty \log\frac{e^x + 1}{e^x - 1}\, \frac{dx}{x} = \frac{\pi^2}{4}$.

3: $\displaystyle\int_0^\infty \frac{dx}{(x + a)(\log^2 x + \pi^2)} = \frac{1}{1 - a} + \frac{1}{\log a}$, where $a > 0$. What is the value of the integral if $a = 1$?

4: $\displaystyle\int_0^\infty \frac{(\log x)^2}{(1 + x^2)^2}\, dx = \frac{\pi^3}{16}$.

5: $\displaystyle\int_0^\infty \frac{(\log x)^n}{1 + x^2}\, dx = 0$ for $n = 2k + 1$, $k = 0, 1, 2, \ldots$.

6: $\displaystyle\int_0^\pi \frac{ax \sin x}{1 - 2a \cos x + a^2}\, dx = \pi \log\left(1 + \frac{1}{a}\right)$.

7: $\displaystyle\int_0^\infty \frac{(\log x)^2}{(1 + x^2)}\, dx = \frac{\pi^3}{8}$.

8: $\displaystyle\int_0^\infty \frac{x^{a-1} - x^{b-1}}{\log x}\, \frac{dx}{1 + x^2} = \log\left[\frac{\tan(\pi a/4)}{\tan(\pi b/4)}\right]$.

9: Given that $\displaystyle\int_0^\infty \frac{\log(1 + x^2)}{1 + x^2}\, dx = \pi \log 2$, deduce that

$$\int_0^1 \frac{\log[x + (1/x)]}{1 + x^2}\, dx = \frac{\pi}{2}\log 2.$$

9.6

The theory of residues has many different uses and we shall now study the summation of infinite series. We commence with a lemma that gives us a bound for $\cot \pi z$ on a square with vertices at $(n + \tfrac{1}{2})(\pm 1 \pm i)$, $n = 0, 1, \ldots$

9.6.1
Lemma

$$|\cot \pi z| \begin{cases} \leq \coth \pi & \text{for } |y| \geq 1 \text{ and } z = x + iy, \\ < 1 & \text{for } x = n + \tfrac{1}{2}, \end{cases}$$

where n is an integer (*positive, negative, or zero*).

PROOF

$$|\cot \pi z|^2 = \frac{2 \cos \pi z \cos \pi \bar{z}}{2 \sin \pi z \sin \pi \bar{z}} = \frac{\cos 2\pi x + \cosh 2\pi y}{\cosh 2\pi y - \cos 2\pi x}$$

$$\leq \frac{\cosh 2\pi y + 1}{\cosh 2\pi y - 1} = \frac{\cosh^2 \pi y}{\sinh^2 \pi y}$$

$$= \coth^2 \pi y \leq \coth^2 \pi \quad \text{for} \quad |y| \geq 1$$

(since $\coth \pi y$ is a decreasing function).

Thus $|\cot \pi z| \leq \coth \pi$ for $|y| \geq 1$ and also

$$|\cot \pi z|^2 = \frac{\cosh 2\pi y - 1}{\cosh 2\pi y + 1} < 1 \quad \text{for} \quad x = n + \tfrac{1}{2}. \qquad \square$$

9.6.2 Summation of Series

If $f(z)$ is an analytic function of a simple kind, it is sometimes possible to use contour integration to evaluate $\sum_n f(n)$. Let C be a closed contour containing the points $z = m, m+1, \ldots, n$ and let $f(z) \in A$ inside C except for (simple) poles at a finite number of points $a_1, \ldots, a_k$, with residues $b_1, \ldots, b_k$, respectively. Consider

$$\int_C \pi \cot \pi z f(z) \, dz.$$

The function $\pi \cot \pi z$ has simple poles at $z = m, m+1, \ldots, n$ with residue 1 at each pole; thus the function $\pi \cot \pi z f(z)$ has residues $f(m), f(m+1), \ldots, f(n)$. By including the residues due to the poles of $f(z)$, we have

$$I = \int_C \pi \cot \pi z f(z) \, dz = 2\pi i \big[f(m) + f(m+1) + \cdots + f(n)$$
$$+ b_1 \pi \cot \pi a_1 + \cdots + b_k \pi \cot \pi a_k \big].$$

In particular, if for example $f(z)$ is a rational function whose poles are not integers and if $f(z) = O\{z^{-2}\}$ at infinity, then by taking C to be the square with corners $(n + \tfrac{1}{2})(\pm 1 \pm i)$, we get that $|I| \to 0$ as $n \to \infty$, since $|f| < M/z^2$ whenever $|z| > r_0$. Choosing $n > r_0$, we obtain

$$\left| \int_{\Gamma_n} f(z) \pi \cot \pi z \, dz \right| \leq \frac{M}{n^2} \coth \pi \cdot 8\left(n + \frac{1}{2}\right) \to 0 \quad \text{as} \quad n \to \infty.$$

Consequently we have

$$\lim_{n \to \infty} \sum_{m=-n}^{n} f(m) = -\pi(b_1 \cot \pi a_1 + \cdots + b_k \cot \pi a_k).$$

If we choose $\pi \csc \pi z$ instead of $\pi \cot \pi z$, we can obtain the expression for sums of the form $\sum_m (-1)^m f(m)$.

EXAMPLE 1. Expansion of a Meromorphic Function. Consider $g(z) = \cot \pi z$. Then $|g(z)| \leq \coth \pi$ on Γ_n, where Γ_n is the square $x = \pm(n + \frac{1}{2})$, $y = \pm(n + \frac{1}{2})$, $n = 1, 2, \ldots$. Let

$$f(z) = \frac{g(z)}{z(z-a)}, \qquad \text{where } a \not\equiv 0 \pmod{1},$$

that is, a is not an integer. Take $n > |a|$. Then on Γ_n we have

$$|f(z)| \leq \frac{\coth \pi}{n(n-|a|)} \quad \text{and} \quad \left| \int_{\Gamma_n} f(z) dz \right| \leq \frac{\coth \pi \cdot 8(n + \frac{1}{2})}{n(n-|a|)},$$

thus $\int_{\Gamma_n} f(z) dz \to 0$ as $n \to \infty$.

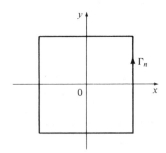

Now, $f(z)$ has simple poles at $z = a, \pm 1, \pm 2, \ldots$, and a pole of order 2 at $z = 0$. Then

$$\operatorname{Res} f(a) = \frac{g(a)}{a} = \frac{\cot \pi a}{a}.$$

The residue of $f(r)$, $r = \pm 1, \pm 2, \ldots$ is given by

$$\operatorname{Res} f(r) = \frac{\cos \pi r}{r(r-a)[d(\sin \pi z)/dz]_{(z=r)}} = \frac{1}{\pi r(r-a)}.$$

Now, as $z \to 0$, we get

$$f(z) = \frac{1 + O(z^2)}{z(z-a)(\pi z + z \cdot O(z^2))}$$

$$= \frac{1}{-a\pi z^2} [1 + O(z^2)] \left[1 + \frac{z}{a} + O(z^2) \right] [1 + O(z^2)],$$

Laurent Series, Singularities

whence Res $f(0) = -1/\pi a^2$. From the theorem of residues it follows that

$$\frac{1}{2\pi i}\int_{\Gamma_n} f(z)dz = \frac{g(a)}{a} - \frac{1}{\pi a^2} + \frac{1}{\pi}\sum_{r=1}^{n}\left[\frac{1}{r(r-a)} + \frac{1}{-r(-r-a)}\right]$$

$$= \frac{\cot \pi a}{a} - \frac{1}{\pi a^2} + \frac{2}{\pi}\sum_{r=1}^{n}\frac{1}{r^2-a^2}.$$

Letting $n \to \infty$, we get

$$\frac{1}{a}\cot \pi a = \frac{1}{\pi a^2} - \frac{2}{\pi}\sum_{r=1}^{\infty}\frac{1}{r^2-a^2},$$

thus

$$\frac{1}{z}\cot \pi z = \frac{1}{\pi z^2} - \frac{2}{\pi}\sum_{r=1}^{\infty}\frac{1}{r^2-z^2}, \quad z \not\equiv 0 \pmod{1}.$$

EXAMPLE 2. Find an expression for the following sum in a closed form:

$$\sum_{n=1}^{\infty}\frac{1}{n^2+a^2}.$$

SOLUTION. Poles of $f(z) = 1/(z^2+a^2)$ occur at $z = \pm ia$; thus the residue of

$$\left.\frac{\pi \cot \pi z}{z^2+a^2}\right|_{ia} = \frac{\pi \cot \pi ia}{2ia}$$

and the residue at $-ia$ becomes $\pi \cot(-\pi ia)/(-2ia)$. Then

$$\sum_{n=-\infty}^{\infty}\frac{1}{n^2+a^2} = \frac{1}{a^2} + 2\sum_{n=1}^{\infty}\frac{1}{n^2+a^2}.$$

Thus, using contours C_n to be squares with vertices at $(n+\frac{1}{2})(\pm 1 \pm i)$, we obtain

$$\frac{1}{a^2} + 2\sum_{n=1}^{\infty}\frac{1}{n^2+a^2} = -\frac{\pi \cot \pi ia}{ia}$$

$$= \frac{i\pi}{a}\frac{e^{-\pi a}+e^{\pi a}}{e^{-\pi a}-e^{\pi a}}i = \frac{e^{\pi a}+e^{-\pi a}}{e^{\pi a}-e^{-\pi a}}\frac{\pi}{a}.$$

Thus,

$$\sum_{n=1}^{\infty}\frac{1}{n^2+a^2} = \frac{1}{2}\left[\frac{\pi}{a}\coth \pi a - \frac{1}{a^2}\right] = \frac{1}{2a^2}[\pi a \coth \pi a - 1].$$

9.6.3
Exercises

Show that:

1: $\displaystyle\sum_{n=-\infty}^{\infty}\frac{1}{(n+a)^2} = \pi^2 \operatorname{cosec}^2 \pi a.$

2: $\sum_{n=0}^{\infty} \dfrac{(-1)^n}{n^2 + a^2} = \dfrac{1}{2a^2}\left[\dfrac{\pi a}{\sinh \pi a} - 1\right].$

3: $\sum_{n=0}^{\infty} \dfrac{1}{(2n+1)^2} = \dfrac{\pi^2}{8}.$

Evaluate

4: $\sum_{n=1}^{\infty} \dfrac{1}{n^4 + a^4}.$ 5: $\sum_{n=1}^{\infty} \dfrac{n^2}{n^4 + a^4}.$

6: Show that $\sum_{n=1}^{\infty} 1/n^2 = \pi^2/6.$ 7: Show that $\sum_{n=1}^{\infty} 1/n^4 = \pi^4/90.$

8: Show that $\sum_{n=0}^{\infty} \dfrac{1}{a + bn^2} = \dfrac{1}{2a} + \dfrac{\pi}{2\sqrt{ab}} \coth\left(\pi \sqrt{\dfrac{a}{b}}\right).$

9: Evaluate $\sum_{n=1}^{\infty} 1/n^6.$

10: Try to evaluate $\sum_{n=1}^{\infty} 1/n^3$; discuss.

11: Prove that $\sum_{n=-\infty}^{\infty} \dfrac{1}{(n+a)^2 + b^2} = \dfrac{\pi}{b} \dfrac{\sinh 2\pi b}{\cosh 2\pi b - \cos 2\pi a}.$

9.7 Simple Properties of Meromorphic Functions: Rouché's Theorem

9.7.1

We now study properties of zeros and poles of meromorphic functions. A great deal of work has been done on this subject and several authoritative treatises exist.*

We consider a few of the simpler theorems as follows.

9.7.2

Theorem *If*

  i. $f(z) \in M$ in a domain Δ,
  ii. D is the interior domain of crJc C,
  iii. $f(z)$ has no zeros or poles on C, and $D \subseteq \Delta$, *then*

$$\dfrac{1}{2\pi i} \int_C \dfrac{f'(z)}{f(z)} dz = N - P,$$

---

*For example, *Distribution of Zeros of Entire Functions*, by B. Ja. Levin, Vol. 5, Translation of Mathematical Monographs, American Mathematical Society, 1964.

where N and P are, respectively, the number of zeros and the number of poles of f in D of multiplicity m.

PROOF. Since $f \in M$ in D, then the number of zeros and poles in $\overline{D} \subset \Delta$ is finite. Also, $f'(z)/f(z) \in A$ in D except at zeros and poles of f. Suppose f has a pole of order p at $z = a$, where $a \in D$. Then there exists $\delta > 0$ such that

$$f(z) = (z-a)^{-p}\varphi(z) \quad \text{for} \quad 0 < |z-a| < \delta,$$

where $\varphi \in A$ and $\varphi \neq 0$ for $|z-a| < \delta$. Thus, for $0 < |z-a| < \delta$, we have

$$\frac{f'(z)}{f(z)} = \frac{-p}{z-a} + \frac{\varphi'(z)}{\varphi(z)},$$

so that $f'(z)/f(z)$ has a simple pole with residue $-p$ at $z = a$.

Similarly, if f has a zero of order n at $z = b$, where $b \in D$, then there exists $\eta > 0$ such that

$$f(z) = (z-b)^n \psi(z) \quad \text{for} \quad |z-b| < \eta,$$

where $\psi \in A$ and $\psi \neq 0$ for $|z-b| < \eta$. Thus, for $0 < |z-b| < \eta$, we have

$$\frac{f'(z)}{f(z)} = \frac{n}{z-b} + \frac{\psi'(z)}{\psi(z)},$$

and it follows that $f'(z)/f(z)$ has a simple pole with residue n at $z = b$. Hence by the residue theorem,

$$\frac{1}{2\pi i} \int_C \frac{f'(z)}{f(z)} dz = \sum n - \sum p = N - P. \qquad \square$$

Corollary 1. *If $P = 0$, then*

$$\frac{1}{2\pi i} \int_C \frac{f'(z)}{f(z)} dz$$

is the number of zeros of the entire function f inside C.

Corollary 2. *If $\varphi(z) \in A$ in and on C and if f has zeros at $a_1, \ldots, a_p$ and poles at $b_1, \ldots, b_q$, whose multiplicity is counted as before, then*

$$\frac{1}{2\pi i} \int_C \frac{f'(z)}{f(z)} \varphi(z) dz = \sum_{\mu=1}^{p} \varphi(a_\mu) - \sum_{\nu=1}^{q} \varphi(b_\nu).$$

PROOF. If $z = a$ is a zero of order m, then in the neighborhood of $z = a$ we have $f(z) = (z-a)^m g(z)$, where $g(z) \neq 0$ and analytic; thus

$$\frac{f'(z)}{f(z)} \varphi(z) = \frac{m\varphi(z)}{z-a} + \frac{g'(z)}{g(z)} \varphi(z).$$

The last term is analytic at $z = a$ and the first term has a simple pole at

$z = a$ with residue $m\varphi(a)$. Likewise, if $z = b$ is a pole of order n, $\dfrac{f'}{f}\varphi$ then $f'(z)\varphi(z)/f(z)$ has a simple pole with residue $n\varphi(b)$. Then by the previous theorem we get that the sum of residues is

$$\frac{1}{2\pi i}\int_C \frac{f'}{f}\varphi\,dz = \sum_{\mu=1}^{p}\varphi(a_\mu) - \sum_{\nu=1}^{q}\varphi(b_\nu). \qquad \square$$

EXAMPLE 1. Consider $f(z) = e^z$; then $f'(z) = e^z$ and thus

$$\frac{1}{2\pi i}\int_C \frac{f'(z)}{f(z)}\,dz = 0$$

for *all* crJc C. Since e^z has no poles, it follows that e^z has no zeros.

EXAMPLE 2. Consider $f(z) = z^n$; then $f'(z) = nz^{n-1}$ and thus

$$\frac{1}{2\pi i}\int_C \frac{f'(z)}{f(z)}\,dz = \frac{1}{2\pi i}\int_C \frac{n}{z}\,dz = \begin{cases} n & \text{if } z = 0 \text{ is inside } C, \\ 0 & \text{if } z = 0 \text{ is outside } C. \end{cases}$$

9.7.3

Theorem (Argument Principle). *Under the conditions of Theorem 9.7.2*

$$N - P = \frac{1}{2\pi}[\arg f(z)]_C,$$

where $[\arg f(z)]_C$ *denotes the increase in* $\arg f(z)$ *as* z *describes* C *in the positive sense, while the value of* $\arg f(z)$ *varies continuously as* z *describes* C.

PROOF. We use the theorem that if $F(z) \in A$ in a (simply-connected) domain D, then

$$F(z_2) - F(z_1) = \int_{z_1}^{z_2} F'(z)\,dz,$$

where $z_1, z_2 \in D$ and the integral is taken along any rectifiable Jordan curve lying in D and joining z_1 to z_2.

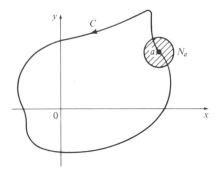

Laurent Series, Singularities

Let a be any point of C. Since $f(z) \in A$ at $z = a$ and $f(a) \neq 0$, there exists a neighborhood $N_a = N(a, \rho)$ of a, in which $f(z) \in A$ and $|f(z) - f(a)| < \frac{1}{2}|f(a)|$. Thus,

$$|f(z) - f(a)| \geq |f(a)| - |f(z)|$$

and

$$|f(z)| \geq |f(a)| - |f(z) - f(a)| > \frac{1}{2}|f(a)| > 0.$$

Hence for $z \in \{|z - a| \leq \rho\}$, there exist branches of $\log f(z)$ that are analytic, according to (8.10.5). Consider any one such branch. We have

$$\frac{d}{dz}[\log f(z)] = \frac{f'(z)}{f(z)}, \quad \forall z \in N_a,$$

that is,

$$\int_{z_1 \atop (k)}^{z_2} \frac{f'(z)}{f(z)} dz = \log[f(z_2)] - \log[f(z_1)],$$

where $z_1, z_2 \in N_a$, lie on C, and the path (k) is contained in N_a. Now each point a of the closed bounded set C is the center of such a neighborhood N_a. Whence, by the Heine–Borel theorem C can be covered by a finite set of overlapping neighborhoods similar to N_a. Thus,

$$\int_C \frac{f'(z)}{f(z)} dz = \{\log[f(z)]\}_C,$$

where $\log f(x)$ denotes a value of the logarithm that varies continuously as z traces out the complete curve C.

Since $\log[f(z)] = \log|f(z)| + i \arg f(z)$ and $[\log|f(z)|]_C = 0$, we have

$$\int_C \frac{f'(z)}{f(z)} dz = i[\arg f(z)]_C,$$

where we consider a value of $\arg f(z)$ that varies continuously as z describes C. Thus,

$$N - P = \frac{1}{2\pi}[\arg f(z)]_C. \qquad \square$$

EXAMPLE 1. Consider $f(z) = z^2$ and $C: |z| = 1$, $\arg z = \theta$, $0 \leq \theta \leq 2\pi$. On C we have

$$z = \operatorname{cis} \theta, \qquad z^2 = e^{2i\theta}, \qquad \text{and} \qquad \arg z^2 = 2\theta.$$

Thus,

$$\frac{1}{2\pi}[\arg z^2]_C = \frac{1}{2\pi} \cdot 4\pi = 2$$

where z^2 has a repeated zero at $z = 0$.

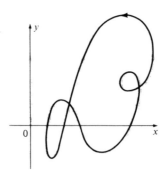

Note: When $R\{f(z)\} > 0$ on C, then

$$-\frac{\pi}{2} < \text{P.V.} \arg f(z) < \frac{\pi}{2}, \quad \forall z \in C.$$

Thus $[\arg f(z)]_C = 0$, since $\arg f(z)$ returns to its original value after description of C.

This note is useful in a number of ways, in particular, for entire functions ($P = 0$) in which we can sometimes determine the position of zeros by evaluating the change in argument as z describes (counterclockwise) a closed rectifiable Jordan curve. If we require information in a quadrant or half-plane, we take an appropriate quadrant or semicircle of radius R and assume that $R \to \infty$. The following example illustrates the method.

EXAMPLE 2. In the right half-plane, find the number of roots of the polynomial

$$P(z) = z^6 + z^5 + 6z^4 + 5z^3 + 8z^2 + 4z + 1.$$

SOLUTION. Let us use the argument principle around C in the following figure, where C is the first quadrant of a circle of radius R.

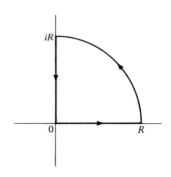

On $y = 0$, $0 \le x \le R$, we have $P(x) > 0$; thus by the previous analysis
$$\arg P(x)\big]_0^R = 0.$$
On $z = Re^{i\theta}$, $0 \le \theta \le \pi/2$, we have $P(z) = R^6 e^{6i\theta}(1 + w)$, where
$$|w| = O(1/R) \quad \text{when } R \to \infty.$$
Thus, $\arg P(Re^{i\theta}) = 6\theta + \arg(1 + w)$, and for large R we get
$$\arg P(Re^{i\theta})\big]_0^{\pi/2} = 3\pi + \delta \quad \text{where } \delta \to 0 \text{ as } R \to \infty.$$
On $x = 0$, $0 < y < R$, we have
$$P(iy) = -(y^6 - 6y^4 + 8y^2 - 1) + i(y^5 - 5y^3 + 4y)$$
$$= -(y^6 - 6y^4 + 8y^2 - 1) + iy(y^2 - 4)(y^2 - 1).$$
As y decreases from R to 0, the function $I\{P(iy)\}$ changes its sign at points 2 and 1. However, if we examine $-(y^6 - 6y^4 + 8y^2 - 1)$, we see that the real part has the following signs for the corresponding values of y:

y	-3	-2	-1	0	1	2	3
$R\{P(iy)\}$	$-$	$+$	$-$	$+$	$-$	$+$	$-$

Consequently, the six roots of the polynomial $-y^6 + 6y^4 - 8y^2 + 1 = 0$ are located in the interval $-3 < y < 3$.

Let the first three zeros of the polynomial be ordered as follows: $\alpha_1 \in (2, 3)$, $\alpha_2 \in (1, 2)$, and $\alpha_3 \in (0, 1)$.

Then for $R > y > \alpha_1$:
$$I\{P(iy)\} > 0, \quad R\{P(iy)\} < 0$$
and $\arg P(iy)$ is in the second quadrant.

For $2 < y < \alpha_1$:
$$I\{P(iy)\} > 0, \quad R\{P(iy)\} > 0$$
and $\arg P(iy)$ is in the first quadrant.

For $\alpha_2 < y < 2$:
$$I\{P(iy)\} < 0, \quad R\{P(iy)\} > 0$$
and $\arg P(iy)$ is in the fourth quadrant.

For $1 < y < \alpha_2$:
$$I\{P(iy)\} < 0, \quad R\{P(iy)\} < 0$$
and $\arg P(iy)$ is in the third quadrant.

For $\alpha_3 < y < 1$:
$$I\{P(iy)\} > 0, \quad R\{P(iy)\} < 0$$
and $\arg P(iy)$ is in the second quadrant.

For $0 < y < \alpha_3$:
$$I\{P(iy)\} > 0, \quad R\{P(iy)\} > 0$$
and $\arg P(iy)$ is in the first quadrant.

Thus as y decreases from R to 0, $\arg P(iy)$ decreases from a value close to (but greater than) $\pi/2$ to zero, that is,

$$\arg P(iy)]_{y=R}^{0} = -(2\pi + \pi/2 + \delta_1) \quad \text{where } \delta_1 = \pi/2 \text{ as } R \to \infty$$

and

$$\arg P(z)]_C = 3\pi + \delta - 3\pi \to 0 \quad \text{as } R \to \infty.$$

Thus the number of zeros in the first quadrant is zero. Since the roots occur in conjugate pairs, there are no roots in the fourth quadrant. Thus there are no roots in the right half-plane. □

We continue with an important theorem that provides a method for determining the number of zeros interior to a crJc.

9.7.4

Theorem (Rouché). *If*

i. *D is the interior domain of a crJc C;*
ii. $f(z), g(z) \in A$ *in* $\Delta \supset \bar{D}$;
iii. $|g(z)| < |f(z)|, \forall z \in C$,
then $f + g$ and f have the same number of zeros in C.

PROOF. Since $\bar{D} \subset \Delta$, the functions f and $f + g$ have only a finite number of zeros in $\bar{D}$. By (iii), f has no zeros on C. Also, since $|f + g| \geq |f| - |g| > 0$ on C, it follows that $f + g$ has no zeros on C.

Suppose f has M zeros, and $f + g$ has N zeros in D. Let $F = (f + g)/f$; then F is meromorphic in Δ as F has $(M - P)$ poles and $(N - P)$ zeros in D. Thus,

$$N - M = \frac{1}{2\pi i} \int_C \frac{f'(z)}{f(z)} dz.$$

Since

$$R\{F(z)\} = 1 + R\left\{\frac{g}{f}\right\} \geq 1 - \frac{|g|}{|f|}, \quad \text{where } |f| \geq |R\{f\}|,$$

we obtain that $R\{F(z)\} > 0$ on C. Thus,

$$N - M = \frac{1}{2\pi i} \int_C \frac{f'(z)}{f(z)} dz = 0$$

by the previous lemma, and $N = M$. Alternatively: Let $\varphi(z) = g(z)/f(z)$, then

$$N' = \frac{1}{2\pi i} \int_C \frac{f' + g'}{f + g} dz = \frac{1}{2\pi i} \int_C \left[\frac{f'}{f} + \frac{\varphi'}{1 + \varphi}\right] dz.$$

Laurent Series, Singularities

Now, since $|\varphi| < 1$, we have

$$N' = N + \frac{1}{2\pi i} \int_C \varphi' \left[\sum_0^\infty \varphi'(-1)^r \right] dz \quad \text{and} \quad \int_C \varphi' \varphi^n dz = \left[\frac{\varphi^{(n+1)}}{n+1} \right]_C = 0,$$

thus $N' = N$. The interchange of summation and integration is justified by uniform convergence of the series. We can improve this result by generalizing as follows. □

9.7.5

Theorem *If we use the first two conditions of Rouché's theorem and assume that $f(z)$ and $g(z)$ are two functions meromorphic in D, then the change in the argument of $f + g$ (when z describes C) is the same as the change in argument of f. Further, the difference between the number of zeros and the number of poles is the same for both functions. Thus,*

$$(N - P)_{f+g} = (N - P)_f.$$

PROOF. Write $F(z) = f(z) + g(z)$, then

$$F(z) = f(z) \left[1 + \frac{g(z)}{f(z)} \right].$$

As z describes C, $|g(z)/f(z)| < 1$. Thus if we write

$$w(z) = 1 + \frac{g(z)}{f(z)},$$

we have $|w(z) - 1| < 1$ on C. Consequently,

$$|\arg w| < \frac{\pi}{2} \quad \text{and} \quad \left[\arg \frac{F}{f} \right]_C = 0$$

or $[\arg(f + g)]_C = [\arg f]_C$, and we have the result. □

One of the particularly useful applications of this form of Rouché's theorem is in discussing the nature of roots of transcendental equations.

EXAMPLE 3. Show that the transcendental equation $\tan z = az$, $a > 0$, has two complex (purely imaginary) roots if $0 < a < 1$, together with infinitely many real and only real roots if $a \geq 1$.

SOLUTION. Set $f = az$, $g = -\tan z$, $F(z) = az - \tan z$. Consider a square with vertices at $z = n\pi(\pm 1 \pm i)$, where we choose n to be an integer large enough, so that $na\pi > 1.005$.

On the vertical sides of the square, $|\tan z| < 1$ (by a simple calculation). On the horizontal sides,

$$\max|\tan z| = \coth n\pi < 1.005 \quad \text{for} \quad n \geq 1$$

(by tables or calculation). Thus, on the sides of the square we have $|az| > |\tan z|$ and consequently

$$(N - P)_F = (N - P)_f \quad \text{or} \quad N_F = N_f - P_f + P_F = 2n + 1.$$

If we plot the graphs of $y = \tan x$, $y = ax$, $-n\pi \leq x \leq n\pi$, we observe that the line has only one intersection with the tangent curve in each of the intervals

$$\left[-\left(k + \frac{1}{2}\right)\pi, -k\pi \right] \quad \text{and} \quad \left[k\pi, \left(k + \frac{1}{2}\right)\pi \right] \quad \text{where } k = 1, 2, \ldots, n - 1.$$

Further, there are three intersections in $(-\pi/2, \pi/2)$ if $a \geq 1$ but only one if $0 < a < 1$. Consequently, if $a \geq 1$, all roots are real; if $0 < a < 1$, there are two complex (and $z = 0$) roots. Since $f(-z) = -f(z)$ and f is real on the real axis, the two roots are of the form $\pm i\beta$. Actually, β is a root of $\tanh y = ay$. □

We continue with some miscellaneous examples illustrating Rouché's theorem.

EXAMPLE 4. Show that $2z^5 + 8z - 1 = 0$ has exactly four roots in $1 < |z| < 2$.

SOLUTION. We show first that there exist five roots in $|z| < 2$. Choose $f = 2z^5$, $g = 8z - 1$.
On $|z| = 2$:

$$|f| = 64 \quad \text{and} \quad |g| \leq 8|z| + 1 = 17,$$

and therefore $|f| > |g|$. Thus, since f has five roots in $|z| < 2$, $f + g = 2z^5 + 8z - 1$ has five roots in $|z| < 2$.

We now show that in $|z| < 1$ there exists one (real) root. Choose $f = 8z - 1$, $g = 2z^5$.
On $|z| = 1$:

$$|g| = 2 \quad \text{and} \quad |f| = |8z - 1| \geq 8|z| - 1 = 7,$$

consequently $|f| > |g|$.

In $|z| = 1$, the zeros of f are $z = 1/8$, that is, there is *one* zero. Thus, in $|z| < 1$, there exists one root of $2z^5 + 8z - 1 = 0$.

Also since $f(z) = 2z^2 + 8z - 1$, $f(0) = -1$, and $f(1) = 9$, the root is real since f changes sign.

The only thing left to show is that there are no roots on $|z| = 1$. Write $f(z) = 2z^5 + 8z - 1$. On $|z| = 1$:

$$|f| = |2z^5 + 8z - 1| \geq 8|z| - 2|z^2| - 1 = 5 > 0.$$

Thus there are exactly four zeros of f in $1 < z < 2$. □

EXAMPLE 5. Find the number of roots of $z^8 - 4z^5 + z^2 - 1 = 0$ in the disk $|z| < 1$.

SOLUTION. Let $f = z^8 - 4z^5$ and $g = z^2 - 1$.
On $|z| = 1$, note that
$$|f| = |z^3 - 4| \geq 4 - |z^3| = 3,$$
$$|g| \leq |z^2| + 1 = 2,$$
and the conditions of Rouché's theorem are satisfied. Thus the number of roots equals the number of roots of f in $|z| < 1$, which equals the number of roots of $z^5(z^3 - 4) = 0$ in the open disk $|z| < 1$.
Consequently we have five roots since $(z^3 - 4) \neq 0$ in $|z| < 1$. □

EXAMPLE 6. Find the number of roots of $e^{z-\lambda} = z$, $\lambda > 1$, in $|z| < 1$.

SOLUTION. Let $f = z$ and $g = -e^{z-\lambda}$.
On $|z| = 1$ we have
$$|f| = 1 \quad \text{and} \quad |g| = e^{x-\lambda} \leq e^{1-\lambda} < 1.$$
Thus the number of zeros of f in $|z| < 1$ equals the number of zeros of $z - e^{z-\lambda}$ in $|z| < 1$. Hence there is exactly one root.
Note that $\varphi(x) = e^{x-\lambda} - x$ is positive for $x = 0$ and negative for $x = 1$. Thus, φ is zero in $(0, 1)$ and the only zero of $e^{z-\lambda} - z$ in $|z| < 1$ is attained at a real and positive z.

The following example illustrates a different technique.

EXAMPLE 7. Show that the equation $z = \lambda - e^{-z}$, $\lambda > 1$, has only one root in the right half-plane and that this root is real.

SOLUTION. Write $f(z) = z - \lambda$, $g(z) = e^{-z}$, and consider $C: |z| = R > 2\lambda$, $x > 0$, and $z = iy$, $-R \leq y \leq R$.
On $|z| = R$:
$$x > 0, \quad |f| \geq |z| - \lambda > \lambda > 1, \quad \text{and} \quad |g| = e^{-x} < 1.$$
On $z = iy$:
$$|f| \geq |\Re\{z - \lambda\}| = \lambda > 1 \quad \text{and} \quad |g| = 1,$$
thus $|f| > |g|$ on C.
Now, f has one zero ($z = \lambda$) in C, therefore $f + g = z - \lambda + e^{-z}$ has only one zero in C.
Further, if $\varphi(z) = z - \lambda + e^{-z}$, then
$$\varphi(0) = 1 - \lambda < 0 \quad \text{and} \quad \varphi(\lambda) = \lambda - \lambda + \frac{1}{e^\lambda} > 0,$$
showing that there is one zero between $z = 0$ and $z = \lambda$.

9.7.6
Exercises

1: Use Rouché's theorem to prove the fundamental theorem of algebra.

2: If $a > e$, show that $e^z = az^n$ has n roots in $|z| < 1$.

3: Show that $e^{k-z}z^n = 1$ has n roots in $|z| < 1$, provided $k > 1$.

4: If n is sufficiently large, show that
$$f_n(z) = 1 + \frac{1}{z} + \frac{1}{2!\,z^2} + \cdots + \frac{1}{n!\,z^n}$$
has all of its zeros in $|z| \leq \rho$, where ρ is arbitrarily small.

5: Prove the following theorem of Hurwitz: Let $f_n(z)$ be a sequence of functions in a domain D bounded by a simple closed contour. Let $f_n(z) \to f(z)$ uniformly. Assume that $f(z) \not\equiv 0$ and let z_0 be an interior point of D. Then z_0 is a zero of $f(z)$ if and only if z_0 is a limit point of zeros of $f_n(z)$, $n = 1, 2, \ldots$, and points that are zeros for infinite n are counted as limit points.

We conclude the chapter with an interesting application of Rouché's theorem which will be useful in later analysis.

9.7.7
Theorem *If $f(z) \in A$ in a domain D containing the closed circle $C = \{z \mid |z - z_0| \leq r\}$ and has a zero of order k at z_0, and if neither f nor f' have any other zeros on C, then there is a neighborhood H of zero, such that for each $\alpha \in H$ the equation $f(z) = \alpha$ has precisely k solutions in C.*
Alternatively: *We may state the theorem as follows: If $f \in A$ in a domain containing z_0 and has a zero of order k there, then there are neighborhoods K of z_0 and H of 0 such that f assumes every value $\alpha \in H$ precisely k times in K.*

PROOF. Let $\gamma = \{|z - z_0| = r\}$ be oriented counterclockwise. Then
$$k = \frac{1}{2\pi i} \int_\gamma \frac{f'(z)}{f(z)}\,dz.$$
If $\alpha \notin f(\gamma)$, then by replacing f by $f - \alpha$ we obtain
$$N(\alpha) = \frac{1}{2\pi i} \int_\gamma \frac{f'(z)}{f(z) - \alpha}\,dz \ldots, \tag{9.5}$$

which is the number of solutions of the equation $f(z) = \alpha$ inside γ. Since $f'(z) \neq 0$ for $z \neq z_0$, there are no multiple solutions except for $\alpha = 0$.

Choose ξ such that $0 < \xi < \rho(0, f(\gamma))$. Then $N(\alpha)$ is continuous on the disk $H = \{|\alpha| < \xi\}$ since the right side of (9.5) is continuous. On the other hand, $N(\alpha)$ is the number of solutions and therefore an integer. Consequently, it must be constant on the disk. But $N(0) = k$, therefore $N(\alpha) = k$ for all $|\alpha| < \xi$, that is, $f(z)$ assumes all those complex numbers α precisely k times. Thus H has the required property. □

10
Conformal Representation

We shall now study mappings induced by the so-called elementary functions. This is by no means an exhaustive list of mappings and their properties, but is a general cross section of methods.

10.1 Linear functions

Write $z = x + iy$ and $w = f(z) = u + iv$, where u, v, x, y are real.

Consider the mapping induced by $w = z + C$, where $C = C_1 + iC_2$. The image of a point (x, y) in the z-plane is the point (u, v) in the w-plane such that

$$u = x + C_1 \quad \text{and} \quad v = y + C_2.$$

Hence the image of a region in the z-plane is a translation of the original image, where shape, size, and orientation are preserved.

Let B(complex) be given by $B = be^{i\beta}$. If we write $z = re^{i\theta}$, then

$$w = Bz = bre^{i(\theta + \beta)}$$

implies that

$$(r, \theta) \Rightarrow (br, \theta + \beta) = (\rho, \varphi).$$

Thus $w = Bz$ is a mapping that rotates the radius vector z about the origin through an angle $\arg \beta$ and magnifies it by $|B| = b$. Hence $w = Bz + C$ consists of a rotation, magnification, and translation.

EXAMPLE. Transform the rectangular region in the diagram, using $w = (1 + i)z + 2 - i$. Since $w = (1 + i)z + 2 - i$, we have

$$\arg(1 + i) = \pi/2 \quad \text{and} \quad |1 + i| = \sqrt{2}\ .$$

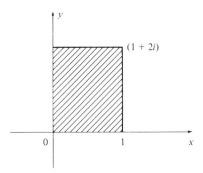

Thus $(0,0) \Rightarrow w = 2 - i$ and $(0, 1) \Rightarrow w = 3$, and our image in the w-plane is depicted in the following diagram.

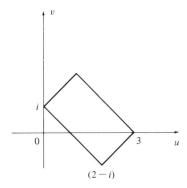

10.2

Consider the mapping $w = z^2$.

Write $w = \rho e^{i\varphi}$ and $z = re^{i\theta}$; then the mapping is

$$\rho e^{i\varphi} = r^2 e^{2i\theta} \quad \text{or} \quad \rho = r^2 \text{ and } \varphi = 2\theta.$$

Consequently an angle in the z-plane becomes doubled in the w-plane. For example, the first quadrant $0 \leq \theta \leq \pi/2, r > 0$, in the z-plane becomes $0 \leq \varphi \leq \pi, \rho > 0$, that is, the upper half of the w-plane.

Further, writing $w = u + iv$ and $z = x + iy$, we obtain

$$w = z^2 \Rightarrow u + iv = x^2 - y^2 + 2ixy$$

or

$$u = x^2 - y^2 \quad \text{and} \quad v = 2xy.$$

Every point on the hyperbola $2xy = k$ has an image on the line $v = k$. Two points $z, -z$ on the hyperbola correspond to each point w on the line.

Also, $u = k$ is the image of $x^2 - y^2 = k$. Note that the image of the domain $x > 0, y > 0, xy < 1$ is the strip $0 < v < 2$.

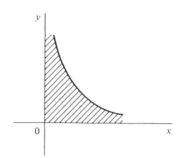

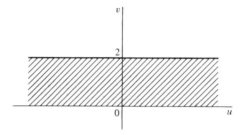

EXAMPLE. Show, using $w = z^2$, that the region in the z-plane bounded by the lines $y = \pm x$, $x = 1$ maps onto a region in the w-plane bounded by the v-axis and $\rho = 2/(1 + \cos\varphi)$.

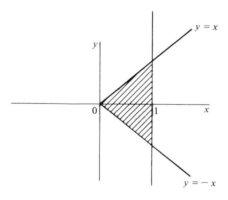

SOLUTION. Given $w = z^2$, then $u = x^2 - y^2$ and $v = 2xy$. Thus
$$y = x \Rightarrow u = 0, \quad v = 2x^2,$$
$$y = -x \Rightarrow u = 0, \quad v = -2x^2.$$
Also,
$$x = 1 \Rightarrow u = 1 - y^2 = 1 - \frac{v^2}{4} \quad \text{or} \quad v^2 = -4(u - 1).$$

Conformal Representation

This is a parabola with a vertex at $u = 1$, $v = 0$ and open to the left. Clearly

$$v^2 = -4(u - 1) \Rightarrow \rho = \frac{2}{1 + \cos\varphi}.$$

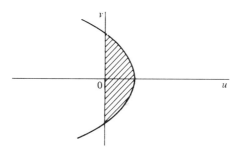

10.3

Consider the mapping $w = z^n$, where n is a natural number. As in 10.2, $\rho = r^n$ and $\varphi = n\theta$. Thus an angle π/n in the z-plane maps onto the upper half of the w-plane. Note also that $w = z^n$, where n is a positive integer, maps circles onto circles.

10.4

We now consider the mapping $w = 1/z$. There is clearly a one-to-one correspondence between the points of the z- and w-planes, except that $z = 0$ and $w = 0$ have no images. Writing $w = \rho e^{i\varphi}$ and $z = re^{i\theta}$, we have

$$\rho e^{i\varphi} = \frac{e^{-i\theta}}{r}.$$

Let $z' = \dfrac{e^{i\theta}}{r}$, then $w = \bar{z}'$. Thus $w = 1/z$ corresponds to the so-called inversion with respect to the unit circle followed by a reflection in the real axis. Note that a point z' on the radius through z has the property $|z'||z| = 1$. Points outside the unit circle are mapped onto points inside the unit circle, and conversely.

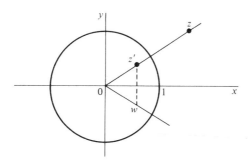

Also,

$$u + iv = \frac{1}{x + iy} \Rightarrow u = \frac{x}{x^2 + y^2}, \quad v = -\frac{y}{x^2 + y^2}$$

or

$$x = \frac{u}{u^2 + v^2}, \quad y = -\frac{v}{u^2 + v^2}.$$

Hence, $w = 1/z$ transforms circles to circles or straight lines. Thus

$$a(x^2 + y^2) + bx + cy + d = 0$$

with real constants implies that

$$d(u^2 + v^2) + bu - cv + a = 0.$$

If $a, d \neq 0$, then both the curve and the image are circles. Circles through $z = 0$ transform to straight lines in the w-plane. Also,

$$x = c(\neq 0) \Rightarrow u^2 + v^2 - \frac{u}{c} = 0.$$

10.4.1 Point at Infinity

Under $w = 1/z$, we say that $z = \infty \Rightarrow w = 0$; that is, points outside an arbitrarily large circle in the z-plane map to points that are arbitrarily close to the origin $w = 0$.

EXAMPLE. Find the image of $x^2 - y^2 = 1$ under the mapping $w = 1/z$.

SOLUTION. We have $x = u/u^2 + v^2$, $y = -v/u^2 + v^2$, thus

$$x^2 - y^2 = \frac{u^2 - v^2}{(u^2 + v^2)^2} = 1 \quad \text{or} \quad u^2 - v^2 = (u^2 + v^2)^2.$$

Using polar coordinates $u = \rho \cos \varphi$, $v = \rho \sin \varphi$ we obtain $\rho^2 = \cos 2\varphi$, (lemniscate of Bernoulli).

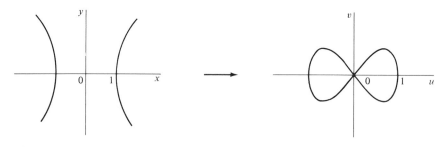

10.5

Consider the mapping $w = e^z$. With the usual polar transformation we get $\rho e^{i\theta} = e^x e^{iy}$, thus

$$\rho = e^x \quad \text{and} \quad \varphi = y.$$

This mapping takes lines $x = k$ into circles $\rho = e^k$. Also, lines $y = k$ map to rays $\varphi = k$. Thus,

$$\begin{cases} c_1 \leq x \leq c_2, \\ c_3 \leq y \leq c_4, \end{cases} \quad \text{becomes} \quad \begin{cases} e^{c_1} \leq \rho \leq e^{c_2}, \\ c_3 \leq \varphi \leq c_4. \end{cases}$$

The mapping is one-to-one if $c_4 - c_3 < 2\pi$; otherwise the angular sector is repeated.

Consider the strip

$$x \leq 0, \quad 0 \leq y \leq \pi.$$

This maps onto

$$0 \leq \rho \leq 1, \quad 0 \leq \varphi \leq \pi.$$

Note, for example, that $z = \pi i \Rightarrow u = -1$.

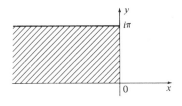

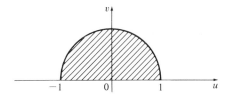

10.6

Let us now consider the map $w = \sin z$. We have

$$\sin z = \sin x \cosh y + i \cos x \sinh y,$$

thus

$$u = \sin x \cosh y \quad \text{and} \quad v = \cos x \sinh y.$$

Consider the semiinfinite strip $-\pi/2 \leqslant x \leqslant \pi/2, y \geqslant 0$. If $x = \pi/2$, then $u = \cosh y$ and $v > 0$; thus $u \geqslant 1$ for all y. If $y = 0$, then $u = \sin x$ and $-1 \leqslant u \leqslant 1$, while $v = 0$. Thus the x-axis maps onto the line segment $-1 \leqslant u \leqslant 1$; however, the map is not one-to-one.

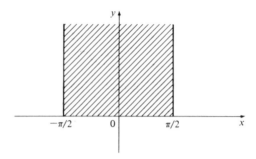

When $-\pi/2 < x < \pi/2$, we have $-1 < u < 1$ and the mapping is one-to-one. Consequently, the semi-infinite strip in the z-plane maps onto the upper half of the w-plane since for $-\pi/2 < x < \pi/2, y > 0$ we have $v > 0$.

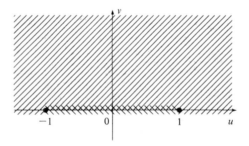

Note: A line segment $-\pi/2 \leqslant x \leqslant \pi/2, y = c$ maps onto the upper half of the ellipse

$$\frac{u^2}{\cosh^2 c} + \frac{v^2}{\sinh^2 c} = 1$$

with foci at $w = \pm 1$.

10.7

When we consider the mapping $w = \cos z$, we observe that this is nothing more than $w = \sin z'$ followed by $z' = z + \pi/2$. Consequently, $w = \cos z$ is the mapping $w = \sin z$ followed by a translation to the right.

10.8

The mapping $w = \sinh z$ corresponds to $w' = \sin z'$, where $z' = iz$ and $w' = iw$, that is, a sine mapping followed by rotating both axes through $\pi/2$.

10.9

Consider the mapping $w = \text{Log } z$. Let $z = re^{i\theta}$, then
$$w = \log r + i(\theta + 2k\pi), \text{ where } k = 0, \pm 1, \ldots,$$
In particular, if $k = 0$, then $u = \log r$ and $v = \theta$. Thus, the angular region
$$\begin{cases} \theta_1 < \arg z < \theta_2, \\ 0 < r < \infty \end{cases} \text{ becomes } \begin{cases} \theta_1 < v < \theta_2, \\ -\infty < u < \infty. \end{cases}$$
For arbitrary k, the angular region corresponds to an infinite collection of infinite strips in the w-plane because of the multivalued nature of the logarithm; however, if $\theta_2 - \theta_1 < 2\pi$, we will not have the plane covered more than once.

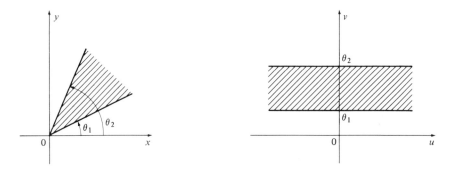

EXAMPLE. Show that $w = z + 1/z$ maps the circle $r = c (\neq 1)$ onto an ellipse. Map the upper semicircle under this transformation.

SOLUTION. Using polar coordinates, we get
$$\rho e^{i\varphi} = re^{i\theta} + \frac{1}{r} e^{-i\theta}.$$

Thus
$$\rho \cos \varphi = \left(r + \frac{1}{r}\right) \cos \theta = u \quad \text{and} \quad \rho \sin \varphi = \left(r - \frac{1}{r}\right) \sin \theta = v.$$

Hence, for $r = c \neq 1$ we have an ellipse:
$$\frac{u^2}{\left(c + \dfrac{1}{c}\right)^2} + \frac{v^2}{\left(c - \dfrac{1}{c}\right)^2} = 1.$$

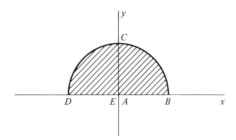

If we wish to examine, say, the upper half of $|z|<1$, let us consider the diagram where E and A represent the origin when it is approached from the left and from the right, respectively. Then,

$$B:\begin{cases} r=1 \\ \theta=0 \end{cases} \Rightarrow B':\begin{cases} u=2 \\ v=0 \end{cases}$$

$$A:\begin{cases} \theta=0 \\ r\to 0^+ \end{cases} \Rightarrow A':\begin{cases} v=0 \\ u\to +\infty \end{cases}$$

$$E:\begin{cases} \theta=0 \\ r\to 0^- \end{cases} \Rightarrow E':\begin{cases} v=0 \\ u\to -\infty \end{cases}$$

$$C:\begin{cases} r=1 \\ \theta=\pi/2 \end{cases} \Rightarrow C':\begin{cases} u=0 \\ v=0 \end{cases}$$

Clearly, for $0<r<1$, $0\leq\theta\leq\pi$, we have $1/r>1$ and

$$\begin{cases} r-\dfrac{1}{r}<0 \\ \sin\theta>0 \end{cases} \Rightarrow v<0 \quad \text{for all } u.$$

Thus, the upper semicircle maps onto the lower half-plane.

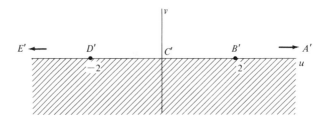

10.10 Mappings and Conformality

Let $w=f(z)=u(x,y)+iv(x,y)$ be a single-valued analytic function of z, where u,v,x,y are real. We now study the structure of mappings of continuous curves through the development of the subsequent four lemmas.

10.10.1

Lemma *If $f(z)$ is continuous on a continuous curve C given by*

$$C: z=z(t)=\theta(t)+i\varphi(t), \quad t\in[a,b],$$

Conformal Representation

where θ and φ are continuous on $[a,b]$, then the mapping of C in the z-plane is a continuous curve Γ in the w-plane given by

$$w = w(t) = u\{\theta(t),\varphi(t)\} + iv\{\theta(t),\varphi(t)\}, \quad t \in [a,b].$$

PROOF. Since u,v are continuous on $[a,b]$, the components of w are continuous since they are continuous functions of continuous functions. □

Note: 1. To a given sense of description of C corresponds a definite sense of description of Γ; in both cases, the sense that corresponds to t increasing from a to b is called the *positive sense*.
2. If C is closed, then so is Γ.

10.10.2

Lemma *If C is rectifiable and lies in a domain D where $f(z)$ is analytic, then Γ is rectifiable.*

PROOF. We may assume that D is simply connected, since each point of C is the center of some circle contained in D, and C can be covered by a finite number of these overlapping circles. Thus C consists of a finite number of segments each lying in a simply connected subdomain of D.

Let $w_0, w_1, \ldots, w_n$ be a set of points (in order) on Γ, then they are mappings of the corresponding points $z_0, z_1, \ldots, z_n$ (in order) lying along the appropriate arc of C:

$$w_r - w_{r-1} = f(z_r) - f(z_{r-1}) = \int_{z_{r-1}}^{z_r} f'(z)\,dz,$$

where $r = 1, \ldots, n$ and $z_{r-1}z_r$ is a path k_r. Now, $f'(z)$ is continuous on the closed set C, consequently there exists M such that $|f'(z)| \leq M$ for $z \in C$. Thus

$$\sum_{r=1}^{n} |w_r - w_{r-1}| = \sum_{r=1}^{n} \left| \int_{z_{r-1}}^{z_r} f'(z)\,dz \right|$$

$$\leq \sum_{r=1}^{n} Ml_r \leq Ml,$$

where l_r is the length of k_r and l is the length of C. Hence $w = u + iv$ is of bounded variation and thus Γ is rectifiable. □

10.10.3

Lemma *If*

i. C has a tangent at $z = z_0$ (making an angle γ with the x-axis),
ii. $f(z)$ is continuous on C and differentiable at $z = z_0$, and
iii. $f'(z_0) \neq 0$, then Γ has a tangent at $w_0 = f(z_0)$ and makes an angle $\gamma + \arg f'(z_0)$ with the u-axis.

Note: The positive direction of tangents correspond to a positive sense of description of C and Γ.

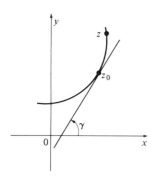 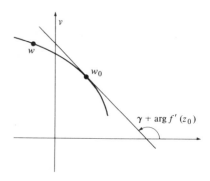

PROOF. Let z be a variable point of C and let w be the corresponding point of Γ. Now,

$$w - w_0 = \frac{w - w_0}{z - z_0}(z - z_0), \quad z \neq z_0,$$

thus

$$\arg(w - w_0) = \arg\left\{\frac{w - w_0}{z - z_0}\right\} + \arg(z - z_0)$$

and

$$\arg(w - w_0) \to \arg\{f'(z_0)\} + \gamma \quad \text{as } z \to z_0$$

through points of C corresponding to values of t less than that corresponding to z_0 (otherwise we add π to the right-hand side).

If $f'(z_0) = 0$, while the first two conditions remain as before, and $f(z)$ is not constant in any neighborhood of z_0, assume that $f \in A$. Then by Taylor's theorem,

$$w - w_0 = (z - z_0)^k [C_0 + C_1(z - z_0) + \cdots], \quad |z - z_0| < \delta, \quad \delta > 0,$$

where $C_0 \neq 0$ and $k = 2, 3, \ldots$. Thus,

$$\arg(w - w_0) = k \arg(z - z_0) + \arg\{C_0 + C_1(z - z_0) + \cdots\}$$
$$\to k\gamma + \arg C_0 \quad \text{as } z \to z_0$$

through points of C between z_0 and $z(b)$. □

Note: Since

$$w - w_0 = \frac{w - w_0}{z - z_0}(z - z_0),$$

then

$$|w - w_0| = \left|\frac{w - w_0}{z - z_0}\right| |z - z_0|,$$

which implies that the image of a small line element $|z - z_0|$ from the point z_0 is magnified by $|f'(z_0)|$ under the mapping $w = f(z)$. Further, the image

Conformal Representation

of a small neighborhood of the point z_0 conforms to the original region in the sense that it is approximately the same shape.

We sum up the three lemmas in the following theorem.

10.10.4
Theorem *If*

i. *two curves C_1 and C_2 pass through z_0 and have tangents at z_0,*
ii. *$f(z)$ is continuous on C_1 and C_2,*
iii. *$f(z)$ is differentiable at z_0,*
iv. *$f'(z_0) \neq 0$, and*
v. *Γ_1 and Γ_2 are the mappings of C_1 and C_2, respectively, then the angle Γ_1 and Γ_2 equals the angle between C_1 and C_2, both sense and magnitude being preserved.*

PROOF. If the two curves C_1 and C_2 pass through z_0 and make angles θ_1 and θ_2 with the positive x-axis, then the angle between the images Γ_1, Γ_2 at w_0 is given by

$$\varphi_2 - \varphi_1 = \theta_2 + \arg f'(z_0) - \theta_1 - \arg f'(z_0) = \theta_2 - \theta_1. \qquad \square$$

10.10.5
Exercises

1: Under the mapping $w = 1/z$, find the images of the curves
   a) $y = kx$ b) $y = x^2$.

2: Find the domain onto which the semicircle $|z| < 1$, $\Re\{z\} > 0$ is mapped by the function $w = z + z^2$.

3: Use the function $w = \frac{1}{2}(z + \frac{1}{z})$ to map the domains
   a) $|z| < R < 1$ b) $R < |z| < 1, I\{z\} > 0$
   c) $I\{z\} > 0$ d) $|z| > 1$.

4: Use the function $w = e^z$ to map the curves and domains
   a) $y = ax + b$
   b) $x < 0, 0 < y < \alpha \leq 2\pi$
   c) $\alpha < y < \beta, 0 \leq \alpha < \beta \leq 2\pi$.

5: Use the function $w = \log z$ to map
   a) $r = Ae^{k\varphi}, A > 0$
   b) $|z| < 1, 0 < \arg z < \alpha \leq 2\pi$
   c) $|z| = R, \arg z = \theta$.

6: Use the function $w = \tan z$ to map
   a) $0 < x < \pi/4$ b) $0 < x < \pi$, $y > 0$.

7: Use the function $w = \cos z$ to map
   a) $0 < x < \pi$
   b) $-\pi/2 < x < \pi/2, y > 0$
   c) The net $x = c, y = c$.

8: Use the function $w = \cosh z$ to map the strip $0 < y < \pi$.

10.11 Definitions Relating to the Point at Infinity

The *finite* complex plane is the set of all finite complex numbers. The *extended* complex (closed) plane consists of all finite complex numbers together with the point $z = \infty$. Geometrically, these two concepts are well illustrated by the stereographic projection; however, we will not spend any time on this topic.

10.11.1

A domain D in the extended plane is a nonempty open set, any two finite points of which can be joined by an open polygon whose points belong to D. If D contains $z = \infty$, then (since D is open) D contains a neighborhood $|z| > R$ of $z = \infty$.

10.11.2

A domain D in the extended plane is called *simply connected* if either the interior (or exterior) domain determined by C consists entirely of points of D, whenever C is a closed Jordan polygon whose points belong to D.

Consider the mapping $w = f(z)$, where $f \in A$ at $z = \infty$. The mapping can be obtained by successive mappings $\xi = 1/z$ and $w = f(1/\xi) = g(\xi)$; then $g \in A$ at $\xi = 0$. If a curve passing through $\xi = 0$ in the ξ-plane is the mapping of a set of points C in the (extended) z-plane, we call C a curve in the z-plane passing through the point at infinity. The angle between two curves meeting at $z = \infty$ in the z-plane is defined to be the angle between their mappings in the ξ-plane at $\xi = 0$. Hence, if $g'(0) \neq 0$, the angle between the mappings in the w-plane (at $w = g(0)$) equals the angle between the mappings in the ξ-plane (at $\xi = 0$) and equals the angle between the curves in the z-plane at $z = \infty$.

EXAMPLES

1. The set $S_1 = \{z \mid a < \Re\{z\} < b, z \neq \infty\}$ is a domain in either the extended or finite plane.
2. The set $S_2 = \{z \mid a < \Re\{z\} < b\}$ is an open set in the extended plane.
3. The sets $S_3 = \{z \mid |z| > r\}$ and $S_4 = \{z \mid r < |z| < \infty\}$ are both domains in the extended plane.
4. The set $S_5 = \{z \mid |z| > r\}$ is simply-connected in the extended plane.
5. The set $S_6 = \{z \mid r < |z| < \infty\}$ is not simply-connected in the extended plane.

We now need Theorem 9.7.7 that will be stated in a slightly different way.

10.12

Theorem *If $w = f(z)$ is such that $f \in A$ at z_0 and $f(z) - f(z_0)$ has a zero of order k at $z = z_0$, then, given an arbitrarily small neighborhood U of z_0, there exists a neighborhood V of $w_0 = f(z_0)$ such that to every value of w in V there corresponds at least k values of z in U.*

In this form, the theorem gives rise to the following corollary.

10.12.1

Corollary *If $w = f(z)$ is such that $f \in A$ at z_0, then the mapping of an arbitrarily small neighborhood of z contains a neighborhood V of w_0, that is, $V \subseteq f(U)$.*

This, in turn, gives rise to the following two theorems.

10.13

Theorem (Open-Mapping Theorem). *If $f(z) \in A$ and is not constant in an open set G, then $f(G)$ is open.*

PROOF. Suppose $w_0 \in f(G)$. Then there exists $z_0 \in G$ such that $w_0 = f(z_0)$. Since G is open, there exists a neighborhood U of z_0 such that $U \subset G$. By corollary 10.12.1, there exists a neighborhood V of w_0 such that $V \subseteq f(U)$, hence $V \subseteq f(G)$, that is, $f(G)$ is open. □

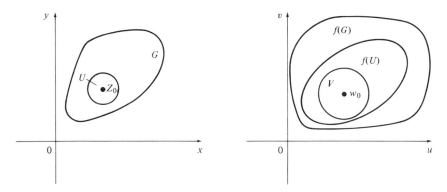

10.14

Theorem *If $f(z) \in A$ and is not constant in a domain D, then $f(D)$ is a domain.*

PROOF. By Theorem 10.13, $f(D)$ is open. Let w_1 and w_2 be any two distinct points of $f(D)$, and z_1, z_2 points of D such that $w_1 = f(z_1)$, $w_2 = f(z_2)$. First assume that $|z_1|, |z_2| < \infty$. Then z_1 and z_2 can be joined by a rectifiable continuous curve that belongs entirely to D. The mapping is a rectifiable continuous curve joining w_1 to w_2, that belongs entirely to $f(D)$.

If $z_1 = \infty$, join w_1 to a point w_3 by a straight line all of whose points are in $f(D)$. Then w_3 can be joined to w_2 by a rectifiable continuous curve. Thus w_1 is joined to w_2 by a rectifiable continuous curve. Thus $f(D)$ is a domain. $\square$

10.15 Simple Conformal Mappings

10.15.1

Definition Let $w = f(z)$ be such that $f \in A$ in a domain D, so that the map $\Delta = f(D)$ is a domain. The mapping of D into Δ is said to be *conformal* if any two curves C_1, C_2 in D (that intersect in a point $z_0 \in D$ and both have tangents at z_0) are mapped into curves Γ_1, Γ_2 in Δ such that the angle between Γ_1 and Γ_2 at $w_0 = f(z_0)$ is equal to the angle between C_1 and C_2 at z_0 (in sense and magnitude). If, in addition, the mapping is one-to-one, it is said to be a *simple (schlicht) conformal* mapping.

10.15.2

Definition If $f(z)$ takes none of its values more than once in a domain D, then $f(z)$ is said to be *univalent (schlicht)* in D.

This is the case if and only if the inverse mapping $z = f^{-1}(w)$ is single-valued for all $w \in \Delta = f(D)$, that is, if and only if the mapping of D on Δ is one-to-one.

We have the result (Theorem 10.10.4) that, if $f(z) \in A$ in a domain D and if $f'(z) \neq 0$ in D, then the mapping of D on $\Delta = f(D)$ is conformal. Let us now examine some related theorems.

10.15.3

Theorem *If $f(z) \in A$ and is univalent in a domain D, then $f'(z) \neq 0$ in D.*

PROOF. Suppose $f'(z) = 0$ at some $z_0 \in D$. Since D is open, there exists a neighborhood U of z_0 such that $U \subset D$. Also, $f'(z_0) = 0$; thus $\{f(z) - f(z_0)\} \in A$ in D and has a zero of order $k \geq 2$ at z_0. Hence, there exists a neighborhood V of $w_0 = f(z_0)$ such that $f(z)$ takes every value in V at least k times in U. This is impossible since $f(z)$ is univalent in D and $U \subseteq D$. $\square$

Note: The converse of Theorem 10.15.3 is not necessarily true, i.e., if $f(z) \in A$ and $f'(z) \neq 0$ in D, it does not follow that $f(z)$ is univalent in D. For example, if $f(z) = e^z$, then $f'(z)$ has no zeros, but $f(z)$ is not univalent in $D = \{z \mid |z| < 2\pi\}$; take $z = \pm i\pi$. However, we have a weaker result that we shall now prove.

10.15.4

Theorem *If $f(z) \in A$ at the point z_0 and $f'(z_0) \neq 0$, then $f(z) \in A$ and is univalent in some neighborhood of z_0.*

PROOF. 1. Suppose that z_0 is finite. Since $f'(z_0) \neq 0$, the difference $f(z) - f(z_0)$ has a zero of order 1 at z_0. Hence there exists $r > 0$, $\eta(r) > 0$ such that $f(z) \in A$ in $|z - z_0| < r$ and takes each value w in the domain $|w - w_0| < \eta$ exactly once in $|z - z_0| < r$.

Since $f(z)$ is continuous at z_0, there exists $\rho > 0$ such that $\rho \leq r$ and $|f(z) - f(z_0)| = |w - w_0| < \eta$ whenever $|z - z_0| < \rho$. (There may be points in $|z - z_0| < r$ such that $w = f(z)$ lies outside $|w - w_0| < \eta$.) Thus each z, such that $|z - z_0| < \rho$, corresponds to one and only one value $w = f(z)$ in $|w - w_0| < \eta$ and this point w corresponds to one and only one point z in $|z - z_0| < \rho$. Hence $f(z) \in A$ and is univalent in $|z - z_0| < \rho$.

2. Suppose that $z_0 = \infty$. By Theorem 9.7.7, given $\epsilon > 0$, there exists $r > 0$, $\eta(r) > 0$, where $0 < r < \epsilon$, such that $f(z) \in A$ for $|z| > 1/r$ and takes each value w, satisfying $|w - w_0| < \eta(r)$ exactly once in $|z| > 1/r$. Since $f(z)$ is continuous at $z = \infty$, there exists $\rho > 0$ such that $\rho \leq r$ and $|f(z) - f(z_0)| = |w - w_0| < \eta$ whenever $|z| > 1/\rho$. Thus, each z, such that $|z| > 1/\rho$, corresponds to one and only one value $w = f(z)$ in $|w - w_0| < \eta(r)$, and this point w corresponds to one and only one point z in $|z| > 1/\rho$. Hence $f(z) \in A$ and is univalent in $|z| > 1/\rho$. □

The next theorem is a direct consequence of Theorem 10.15.3.

10.15.5

Theorem *If $f(z) \in A$ and is univalent in a domain D, then the mapping of D on $\Delta = f(D)$ is a simple conformal mapping.*

10.15.6

Theorem *If $w = f(z)$ and $f(z) \in A$ and is univalent in a domain D that does not contain the point $z = \infty$, then the inverse function $z = f^{-1}(w) \in A$ and is univalent in $\Delta = f(D)$.*

PROOF. Since the mapping of D on Δ is one-to-one, $f^{-1}(w)$ is single-valued and univalent in Δ.

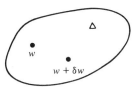
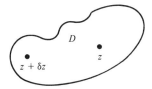

Take w (a fixed point of Δ) and $w + \delta w$ (a variable point). Let $z = f^{-1}(w)$ and $z + \delta z = f^{-1}(w + \delta w)$. Then

$$\frac{f^{-1}(w + \delta w) - f^{-1}(w)}{\delta w} = \frac{\delta z}{\delta w}, \quad \delta w \neq 0.$$

Since $f(z) \in A$ and is univalent in D, we get $\delta w / \delta z \to f'(z) \neq 0$ as $\delta z \to 0$ in any manner. Hence $\delta z \to 0$ and $\delta w \to 0$ together and

$$\frac{\delta z}{\delta w} \to \frac{1}{f'(z)} \quad \text{as} \quad \delta z \to 0 \quad \text{and} \quad \delta w \to 0.$$

Thus,

$$\lim_{\delta w \to 0} \frac{f^{-1}(w + \delta w) - f^{-1}(w)}{\delta w} = \frac{1}{f'(z)} \quad \text{for all} \quad z \in D. \qquad \square$$

10.15.7

Corollary (Inverse-Function Theorem). *If $f(z) \in A$ at z_0 (finite) and $f'(z_0) \neq 0$, the relation $w = f(z)$ defines z as an analytic function $f^{-1}(w)$ in some neighborhood of the point $w_0 = f(z_0)$.*

PROOF. Let U be the domain $|z - z_0| < \rho$ in Theorem 10.15.6. Then $f(z) \in A$ and is univalent in U. Hence by Theorem 10.15.5, we get that $f^{-1}(w) \in A$ and is univalent in $f(U)$. Since $f(U)$ is open (by Theorem 10.13) and $w_0 \in f(U)$, then $f^{-1}(w) \in A$ in a neighborhood of w_0. $\qquad \square$

We have seen that if $f(z) \in A$ and $f'(z_0) \neq 0$, then the mapping $w = f(z)$ is conformal at z_0. We now consider the converse problem.

10.15.8

Theorem *Suppose that $w = f(z)$ and defines a one-to-one mapping $f: S \to T$. If f has continuous partial derivatives of at least order 1 and if f maps S onto T in such a way as to preserve the angles, then $f(z) \in A$.*

PROOF. Let $z_0 \in S$ and $w_0 = f(z_0)$. Consider any neighboring point $z = z_0 + \rho e^{i\alpha}$, where ρ is small. Using the Taylor expansion about (x_0, y_0), we get

$$w - w_0 = \rho(f_x \cos\alpha + f_y \sin\alpha) + o(\rho)$$

and

$$\frac{w - w_0}{z - z_0} = \frac{f_x \cos\alpha + f_y \sin\alpha}{e^{i\alpha}} + o(1)$$

$$= \frac{1}{2}\left(f_x + \frac{1}{i} f_y\right) + \frac{1}{2}\left(f_x - \frac{1}{i} f_y\right) e^{-2i\alpha} + o(1).$$

If the mapping preserves angles, then in the limit $\arg[(w - w_0)/(z - z_0)]$

must be independent of α as $z \to z_0$, that is,
$$f_x = -if_y,$$
which is the complex form of the Cauchy–Riemann equations. Hence $f(z) \in A$. □

If the mapping $w = f(z)$ has uniform magnification at z independent of direction, then in the limit as $z \to z_0$, the modulus $|(w - w_0)/(z - z_0)|$ must be independent of α. This implies that either
$$f_x = -if_y \quad \text{or} \quad f_x = if_y.$$
The first equation implies that $w = f(z) \in A$. The second equation implies that $w = f(\bar{z})$, where $f(\bar{z}) \in A$. It is only necessary to consider the first equation, since the mapping $w = f(\bar{z})$ is equivalent to $w = f(\xi)$ followed by $\xi = \bar{z}$, a reflection in the real axis.

Several more theorems can be proved, however, we will only state the results here since the proofs are somewhat tedious.

10.15.9

Theorem (Darboux). *If*

i. D_i *is the interior domain bounded by a crJc C,*
ii. $f(z) \in A$ *in a domain containing $\overline{D}_i$,*
iii. $f(z)$ *is univalent on the curve C (that is, f takes each value $w \in f(C)$ once and only once on C),*

then the transformation $w = f(z)$ maps C onto a crJc Γ, while D_i is mapped onto the interior domain Δ_i bounded by Γ. Further, $f(z) \in A$ and is univalent in D_i, and also, as z describes C_+, its map $w = f(z)$ describes Γ_+. (The $+$ sign refers to the sense of description of the curves.)

Note: We may replace D_i by D_e, the exterior domain of the extended plane bounded by C. Provided all the conditions of Theorem 10.15.9 hold, the conclusion of that theorem follows, except that, as z describes C_+, the point $w = f(z)$ describes Γ_-.

Finally, we state the following theorem.

10.15.10

Theorem *If $f(z) \in A$ and is univalent in a simply connected domain D of the extended plane, then the mapping $\Delta = f(D)$ of D is also simply connected.*

10.16 Bilinear Transformation

We now study one of the most important of all mappings, the *bilinear mapping* (or *transformation*). This mapping may also be called a *Möbius transformation* or *linear fractional transformation*, and we define it as follows:

$$w = f(z) = \frac{az + b}{cz + d}, \quad \text{where} \quad ad - bc \neq 0. \tag{10.1}$$

Convention:

$$w = \infty \quad \text{iff} \quad z = -\frac{d}{c} \quad \text{and} \quad z = \infty \quad \text{iff} \quad w = \frac{a}{c}.$$

Clearly, if $ad = bc$, then Equation (10.1) reduces to a constant, since

$$f(z) = \frac{a}{c} + \frac{bc - ad}{c(cz + d)} \quad \text{for} \quad c \neq 0.$$

The term "bilinear" is justified to some extent by the fact that Equation (10.1) becomes $(zw)c + dw - az - b = 0$, which is linear with regard to w and z.

If $w = (az + b)/(cz + d)$, then $z = (b - dw)/(cw - a)$, that is,

$$f^{-1}(w) = \frac{-d(w) + b}{cw + (-a)} = \frac{Aw + B}{Cw + D},$$

where

$$AD - BC = (-d)(-a) - (b)(c) = ad - dc \neq 0.$$

Further,

$$f'(z) = \frac{ad - bc}{(cz + d)^2}, \quad \text{where} \quad z \neq -\frac{d}{c}.$$

Thus we have the following properties of the bilinear transformation:

1. $w = f(z)$ gives a one-to-one mapping of the extended z-plane onto the extended w-plane:

$$z = -\frac{d}{c} \text{ is mapped on to } w = \infty,$$

$$z = \infty \text{ is mapped on to } w = \frac{a}{c}.$$

2. $f(z) \in A$ and is univalent in any domain that does not contain $z = -d/c$.
3. $z = -b/a$ is mapped on to $w = 0$, and a direction at $z = -b/a$ is rotated through a positive angle

$$\arg\left[f'\left(-\frac{b}{a}\right)\right] = \arg \frac{a^2}{ad - bc}.$$

4. The inverse mapping $z = f^{-1}(w) = (b - dw)/(cw - a)$ is a bilinear mapping.

5. The product of two bilinear maps is a bilinear map since, if

$$z_1 = \frac{az+b}{cz+d}, \quad ad-bc \neq 0 \quad \text{and} \quad w = \frac{\alpha z_1 + \beta}{\gamma z_1 + \delta}, \quad \alpha\delta - \beta\gamma \neq 0,$$

that is,

$$w = F(z_1) = F[f(z)] = \frac{\alpha(az+b) + \beta(cz+d)}{\gamma(az+b) + \delta(cz+d)} = \frac{Az+B}{Cz+D},$$

then

$$AD - BC = \begin{vmatrix} A & B \\ C & D \end{vmatrix} = \begin{vmatrix} \alpha & \beta \\ \gamma & \delta \end{vmatrix} \begin{vmatrix} a & b \\ c & d \end{vmatrix} \neq 0.$$

6. Under $w = f(z)$, the map of a circle is a circle or straight line (and a line is a circle or straight line). Let us examine

$$\left| \frac{z-p}{z-q} \right| = \lambda. \tag{10.2}$$

If $\lambda = 1$, then $|z-p|^2 = |z-q|^2$, that is,

$$|z|^2 - pz - p\bar{z} + p^2 = |z|^2 - qz - q\bar{z} + q^2$$

or

$$p^2 - q^2 + 2x(q-p) = 0.$$

Equation (10.2) represents a straight line (the perpendicular bisector of the line segment pq).

If $\lambda \neq 1$ and $\lambda > 0$, a short calculation shows that (10.2) represents a circle.

Consequently, the equation of any circle or line in the z-plane can be written in the form (10.2) with $p \neq -d/c$, $q \neq -d/c$. The transformation of Equation (10.2) yields the equation

$$\left| \frac{(b-dw) - p(cw-a)}{(b-dw) - q(cw-a)} \right| = \lambda,$$

or

$$\frac{|d+cp| \left| w - \frac{ap+b}{cp+d} \right|}{|d+cq| \left| w - \frac{aq+b}{cq+d} \right|} = \lambda,$$

that is,

$$\left| \frac{w-p'}{w-q'} \right| = \lambda',$$

where p', q' are transformations of p, q, respectively. Hence,
$$p' \neq q' \quad \text{and} \quad \lambda' = \lambda \left| \frac{d + cq}{d + cp} \right|.$$

Thus, the mapping of (10.2) is a circle or straight line; it is a straight line iff
$$\lambda \left| \frac{d + cq}{d + cp} \right| = 1 \quad \text{or} \quad \left| \left(-\frac{d}{c} - p \right) \Big/ \left(-\frac{d}{c} - q \right) \right| = \lambda,$$
that is, iff the point $z = -d/c$ lies on (10.2).

If we assume that (10.2) is a circle C and that $z = -d/c$ does not lie on C, then C is mapped into a circle Γ.

If $-d/c$ lies *outside* C, that is, $-d/c \in D_e$, then, by Darboux's theorem, D_i (interior of C) maps onto Δ_i (interior of Γ) and C_+ corresponds to Γ_+. Since the mapping is one-to-one, the exterior domain D_e of C maps onto the exterior domain Δ_e of Γ.

If $-d/c$ lies *inside* C, that is, $-d/c \in D_i$, then $D_e \to \Delta_i$ and $D_i \to \Delta_e$, also C_+ corresponds to Γ_-.

10.16.1 Inversion with Respect to a Circle

Given a circle C with center z_0 and radius r, let z_1 be a point inside the circle and z_2 a point outside the circle such that z_0, z_1, z_2 are collinear and
$$|z_0 - z_1||z_0 - z_2| = r^2,$$
then z_1 and z_2 are said to be *inverse points* with respect to the circle.

If we write
$$z_1 = z_0 + \rho e^{i\lambda}, \quad z_2 = z_0 + \frac{r^2}{\rho} e^{i\lambda},$$
then it is a simple task to show that
$$\left| \frac{z - z_1}{z - z_2} \right| = \frac{\rho}{r}$$
is a circle with center z_0 and radius r, while z_1 and z_2 are inverse points.

The bilinear transformation has many properties, however we mention only one further result.

10.16.2 Fixed Points

We call $z = \alpha$ a *fixed point* of the mapping $w = f(z)$ if $z = \alpha$ maps onto $w = \alpha$. Thus, for example, the fixed point of the mapping $w = (6z - 9)/z$ occurs for $w = z$, or
$$z^2 - 6z + 9 = 0 \Rightarrow z = 3.$$

Note: Every bilinear map (not the identity) has at most two fixed points.

Conformal Representation

EXAMPLE. If the point $z = 0$ is to be a fixed point of
$$w = \frac{az + b}{cz + d},$$
then $w = b/d = 0 \Rightarrow b = 0$. Thus,
$$w = \frac{az}{cz + d} \quad \text{or} \quad w = \frac{z}{pz + q}, \quad q \neq 0.$$

We now prepare to solve several problems illustrating the power of the bilinear transformation. The concept of *points inverse with respect to a circle* is often used, and we state the definition and develop a few consequences.

10.16.3

Definition We say that two points p and q are *symmetric* with respect to a circle Γ, if every circle or straight line passing through p and q intersects Γ orthogonally.

We leave it as an exercise to show that two finite points p, q are symmetric with respect to a circle Γ, iff p and q are inverse points with respect to Γ.

Now, we know that bilinear mappings are conformed, thus orthogonal circles or straight lines are mapped into orthogonal circles or straight lines. Hence we have the following theorem.

10.16.4

Theorem *Let Γ be a circle or straight line. Then, under a bilinear mapping, points symmetric with respect to Γ are mapped onto points symmetric with respect to the image of Γ (and conversely).*

PROBLEM 1. Find all bilinear transformations of $|z| < 1$ into $|w| < 1$.

SOLUTION. $w = (az + b)/(cz + d)$ maps $|z| < 1$ onto $|w| < 1$, provided that $|z| = 1$ is mapped onto $|w| = 1$ and $z = -d/c$ lies outside $|z| = 1$. Thus, let us assume that $|d| > |c|$ and, in particular, that $d \neq 0$.

Since $z = -b/a \to w = 0$, we suppose that $|b| < |a|$ and, in particular, that $z\bar{z} = 1$ is mapped onto
$$\left[\frac{b - dw}{cw - a} \right] \left[\frac{\bar{b} - \bar{d}\bar{w}}{\bar{c}\bar{w} - \bar{a}} \right] = 1,$$
that is,
$$(|d|^2 - |c|^2)w\bar{w} + (\bar{a}c - \bar{b}d)w + (a\bar{c} - b\bar{d})\bar{w} + |b|^2 - |a|^2 = 0.$$

This is the circle $|w| = 1$, provided
$$|d|^2 - |c|^2 = |a|^2 - |b|^2 \tag{10.3}$$

and
$$\bar{a}c - \bar{b}d = 0. \tag{10.4}$$

Since $a \neq 0$, $d \neq 0$, Equation (10.4) implies $\bar{b}/\bar{a} = c/d = -\bar{\gamma}$, where we have $|\gamma| < 1$ since $|d| > |c|$. Substituting in (10.3), we get

$$|d|^2(1 - |\gamma|^2) = |a|^2(1 - |\gamma|^2),$$

hence $|a| = |d|$. And since $|a| \neq 0$, we write $a/d = e^{i\lambda}$, where λ is real. Thus, the required transformation is

$$w = \frac{a(z - \gamma)}{d(1 - \bar{\gamma}z)}$$

or

$$w = \frac{e^{i\lambda}(z - \gamma)}{1 - \bar{\gamma}z}, \quad \lambda \text{ real}, |\gamma| < 1.$$

Note: With $w = \dfrac{e^{i\lambda}(z - \gamma)}{1 - \bar{\gamma}z}$, we have

$$\frac{dw}{dz} = \frac{e^{i\lambda}}{1 - |\gamma|^2} \quad \text{for} \quad z = \gamma.$$

Thus, since $|\gamma| < 1$, for $z = \gamma$ we get $\arg\{dw/dz\} = \lambda$. Consequently, there exists one and only one bilinear transformation of $|z| < 1$ into $|w| < 1$, in which a given point $z = \gamma$ inside $|z| < 1$ is mapped into $w = 0$ and a direction at $z = \gamma$ is rotated through a given angle λ.

If, in particular, $z = 0$ corresponds to $w = 0$, then $\gamma = 0$ and $w = e^{i\lambda}z$. Also, if $dw/dz > 0$ at $z = 0$, then $w = z$.

PROBLEM 2. Find a conformal mapping that takes $\{|z| < 1\} \cap \{|z - 1| < 1\}$ onto $|w| < 1$.

SOLUTION. Let us illustrate the common part of $|z| < 1$ and $|z - 1| < 1$.
Let the circles intersect at P and Q; thus if P is the point $z = \alpha$, then Q

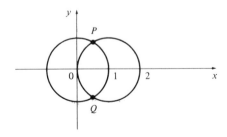

Conformal Representation

is the point $z = \bar{\alpha}$. A simple calculation shows that

$$\alpha = \cos \frac{\pi}{3} + i \sin \frac{\pi}{3} \equiv \operatorname{cis} \frac{\pi}{3} \quad \text{and} \quad \bar{\alpha} = \operatorname{cis}\left(-\frac{\pi}{3}\right).$$

We map from the z-plane to the z_1-plane by choosing

$$z_1 = \frac{z - \bar{\alpha}}{z - \alpha}.$$

This is chosen because it takes:

$$z = \alpha \to z_1 = \infty,$$
$$z = \bar{\alpha} \to z_1 = 0,$$
$$z = 0 \to z_1 = \operatorname{cis}\left(-\frac{2\pi}{3}\right) \equiv R,$$
$$z = 1 \to z_1 = \operatorname{cis}\left(\frac{2\pi}{3}\right) \equiv S.$$

It is also a bilinear transformation, therefore it maps $|z| = 1$ and $|z - 1| = 1$ into circles or straight lines. The two given circles are clearly mapped into lines through the origin and $\operatorname{cis}(\pm 2\pi/3)$, since the point $z = \alpha$ (on both) goes to $z_1 = \infty$.

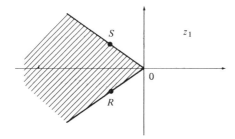

The intersection of the two circles maps onto the left half-wedge of angle 120°. We can check, for example, that the point $z = \tfrac{1}{2} \to z_1 = -1$. The idea is now to send this wedge into a half-plane and then use a bilinear transformation to take the half-plane to a unit circle. We proceed as follows.

Write $z_2 = z_1^{3/2}$. Then, with $z_2 = \rho e^{i\varphi}$ and $z_1 = R e^{i\varphi'}$, we have

$$\rho = R^{3/2}, \qquad \varphi = \frac{3}{2} \varphi',$$

Since $R \geqslant 0$, then $\rho \geqslant 0$ and for $2\pi/3 \leqslant \varphi' \leqslant 4\pi/3$ we have $\pi \leqslant \varphi \leqslant 2\pi$. Thus we have the wedge mapped onto the lower half of the z_2-plane.

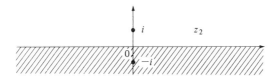

If we imagine the real z_2-axis to be a circle of infinite radius, let us choose a bilinear transformation to map this "circle" into a unit circle as follows. Since i and $-i$ are inverse points (in the z_2-plane) with respect to the line $I(z_2) = 0$, their images under a bilinear map must be symmetric with respect to the image of $I(z_2) = 0$. Choose

$$w = \frac{z_2 + i}{z_2 - i}$$

then

$$\begin{aligned} z_2 = \pm\infty & \to w = 1, \\ z_2 = -i & \to w = 0, \\ z_2 = i & \to w = \infty, \\ z_2 = 0 & \to w = -1. \end{aligned}$$

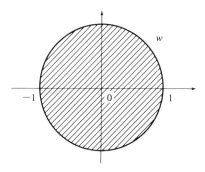

Thus the lower half of the z_2-plane maps onto the unit circle $|w| \leq 1$. It is a circle (and not a straight line), since $w = 0, \infty$ are inverse points.

Finally, the required mapping is

$$w = \frac{z_1^{3/2} + i}{z_1^{3/2} - i} = \frac{(z - \bar{\alpha})^{3/2} + i(z - \alpha)^{3/2}}{(z - \bar{\alpha})^{3/2} - i(z - \alpha)^{3/2}}, \quad \text{where} \quad \alpha = \frac{1 + \sqrt{3}\, i}{2}.$$

PROBLEM 3. Map conformally the region between $|z| = 1$ and $|z - \tfrac{1}{2}| = \tfrac{1}{2}$ onto an upper half-plane.

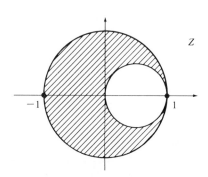

Conformal Representation

SOLUTION. Choose $z_1 = z/(z-1)$, then the real axis in the z-plane maps onto the real axis in the z_1-plane, since if $y = 0$, then

$$I(z_1) = \frac{-y}{(x-1)^2 + y^2} = 0.$$

Also, $z = 1 \to z_1 = \infty$, therefore, by properties of the bilinear transformation, both given circles map onto straight lines in the z_1-plane. (See figure.) Further, the two circles $|z| = 1$ and $|z - \tfrac{1}{2}| = \tfrac{1}{2}$ become lines perpendicular to the real z_1-axis passing through $z_1 = \tfrac{1}{2}, 0$, respectively. (Why?) Now choose $z_2 = iz_1$, then

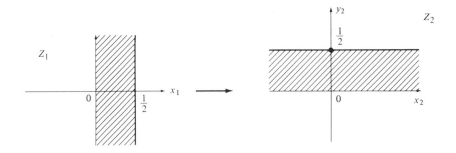

We now wish to expand the infinite strip in the z_2-plane to fill the whole upper half-plane. Choose $w = e^{2\pi z_2}$. Then if we write

$$w = \rho e^{i\varphi} \quad \text{and} \quad z_2 = x_2 + iy_2,$$

we get

$$\rho = e^{2\pi x_2} \quad \text{and} \quad \varphi = 2\pi y_2.$$

Thus, if $0 \le y_2 \le \tfrac{1}{2}$, then $0 \le \varphi \le \pi$ and if $-\infty \le x_2 \le \infty$ then $0 \le \rho \le \infty$, and we have the required result. Consequently the required mapping is $w = e^{2\pi i z/(z-1)}$.

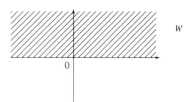

PROBLEM 4. Find a conformal map to take the exterior of $y^2 = 2px$ ($p > 0$) onto the disk $|w| < 1$, such that $z = 0, -p/2$ map onto $w = 1, 0$, respectively.

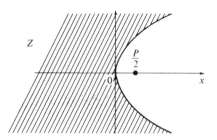

SOLUTION. Let us first look at the graph of $y^2 = 2px$.
Choose $z_1 = z - (p/2)$ which is first a lateral shift of the original graph. We obtain

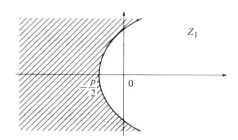

Now choose a branch of $\sqrt{z_1}$, $z_2 = \sqrt{z_1}$, and write

$$z_2 = x_2 + iy_2 \quad \text{and} \quad z_1 = x_1 + iy_1.$$

Then

$$x_1 = x_2^2 - y_2^2 \quad \text{and} \quad y_1 = 2x_2 y_2.$$

The parabola in the z_1-plane is given by

$$y_1^2 = 2p\left(x_1 + \frac{p}{2}\right) \quad \text{and} \quad x_1 = \frac{y_1^2}{4y_2^2} - y_2^2;$$

thus,

$$y_1^2 = 4y_2^2(x_1 + y_2^2) \quad \text{and} \quad y_1^2 = 2p\left(x_1 + \frac{p}{2}\right) \rightarrow y_2 = \sqrt{\frac{p}{2}}.$$

Also,

$$z = 0 \rightarrow z_1 = -\frac{p}{2} \rightarrow z_2 = i\sqrt{\frac{p}{2}}$$

and

$$z = -\frac{p}{2} \rightarrow z_1 = -p \rightarrow z_2 = i\sqrt{p}.$$

Consequently we have the following half-plane bounded by the line $y_2 = \sqrt{p/2}$.

Conformal Representation

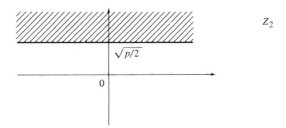

We want to map this half-plane onto a unit disk, so that

$$z_2 = i\sqrt{\frac{p}{2}} \to w = 1 \quad \text{and} \quad z_2 = i\sqrt{p} \to w = 0.$$

Choose

$$w = \frac{az_2 + b}{cz_2 + d}, \quad \text{where} \quad ad - bc \neq 0.$$

We want to take the line $y_2 = \sqrt{p/2}$ onto $|w| = 1$. Now,

$$1 = \left|\frac{az_2 + b}{cz_2 + d}\right|,$$

therefore, as $z_2 \to \infty$, we get $|a/c| \to 1$ and a and c both differ from zero. Setting $a/c = \pm 1$, we may write:

$$w = \pm \frac{z_2 - z'}{z_2 - z''},$$

and since $w = 0, \infty$ are symmetric with respect to $|w| = 1$, we must choose z', z'' symmetric with respect to the line $y_2 = \sqrt{p/2}$. Now, $z_2 = i\sqrt{p} \to w = 0$, therefore the point symmetric to it (with respect to $y_2 = \sqrt{p/2}$) is

$$z_2 = i\left[\sqrt{\frac{p}{2}} - \left(\sqrt{p} - \sqrt{\frac{p}{2}}\right)\right]$$

and we want the latter point to go to $w = \infty$. Consequently we have:

$$w = \pm \frac{z_2 - i\sqrt{p}}{z_2 - i(\sqrt{2p} - \sqrt{p})}.$$

But we want $z_2 = i\sqrt{p/2} \to w = 1$, consequently we must take the minus sign. (Why?) Finally resubstituting, we obtain

$$w = -\frac{\sqrt{z - (p/2)} - i\sqrt{p}}{\sqrt{z - (p/2)} - i(\sqrt{2p} - \sqrt{p})}.$$

10.16.5
Exercises

1: Find a bilinear transformation that carries $|z| = 1$ and $|z - \frac{1}{4}| = \frac{1}{4}$ into concentric circles. Find the ratio of the radii.

   [Hint: choose points that are inverse with respect to both circles.]

2: Mappings that preserve angles but not sense are called *isogonal* mappings. Give several examples of such mappings.

3: Show that if two mappings are conformal, their sum may not be conformal.

4: We define the *cross ratio* of four points z_1, z_2, z_3, z_4 to be the expression
$$(z_1, z_2, z_3, z_4) \equiv \frac{(z_1 - z_4)(z_3 - z_2)}{(z_1 - z_2)(z_3 - z_4)}.$$
Show that if a bilinear transformation takes four distinct finite points z_1, z_2, z_3, z_4 of the z-plane respectively into four corresponding finite points w_1, w_2, w_3, w_4 of the w-plane, then the cross ratio of the four distinct points is invariant.

5: Show that under the bilinear transformation $w = (2z + 3)/(z - 4)$, the circle $C: |z - 2i| = 2$ is mapped onto the circle Γ:
$$\left| w - \left[-\frac{6 + 11i}{8} \right] \right| = \frac{11}{8}.$$

6: Show that four noncollinear points z_1, z_2, z_3, z_4 lie on a circle if and only if their cross ratio is real.

7: Find all bilinear transformations that map the upper half-plane $I(z) > 0$ onto the interior of the circle $|w| = \rho$.

8: Show that if $|c| = |d|$, then the bilinear transformation
$$w = \frac{az + b}{cz + d}, \quad ad - bc \neq 0$$
maps the unit circle in the z-plane into a straight line.

9: Show that if $f(z) \in A$ in $|z| < 1$ and if $R\{f(z)\} > 0$ and if $f(0) = \alpha > 0$, then $|f'(0)| \leq 2\alpha$.

10: Find a bilinear transformation that maps the points $z_1 = 2$, $z_2 = i$, $z_3 = -2$ onto $w_1 = 1$, $w_2 = i$, $w_3 = -1$, respectively.

10.17 Riemann's Mapping Theorem

We now study a very general theorem called Riemann's mapping theorem which basically deals with the mapping of one simply connected domain onto another. The proof requires concepts from the theory of normal families, however, we shall not study the structure of normal families in detail.* We prove that any simply connected domain of the plane, other than the plane itself, can be mapped conformally onto a unit circle. This implies that any two such domains can be mapped conformally onto each other, since we use the unit disk as the intermediate step. The exact statement of the theorem we shall prove is as follows.

10.17.1

Theorem *Given any simply connected domain Ω, which is not the whole plane, and a point $z_0 \in \Omega$, there exists a unique function $f(z)$, analytic in Ω and normalized by the conditions $f(z_0) = 0$, $f'(z_0) > 0$, such that $f(z)$ defines a one-to-one mapping of Ω onto the disk $|w| < 1$.*

Note: The mapping is not possible unless Ω is simply connected. To prove this, assume that $w = f(z)$ maps Ω onto $|w| < 1$ and that $f(z)$ is analytic and univalent in Ω. Then $z = f^{-1}(w)$ maps $|w| < 1$ onto Ω and $f^{-1}(w)$ is analytic and univalent in $|w| < 1$. Thus, since $|w| < 1$ is simply connected, then Ω is also simply connected.

Before proceeding, we need to examine the concept of local uniform (almost uniform) convergence.

10.17.2

Definition A sequence of functions $\{f_n(z)\}$, defined on a set S of the complex plane, is said to converge *locally uniformly* (almost uniformly) on S, if it converges uniformly on every compact subset of S.

For example, the sequence $f_n(z) = z^n$ converges locally uniformly to zero on $|z| < 1$, but it does not converge locally uniformly on $|z| \leq 1$, since $\lim_{n \to \infty} f_n(z) = 1$ at $z = 1$. More precisely, we have the following. Let $z_0 \in G$. Let $\rho > 0$ be such that $S = \overline{N(z_0, \rho)} \subset G$. Then S is compact. A sequence of functions $\{f_n\}$ converges locally uniformly to f on G, if to every $\epsilon > 0$ there exists $N(\epsilon)$ such that $|f_n(z) - f(z)| < \epsilon$ for all $n > N(\epsilon)$ and all $z \in S$.

10.17.3

Lemma *If a sequence of functions $\{f_n\}$ is such that every point of the region S has a neighborhood in which $\{f_n\}$ converges locally uniformly, then $\{f_n\}$ converges locally uniformly in S.*

---

*See, for example, L. V. Ahlfors, *Complex Analysis*, 2nd ed., New York: McGraw-Hill Book Co., 1966, or Z. Nehari, *Conformal Mapping*, New York: McGraw-Hill Book Co., 1952.

PROOF. Let $\{f_n\}$ converge pointwise to f in S and let A be a compact subset of S. Corresponding to $\epsilon > 0$, there exist points $z_1, z_2, \ldots, z_j$ in A, positive numbers $r_1, r_2, \ldots, r_j$, and integers $M_1, M_2, \ldots, M_j$ such that the neighborhoods $N(z_i, r_i)$ cover A and $n > M_i$ implies that $|f_n(z) - f(z)| < \epsilon$ for $z \in N(z_i, r_i)$. Thus if $n > \max_{1 \leq i \leq j} M_i$, it follows that $|f_n(z) - f(z)| < \epsilon$ for all $z \in A$. □

Functions that are *locally bounded* are defined in a similar way.
It is not difficult to prove the following lemma, which is left as an exercise.

10.17.4

Lemma *If $\{f_n\}$ is a sequence of functions continuous on an open set $S \subset \mathbb{C}$ and if the sequence converges locally uniformly to f on S, then f is continuous on S.*

We prove the following result which we shall need later on.

10.17.5

Theorem *If $\{f_n\}$ is a sequence of functions converging locally uniformly to f on a set S, and if $f_n \in A$ for each n, then the limit function $f \in A$ in S. Further, the sequence $\{f_n'\}$ converges locally uniformly to f' on S.*

PROOF. Corresponding to a point $z_0 \in S$, choose $\delta > 0$ that $\overline{N(a,\delta)} \subset S$. Let C be the boundary of $N(a,\delta)$ oriented positively and let $z \in N(a,\delta)$. Since $\{f_n\}$ converges uniformly to f on C and since $|\xi - z|^{-1}$ is bounded with respect to $\xi \in C$, then $f_n(\xi)(\xi - z)^{-1}$ converges uniformly to $f(\xi) \cdot (\xi - z)^{-1}$ with respect to $\xi \in C$. Thus, by 5.6.1 and 6.2.3,

$$f(z) = \lim_{n \to \infty} f_n(z) = \lim_{n \to \infty} \frac{1}{2\pi i} \int_C \frac{f_n(\xi)}{\xi - z} d\xi = \frac{1}{2\pi i} \int_C \frac{f(\xi)}{\xi - z} d\xi.$$

Consequently, by 5.7, $f \in A$ in $N(a,\delta)$ and hence in S. We can use the fact that

$$f'(z) = \frac{1}{2\pi i} \int_C \frac{f(\xi)}{(\xi - z)^2} d\xi$$

to prove similarly that $\{f_n'\}$ converges uniformly to f' on $N(a, \delta/2)$. Now, for $z \in N(a, \delta/2)$ and $\xi \in C$, we have $|z - \xi|^2 < (\delta/2)^2$. Thus for sufficiently large n we get

$$|f_n(\xi) - f(\xi)| < \frac{\epsilon \delta}{4},$$

then

$$|f_n'(z) - f'(z)| = \frac{1}{2\pi} \left| \int_C \frac{f_n(\xi) - f(\xi)}{(\xi - z)^2} d\xi \right| \leq \frac{1}{2\pi} \frac{\epsilon \delta}{4} \left(\frac{\delta}{2}\right)^{-2} 2\pi\delta = \epsilon.$$

The local uniform convergence of $\{f_n'\}$ on S follows from 10.17.3. □

Conformal Representation

We now prove a result that uses the important Cantor diagonal argument.

10.17.6

Lemma *Suppose $f_1(z), f_2(z), \ldots$ is a sequence of functions analytic in a domain D and locally uniformly bounded on D. Then there exists a subsequence $f_{n1}, f_{n2}, \ldots$, that converges locally uniformly to an analytic function $f(z)$, the convergence being uniform on any closed subset of D.*

PROOF. Let $\xi_1, \xi_2, \ldots$, be an enumeration of all the rational points in D. Since $\{f_n(\xi_1)\}$ is a bounded sequence, there is a convergent subsequence, i.e., there exist integers n_{1i} such that $n_{11} < n_{12} < n_{13} < \cdots$ and such that $f_{n_{1i}}(\xi_1)$ approaches a limit as $i \to \infty$. Similarly, there exists a subsequence of integers in $n_{11} < n_{12} < n_{13} < \cdots$, namely,

$$n_{21} < n_{22} < n_{23} < \cdots,$$

such that $f_{n_{2i}}(\xi_2)$ approaches a limit as $i \to \infty$. Continuing the process, we can find integers n_{jk} such that

$$\begin{aligned} n_{11} &< n_{12} < n_{13} < \cdots, \\ n_{21} &< n_{22} < n_{23} < \cdots, \\ &\vdots \\ n_{j1} &< n_{j2} < n_{j3} < \cdots, \end{aligned}$$

and each row is contained in the previous row, and such that $f_{n_{jk}}(\xi_j)$ tends to a limit as $k \to \infty$.

Now consider the function $g_i = f_{n_{ii}}$ ($i = 1, 2, \ldots$). This sequence is a subsequence of the original one and is such that $g_i(\xi_j)$ approaches a limit as $i \to \infty$ for $j = 1, 2, 3, \ldots$. We want to prove that $g_i(z)$, $i = 1, 2, \ldots$, is convergent for all $z \in D$, that the limit function $g(z)$ is analytic on D, and that the convergence is uniform on any subset $E \subset D$.

We have shown that there is a sequence $n_{11} < n_{22} < n_{33} < \cdots$ such that

$$g_i(\xi) = f_{n_{ii}}(\xi)$$

tends to a limit as $i \to \infty$ for any rational point ξ. Also if $z_0 \in D$, there exists $\rho > 0$ such that $\overline{N(z_0, 2\rho)} \subset D$. We will show that the sequence $g_i(z)$ is uniformly convergent as $i \to \infty$ for $z \in N(z_0, \rho)$.

Let $\epsilon > 0$ be given; we must show that there exists a constant $N = N(\epsilon)$ such that

$$|g_i(z) - g_j(z)| < \epsilon, \forall z \in N(z_0, \rho) \quad \text{and} \quad i, j > N.$$

First we use the hypothesis that there exists a constant M such that

$$|g_i(z)| \leq M \quad \text{for} \quad z \in \overline{N(z_0, 2\rho)}.$$

Hence, if $z \in N(z_0, \rho)$, we have for $i = 1, 2, 3, \ldots$:

$$|g_i'(z)| = \left| \frac{1}{2\pi i} \int_{|\xi - z_0| = 2\rho} \frac{g_i(z)}{(\xi - z)^2} d\xi \right| \leq \frac{M 4\pi \rho}{2\pi \rho^2} = k,$$

that is, $g_i'(z)$ is also uniformly bounded on $N(z_0,\rho)$. If $z_1, z_2 \in N(z_0,\rho)$, we have the Lipschitz condition:

$$|g_i(z_1) - g_i(z_2)| = \left|\int_{z_1}^{z_2} g_i'(z)dz\right| \leq k|z_2 - z_1|, \quad i = 1,2,3,\ldots \quad (10.5)$$

For each $z \in N(z_0,\rho)$ let $\Delta(z) = N(z, \epsilon/3k)$. Then $N(z_0,\rho)$ can be covered by a finite number of $\Delta(z)$, that is, there exists an integer n and $z_1, z_2, \ldots, z_n$ such that $N(z_0,\rho) \subset \bigcup_{p=1}^{n} \Delta(z_p)$.

Since $\Delta_p \equiv \Delta(z_p) \cap N(z_0,\rho)$ is open and does not equal ϕ, it contains a rational point η_p. Now we can choose N, so that

$$|g_i(\eta_p) - g_j(\eta_p)| < \frac{\epsilon}{3} \quad \text{for} \quad 1 \leq p \leq n \quad \text{and} \quad i,j > N. \quad (10.6)$$

Let $z \in N(z_0,\rho)$ and $i, j > N$; then there exists some p such that $1 \leq p \leq n$ and $z \in \Delta_p$. Thus $|z - \eta_p| < \epsilon/3k$ and, by Equations (10.5) and (10.6), we have

$$|g_i(z) - g_j(z)| \leq |g_i(z) - g_i(\eta_p)| + |g_i(\eta_p) - g_j(\eta_p)| + |g_j(\eta_p) - g_j(z)|$$
$$< \frac{\epsilon}{3} + \frac{\epsilon}{3} + \frac{\epsilon}{3}$$

This completes the proof. □

We also require the following lemma.

10.17.7

Lemma *If $f_1, f_2, \ldots$ is a sequence of analytic functions that are simple (univalent) on a domain Ω and converge locally uniformly to $f(z)$ in Ω, then either $f(z)$ is simple or $f(z)$ is constant.*

PROOF. Suppose $f(z)$ is not constant and there exists $z_0, z_1 \in \Omega$ such that $z_0 \neq z_1$ and $f(z_0) = f(z_1) = w_0$. Since the zeros of $f(z) - w_0$ are isolated, there exists $\delta > 0$ such that $f(z) \neq w_0$ for all z such that

$$0 < |z - z_0| \leq \delta \quad \text{or} \quad 0 < |z - z_1| \leq \delta$$

(deleted neighborhoods of z_0 and z_1 containing no zeros). Let Γ_i denote the boundaries of the two disks $N(z_0,\delta)$, $N(z_1,\delta)$ that can be assumed to be disjoint and contained in Ω. Let

$$m = \inf\{|f(z) - w_0| : z \in \Gamma_0 + \Gamma_1\}.$$

Then $m > 0$ and $|f(z) - w_0| \geq m$ for $z \in \Gamma_0 + \Gamma_1$. Since $f_n(z) \to f(z)$ locally uniformly in Ω, there exists $N > 0$ such that

$$|f_n(z) - f(z)| < m \quad \text{for} \quad \forall n > N \quad \text{and} \quad z \in \Gamma_0 + \Gamma_1.$$

Therefore, by Rouché's theorem, $[f(z) - w_0] + [f_n(z) - f(z)]$ has the same number of zeros within Γ_i, $i = 0, 1$, as $f(z) - w_0$. Hence, since the disks $N(z_0,\delta)$ and $N(z_1,\delta)$ are disjoint, $f_n(z) - w_0$ has at least two zeros within Ω for $n > N$. This contradicts the hypothesis that $f_n(z)$ is simple in Ω. □

Note: The sequence $f_n(z) = z/n$ converges uniformly for $|z| < 1$; the functions are simple but $\lim_{n \to \infty} f_n(z) = 0$ is a constant.

Finally, we need the following lemma, that is the result obtained in Problem 1 of Chapter 10.

10.17.8

Lemma *If $w = f(z)$ maps the unit disk $|z| \leq 1$ conformally onto $|w| \leq 1$, preserving the origin and the direction of the real axis at the origin, then $w = z$.*

PROOF OF RIEMANN'S THEOREM. We are now in a position to prove Riemann's theorem that any simply connected domain Ω that is not the entire plane can be mapped conformally onto the unit disk. We will prove in fact that if z_0 is an arbitrary point of Ω, then there exists a function $f(z)$ that maps Ω conformally onto the unit disk and is such that $f(z_0) = 0$ and $f'(z_0) > 0$.

Such a function, if it exists, is unique. For if f_1, f_2 are two such functions, then $h(w) f_1[f_2^{-1}(w)]$ maps the unit disk onto itself and

$$h(0) = 0, \quad h'(0) = \frac{f_1'(z_0)}{f_2'(z_0)} > 0$$

that is, the origin and the direction of the real axis at the origin is preserved. Hence, by Lemma 10.17.8, $h(w) \equiv w$ for $|w| < 1$ and $f_1 \equiv f_2$ on Ω.

The sought function $f(z)$ must have the following properties:

1. $f(z)$ is analytic and univalent on Ω,
2. $|f(z)| \leq 1$ on Ω,
3. $f(z_0) = 0$; $f'(z_0) > 0$.

Let F be the class of all functions satisfying all three conditions. We will prove that there is a function $f \in F$ that has the additional property

4. f maps Ω one-to-one and conformally onto the unit disk. First we prove that $F \neq \phi$. Since Ω is not the whole plane, there exists $a \notin \Omega$ and, by 8.9.6, there exists a function $P(z)$ that is analytic and single valued on Ω and is such that $e^{P(z)} = z - a$ on Ω and $P(z_0) = \log(z_0 - a)$. If $z_1, z_2 \in \Omega$ and $z_1 \neq z_2$, then $z_1 - a \neq z_2 - a$ and therefore $P(z_1) - P(z_2)$ is not an integral multiple of $2\pi i$. Hence,

$$q(z) = \frac{1}{P(z) - P(z_0) - 2\pi i}$$

is univalent and analytic on Ω. Since $P'(z)(z - a) = 1$ on Ω, $P'(z) \neq 0$ and therefore $P(\Omega)$ is open. Note, a cannot be the point at infinity. Hence there exists $\epsilon > 0$ such that $N(\omega_0, \epsilon) \subset P(\Omega)$, where $\omega_0 = P(z_0)$. By the remark above, it follows that $N(\omega_0 + 2\pi i, \epsilon) \cap P(\Omega) = \phi$, that is, $|\omega - \omega_0 - 2\pi i| > \epsilon$ for all $\omega \in P(\Omega)$. Hence $|q(z)| < 1/\epsilon$ on Ω. Now put

$$r(z) = \frac{\epsilon}{2}[q(z) - q(z_0)]\frac{|q'(z_0)|}{q'(z_0)},$$

then $r(z) \in F$:

$$q'(z) = \frac{-P'(z)}{[P(z) - P(z_0) - 2\pi i]^2} \neq 0,$$

$q(z)$ is univalent and $r(z)$ satisfies conditions (1) to (3).

Let $d = \text{l.u.b.}_{f \in F} f'(z_0)$. We will show that there is some $f \in F$ such that $f'(z_0) = d$ (we do not presume that d is finite, but our assertion implies that d is finite). Suppose this is not the case. Then there exists a sequence of functions $\{f_n\} \in F$ such that $f_n'(z_0) \to d$ (which may be infinity) as $n \to \infty$.

Since $|f_n(z)| \leq 1$ for all n, and $z \in \Omega$, it follows from 10.17.6 that there is a subsequence which is locally uniformly convergent on Ω. We might as well assume that the original sequence is locally uniformly convergent on Ω with a limit function $f(z)$. The function f is analytic on Ω and is either univalent on Ω or constant. Now the sequence $f_1', f_2', \ldots$ is locally uniformly convergent on Ω with limit f', hence $f(z_0) = 0$ and $f'(z_0) = d > 0$. Thus f is not constant in Ω and is therefore univalent. Also, $|f(z)| \leq 1$ on Ω, therefore $f \in F$. We will show that f is indeed the function we are seeking, i.e., it also satisfies condition (4).

Suppose there exists some number w such that $|w| < 1$ and $f(z) \neq w$ for any $z \in \Omega$. Then $w \neq 0$. Therefore there exists a function $F(z)$ analytic in Ω, such that

$$e^{F(z)} = \frac{f(z) - w}{-1 + \bar{w}f(z)} \quad \text{for} \quad z \in \Omega \quad \text{and} \quad F(z_0) = \log w.$$

Since

$$\left| \frac{f(z) - w}{\bar{w}f(z) - 1} \right| < 1$$

(because $(\xi - w)/(-1 + \bar{w}\xi)$ maps the unit circle onto the unit circle), it follows that $R\{F(z)\} < 0$. Hence,

$$G(z) = \frac{F(z) - F(z_0)}{F(z) + \overline{F(z_0)}}$$

is analytic in Ω and $|G(z)| < 1$. (The denominator does not vanish since $R\{\overline{F(z_0)}\} < 0$.) Both $F(z)$ and $G(z)$ are univalent on Ω. Also,

$$G'(z_0) = \frac{F'(z_0)[F(z_0) + \overline{F(z_0)}]}{[F(z_0) + \overline{F(z_0)}]^2} = \frac{f'(z_0)(w\bar{w} - 1)}{2R[F(z_0)]w}.$$

Now put $g(z) = G(z)|G'(z_0)|/G'(z_0)$. Then $g \in F$ and

$$g'(z_0) = |G'(z_0)| = \frac{d(|w|^2 - 1)}{2|w|\log|w|}.$$

But for $0 < t < 1$,
$$\frac{t^2-1}{2t\log t} > 1,$$
therefore $g(z) \in F$ and $g'(z_0) > d$. The contradiction proves the theorem. □

Note: Let
$$E(t) = 2\log t + \frac{1}{t} - t, \qquad t > 0,$$
$$E'(t) = \frac{2}{t} - \frac{1}{t^2} - 1 = -\frac{1}{t^2}(t^2 - 2t + 1) = -\frac{(t-1)^2}{t^2};$$
therefore $E'(t) < 0$ for $0 < t < 1$. Since $E(1) = 0$, it follows that $E(t) > 0$ for $0 < t < 1$. Hence for $0 < t < 1$ we have
$$2\log t > t - \frac{1}{t} \quad \text{or} \quad \frac{t^2-1}{2t\log t} > 1, \quad (\log t < 0).$$

The theorem establishes a one-to-one correspondence between the points of two simply connected domains D_1 and D_2, however it gives no information about the boundaries C_1 and C_2 of these domains. The theorem does not state that the mapping is continuous on the closure $\overline{D}$ of D or that it establishes a one-to-one correspondence between $\overline{D}_1$ and $\overline{D}_2$. Considerable pathology can be introduced to show that boundary points of D can be "inaccessible" in a sense and the mappings are not continuous at these points.* The following result which we do not prove here, can be found in Nehari's book quoted below.

10.17.9

Theorem *If D_1 and D_2 are domains bounded by closed Jordan curves C_1 and C_2 respectively, then the conformal map $D_1 \to D_2$ is continuous in $\overline{D}_1$ and establishes a one-to-one correspondence between the points of C_1 and C_2.*

10.18 Further Properties of Univalent Functions: Koebe's Constant

We conclude this chapter with a study of several more properties of univalent functions. The class F of functions univalent in $|z| < 1$, with $f(0) = 0$ and $f'(0) = 1$, has been studied in great detail.[†]

For functions $f \in F$, the map of the unit circle is subject to certain limitations. We now obtain several simple properties of the map.

---

*See Z. Nehari, *Conformal Mapping*. New York: McGraw-Hill Book Company, p. 178, 1952.

[†]See W. K. Hayman, *Multivalent Functions*. Cambridge University Press, 1958.

10.18.1

Theorem *For any $f \in F$, no boundary point of the map of the unit circle is nearer to the origin than the point $\frac{1}{4}$.*

We require two lemmas to prove the theorem.

10.18.2

Lemma *Let $w = z + a_1/z + a_2/z^2 + \cdots$ be univalent in $|z| < 1$ and analytic, except at infinity. Then*

$$\sum_{n=1}^{\infty} n|a_n|^2 \leq 1.$$

PROOF. Since w is univalent, any circle $|z| = r > 1$ corresponds to a simple closed curve in the w-plane that encloses area $A \geq 0$:

$$A = \int_0^{2\pi} u(\theta) v'(\theta) \, d\theta.$$

Thus

$$A = \int_0^{2\pi} \frac{w(\theta) + \overline{w}(\theta)}{2} \cdot \frac{w'(\theta) - \overline{w}'(\theta)}{2i} \, d\theta$$

$$= \frac{1}{4} \int_0^{2\pi} \left[re^{i\theta} + re^{-i\theta} + \sum_{n=1}^{\infty} \frac{a_n e^{-in\theta} + \bar{a}_n e^{in\theta}}{r^n} \right]$$

$$\times \left[re^{i\theta} + re^{-i\theta} - \sum_{n=1}^{\infty} \frac{na_n e^{-in\theta} + n\bar{a}_n e^{in\theta}}{r^n} \right] d\theta$$

$$= \frac{\pi}{2} \left[\left(r + \frac{\bar{a}_1}{r}\right)\left(r - \frac{a_1}{r}\right) + \left(r + \frac{a_1}{r}\right)\left(r - \frac{\bar{a}_1}{r}\right) - \sum_{n=2}^{\infty} \frac{2na_n\bar{a}_n}{r^{2n}} \right]$$

$$= \pi r^2 - \pi \sum_{n=1}^{\infty} n|a_n|^2 r^{-2n}.$$

Now $A \geq 0$, thus

$$\sum_{n=1}^{\infty} n|a_n|^2 r^{-2n} \leq r^2,$$

and letting $r \to 1$, we have the result. □

10.18.3

Lemma *If $w = f(z) = z + a_2 z^2 + \cdots$ is univalent in $|z| < 1$, then $|a_2| \leq 2$.*

PROOF. The function

$$F(z) = \sqrt{f(z^2)} = z + \frac{a_2 z^3}{2} + \cdots$$

is univalent in $|z| < 1$, since it is analytic [$f(z^2)$ vanishes only at $z = 0$], and if $F(z_1) = F(z_2)$ then $f(z_1^2) = f(z_2^2)$. However, since $f(z)$ is univalent, $z_1^2 = z_2^2$, that is, $z_1 = \pm z_2$. But $F(z)$ is an odd function, so that $z_1 = -z_2$, which implies that $F(z_1) = -F(z_2)$. Thus the only solution of $F(z_1) = F(z_2)$ is $z_1 = z_2$ and consequently $F(z)$ is univalent. It follows that

$$\left[F\left(\frac{1}{z}\right)\right]^{-1} = z - \frac{a_2}{2z} + \cdots$$

is univalent in $|z| > 1$. Hence, by Lemma 10.18.2,

$$\tfrac{1}{4}|a_2|^2 + \cdots \leqslant 1 \quad \text{and} \quad |a_2| \leqslant 2.$$

PROOF OF THEOREM 10.18.1. Let $w = f(z) = z + a_2 z^2 + \cdots$ be such that $f \in F$. Let c be a value not attained by w in the unit circle, that is, c is a point outside the map of the unit circle. Then

$$\frac{cf(z)}{c - f(z)} = z + \left(a_2 + \frac{1}{c}\right)z^2 + \cdots$$

is analytic and univalent in $|z| < 1$. Hence,

$$\left|a_2 + \frac{1}{c}\right| \leqslant 2 \quad \text{or} \quad \left|\frac{1}{c}\right| \leqslant 2 + |a_2| \leqslant 4,$$

and thus $|c| \geqslant \tfrac{1}{4}$. □

The constant $\tfrac{1}{4}$ is sometimes called *Koebe's constant* and $|c| = \tfrac{1}{4}$ for the extremal function $w = z/(1-z)^2$. Clearly, $a_2 = 2$ and $z/(1-z)^2$ is univalent, since the only solution of

$$\frac{z}{(1-z)^2} = \frac{z'}{(1-z')^2}, \quad |z| < 1, \quad |z'| < 1$$

is $z = z'$.

10.18.4 Bieberbach Conjecture

We have shown that for functions

$$f(z) = z + a_2 z^2 + a_3 z^3 + \cdots$$

belonging to the class F, we have $|a_2| \leqslant 2$. It was conjectured by L. Bieberbach in 1916 that $|a_n| \leqslant n$ if $f(z) \in F$. The proof of this conjecture has turned out to be singularly difficult, and to this day very little progress has been made. In fact, it has only been verified for a very few values of n. In 1923, J. E. Littlewood succeeded in showing that* $|a_n| < en$.

Further improvements have been made by I. E. Bazilyevich who proved

---

*See E. Hille, *Analytic Function Theory*, Chelsea Publishing Company, p. 355, 1973.

in 1951 that

$$|a_n| \leq \frac{en}{2} + 1.51,$$

and for special classes of univalent functions the full conjecture has been verified. For example, $|a_n| \leq n$ for $f(z) \in F$ with real coefficients. The conjecture, if true for all n, is best possible since the function

$$f(z) = \frac{z}{(1 + e^{i\beta}z)^2}, \quad \text{where } \beta \text{ is real,}$$

has the property that $|a_n| = n$.

10.18.5
Exercises

1: Show that the function $f(z) = z/(1-z)^3$ is univalent for $|z| < \frac{1}{2}$, but not in any larger circle with center at the origin.

2: Show that the function $f(z) = z + a_2 z^2 + a_3 z^3 + \cdots$ is univalent for $|z| < 1$, if $\sum_{n=2}^{\infty} n|a_n| \leq 1$.

3: Prove that if the function $f(z)$ satisfies the following conditions:

   i. $f(z)$ is schlicht in $|z| < 1$,
   ii. $f(0) = 0$,
   iii. $f'(0) = 1$,

   and if the points w_1, w_2 lie on the boundary of the domain D into which this function f maps the unit disk and where $\arg w_2 = \arg w_1 + \pi$, then $|w_1| + |w_2| \geq 1$.

   [*Hint*: Apply Koebe's theorem to the function $f(z)/[1 - f(z)/w_1]$.

11
Infinite Products, Expansion of Functions, Mittag–Leffler Theorem, and Gamma Function

11.1

In order to study the so-called nonelementary functions, we need to set up the structure of infinite product representation. We now turn our attention to theorems dealing with the convergence and divergence of these products.

11.2 Infinite Products

Consider $(1 + a_1)(1 + a_2) \cdots$ containing an infinite number of factors, that is, $\prod_{i=1}^{\infty}(1 + a_i)$, where no $a_i = -1$. Then let us write

$$P_n = \prod_{m=1}^{n} (1 + a_m).$$

11.2.1

Definition The infinite product converges if there exists a finite L (not zero), such that

$$\lim_{n \to \infty} P_n = L.$$

If $P_n \to 0$, the product is said to *diverge to zero*. For convergent products it is necessary that $a_n \to 0$, since

$$P_n = P_{n-1} + a_n P_{n-1},$$

but clearly not sufficient, since $\prod_{n=1}^{\infty}[1 + (1/n)]$ diverges, yet $a_n \to 0$.

Examples of some infinite products:

Infinite Product	Partial Product	Behavior of Infinite Product
$\prod_{k=1}^{\infty}\left(1+\dfrac{1}{k}\right)$	$P_n = \dfrac{2}{1}\cdot\dfrac{3}{2}\cdots\dfrac{n+1}{n} = n+1$	Diverges
$\prod_{k=1}^{\infty}\left(1-\dfrac{1}{k+1}\right)$	$P_n = \dfrac{1}{2}\cdot\dfrac{2}{3}\cdots\dfrac{n}{n+1} = \dfrac{1}{n+1}$	Diverges to 0
$\prod_{k=1}^{\infty}\left[1-\dfrac{1}{(k+1)^2}\right]$	$P_n = \dfrac{3}{4}\cdot\dfrac{8}{9}\cdots\dfrac{n^2-1}{n^2} = \dfrac{n+1}{2n}$	Converges to $\tfrac{1}{2}$

The following theorems give criteria for convergence or divergence of infinite products in terms of behavior of an associated series.

11.2.2

Theorem *If $a_n \geq 0$ for all n, then $\prod_{n=1}^{\infty}(1+a_n)$ and $\sum_{n=1}^{\infty} a_n$ converge or diverge together.*

PROOF. We note that P_n is a nondecreasing function of n (since each term is greater than or equal to one). Thus P_n either converges or approaches infinity. We have

$$a_1 + a_2 + \cdots + a_n \leq (1+a_1)(1+a_2)\cdots(1+a_n) \leq e^{a_1+a_2+\cdots+a_n},$$

and if P_n is bounded, then $\sum_{i=1}^{n} a_i$ is bounded. Also, if $\sum_{i=1}^{n} a_i$ is bounded, $\exp[\sum_{i=1}^{n} a_i]$ is bounded, and thus P_n is bounded. □

If $a_n \leq 0$ for all n, write $a_n = -b_n$ and consider $\prod_{n=1}^{\infty}(1-b_n)$.

11.2.3

Theorem *If $b_n \geq 0$, $b_n \neq 1$ for all n, and $\sum_{n=1}^{\infty} b_n$ converges, then $\prod_{n=1}^{\infty}(1-b_n)$ converges.*

PROOF. Since $\sum_{n=1}^{\infty} b_n$ converges, there exists N such that

$$b_N + b_{N+1} + \cdots + b_n < \frac{1}{2}$$

and $b_n < 1$ for $n \geq N$. Thus,

$$(1-b_N)(1-b_{N+1}) \geq 1 - b_N - b_{N+1}$$

and

$$(1-b_N)(1-b_{N+1})(1-b_{N+2}) \geq (1-b_N-b_{N+1})(1-b_{N+2})$$
$$\geq 1 - b_N - b_{N+1} - b_{N+2}.$$

Infinite Products

Therefore,
$$(1 - b_N)(1 - b_{N+1}) \cdots (1 - b_n) \geq 1 - b_N \cdots - b_n > \frac{1}{2}$$

and P_n/P_{n-1} is decreasing [since $(1 - b_n) < 1$ for $n > N$] and has a positive lower limit. Therefore, P_n/P_{n-1} tends to a positive limit, and P_n converges since $P_{n-1} \neq 0$. □

11.2.4

Theorem *If $0 \leq b_n < 1$ for all n and $\sum_{n=1}^{\infty} b_n$ diverges, then $\prod_{n=1}^{\infty}(1 - b_n)$ diverges to zero.*

PROOF. If $0 \leq b_n < 1$, then $1 - b \leq e^{-b}$. Thus,
$$(1 - b_1)(1 - b_2) \cdots (1 - b_n) \leq e^{-b_1 - b_2 - \cdots - b_n}$$

and the right-hand side approaches zero; hence the result. □

11.2.5

Definition The product $\prod_{n=1}^{\infty}(1 + a_n)$ is said to be *absolutely convergent* if $\prod_{n=1}^{\infty}(1 + |a_n|)$ converges. Thus a necessary and sufficient condition that the product should be absolutely convergent is that $\sum_{n=1}^{\infty} |a_n|$ should converge.

11.2.6

Lemma *An absolutely convergent product is convergent.*

PROOF. As before, write
$$P_n = \prod_{m=1}^{n}(1 + a_m) \quad \text{and} \quad q_n = \prod_{m=1}^{n}(1 + |a_m|).$$

Then
$$P_n - P_{n-1} = (1 + a_1) \cdots (1 + a_{n-1}) a_n,$$

and
$$q_n - q_{n-1} = (1 + |a_1|) \cdots (1 + |a_{n-1}|)|a_n|.$$

Thus,
$$|P_n - P_{n-1}| \leq q_n - q_{n-1}.$$

If $\prod_{n=1}^{\infty}(1 + |a_n|)$ converges, q_n approaches a limit and $\sum_{n=1}^{\infty}(q_n - q_{n-1})$ converges; thus $\sum_{n=2}^{\infty}(P_n - P_{n-1})$ converges, that is, P_n approaches a (nonzero) limit. □

We see that this limit cannot be zero, since if $\sum_{n=1}^{\infty} |a_n|$ converges and $1 + a_n \to 1$, the series $\sum_{n=1}^{\infty} |a_n/(1 + a_n)|$ converges. Thus the product

$$\prod_{m=1}^{\infty}\left[1 - \frac{a_m}{1 + a_m}\right]$$

approaches a limit; but this product is $1/P_n$, thus
$$\lim_{n\to\infty} P_n \neq 0.$$

11.2.7

Theorem *A necessary and sufficient condition for the convergence of the infinite product $\prod_{n=1}^{\infty}(1 + a_n)$, where a_n is complex, is the convergence of $\sum_{n=1}^{\infty} \log(1 + a_n)$, where we take the principal value of each logarithm.*

PROOF. Write $S_n = \sum_{r=1}^{n} \log(1 + a_r)$. To establish sufficiency, we have that $P_n = \exp(S_n)$ and since the exponential function is continuous, $S_n \to S$, which implies that $P_n \to e^S$.

To establish necessity, we have that $S_n = \log P_n + 2q_n\pi i$, where q_n is an integer. Since the principal value of the logarithm of the product is not necessarily the sum of the principal values of its factors, q_n is not necessarily zero. We show that q_n is constant for all $n > N$, and from this necessity follows.

Write the principal values α_n, β_n of the arguments of $(1 + a_n)$ and P_n, respectively. If the infinite product converges, then $\alpha_n \to 0$ and $\beta_n \to \beta$ as $n \to \infty$. The integer q_n is then given by
$$\alpha_1 + \alpha_2 + \cdots + \alpha_n = \beta_n + 2\pi q_n$$
and
$$2(q_{n+1} - q_n)\pi = \alpha_{n+1} - (\beta_{n+1} - \beta_n) \to 0 \quad \text{where} \quad n \to \infty.$$
However, since q_n is an integer, this implies that $q_n = q$ for all sufficiently large n. Thus, if P_n tends to the finite nonzero limit P as $n \to \infty$, it follows that
$$S_n \to \log P + 2q\pi i,$$
and the condition is necessary. □

11.2.8

Given a sequence of functions of a complex variable $\{U_n(z)\}$, where $n = 1, 2, \ldots$, we give definitions of an infinite product of functions similar to the definitions of 11.2. Thus, for any m we write
$$Q_n(z) = \prod_{m=1}^{n} [1 + U_m(z)], \quad \text{where} \quad U_m(z) \neq -1$$
and we say that the infinite product converges if there exists a finite (nonzero) M such that
$$\lim_{n\to\infty} Q_n(z) = M.$$

We say that the infinite product $\prod_{n=1}^{\infty}[1 + U_n(z)]$ converges uniformly, if the partial product
$$Q_n(z) = \prod_{m=1}^{n} [1 + U_m(z)]$$
converges uniformly in some domain to a limit which is never zero.

11.2.9

Theorem *The product $\prod_{n=1}^{\infty}[1 + U_n(z)]$ is uniformly convergent in a region where the series $\sum_{n=1}^{\infty}|U_n(z)|$ converges uniformly.*

PROOF. Let M be an upper bound of the sum $\sum_{n=1}^{\infty}|U_n(z)|$ in the considered region. Then

$$[1 + |U_1(z)|][1 + |U_2(z)|] \cdots [1 + |U_n(z)|] < e^{|U_1(z)| + \cdots + |U_n(z)|} \leq e^M.$$

Let

$$P_n(z) = \prod_{m=1}^{n} [1 + |U_m(z)|].$$

Then

$$P_n(z) - P_{n-1}(z) = [1 + |U_1(z)|] \cdots [1 + |U_{n-1}(z)|]|U_n(z)| < e^M |U_n(z)|,$$

hence $\sum_{n=2}^{\infty}[P_n(z) - P_{n-1}(z)]$ converges uniformly since

$$\sum_{n=2}^{\infty} [P_n(z) - P_{n-1}(z)] < e^M \sum_{n=1}^{\infty} |U_n(z)|$$

and $\sum_{n=1}^{\infty}|U_n(z)|$ converges uniformly. The rest follows from arguments similar to 11.2.6. □

11.2.10

Exercises

Prove the following equalities:

1: $\displaystyle\prod_{n=2}^{\infty} \frac{n^3 - 1}{n^3 + 1} = \frac{2}{3}.$

2: $\displaystyle\prod_{n=1}^{\infty} \left[1 + \frac{(-1)^{n+1}}{n}\right] = 1.$

3: Show that $\displaystyle\prod_{n=1}^{\infty} \cos \frac{x}{2^n} = \frac{\sin x}{x}$ and hence

$$\frac{2}{\sqrt{2}} \frac{2}{\sqrt{2 + \sqrt{2}}} \frac{2}{\sqrt{2 + \sqrt{2 + \sqrt{2}}}} \cdots = \frac{\pi}{2}.$$

4: Show that inside the unit circle we have $\prod_{n=0}^{\infty}(1 + z^{2^n}) = 1/(1 - z)$.

5: Show that $\displaystyle\prod_{n=1}^{\infty}\left[1 + \frac{(-1)^{n+1}}{n^z}\right]$ converges in $R(z) > \frac{1}{2}$ and converges absolutely in $R(z) > 1$.

6: Let $\{p_n\}$ be the sequence of primes $2, 3, 5, \ldots$. Then, define $\zeta(z) = \sum_{n=1}^{\infty} n^{-s}$, where $n^{-s} = e^{-s \log n}$. Prove that $1/\zeta(z)$

$= \prod_{n=1}^{\infty}(1 - p_n^{-s})$. Also, prove that the series $\sum_{n=1}^{\infty} 1/p_n$, where p_n are prime numbers, diverges. What can you say about the convergence or divergence of $\sum_{n=1}^{\infty} 1/p_n^2$?

We now study several theorems that give us a practical method for expanding entire or meromorphic functions.

11.3

Theorem *Let $f(z)$ be a function whose only singularities, except at infinity, are poles. Suppose all poles are simple. Let them be $a_1, a_2, \ldots$ and be ordered, so that $0 < |a_1| \leq |a_2| \leq \cdots$, let the residues at the poles be $b_1, b_2, \ldots$. Suppose there is a sequence of contours C_n, such that C_n includes $a_1, a_2, \ldots, a_n$, but no other poles. Let the minimum distance R_n of C_n from the origin approach infinity as $n \to \infty$, while the length L_n of C_n is $O(R_n)$ and such that $f(z) = o(R_n)$ on C_n. This last condition is satisfied if, for example, $f(z)$ is bounded on all C_n. Then*

$$f(z) = f(0) = \sum_{n=1}^{\infty} b_n \left[\frac{1}{z - a_n} + \frac{1}{a_n} \right]$$

for all z except the poles.

PROOF. Consider

$$I = \frac{1}{2\pi i} \int_{C_n} \frac{f(w)}{w(w - z)} dw,$$

where z is inside C_n. The integrand has poles at a_m, 0, and $w = z$, with residues $b_m/a_m(a_m - z)$, $-f(0)/z$, $f(z)/z$, respectively. Then,

$$I = 2\pi i \sum \text{Residues} = \sum_{m=1}^{n} \frac{b_m}{a_m(a_m - z)} - \frac{f(0)}{z} + \frac{f(z)}{z}$$

and, by the conditions imposed,

$$|I| \leq \frac{1}{2\pi} \frac{L_n}{R_n(R_n - |z|)} \cdot \max_{C_n} |f(w)| \to 0 \quad \text{as} \quad n \to \infty.$$

Thus,

$$f(z) = f(0) = -\lim_{n \to \infty} \sum_{m=1}^{n} \frac{zb_m}{a_m(a_m - z)}$$

$$= f(0) + \sum_{i=1}^{\infty} b_i \left[\frac{1}{z - a_i} + \frac{1}{a_i} \right].$$

Note, that the series converges uniformly inside any closed contour containing all the poles. □

EXAMPLE. Consider $f(z) = \pi \cot \pi z - 1/z$, where $z \neq 0$ and $f(0) = 0$. The poles of

$$f(z) = \frac{\pi z \cos \pi z - \sin \pi z}{z \sin \pi z}$$

Infinite Products

are simple and occur at $z = n$, where $n = \pm 1, \pm 2, \ldots$. Then,

$$\text{Res } f(n) = \frac{\pi z \cos \pi z - \sin \pi z}{z \pi \cos \pi z + \sin \pi z}\bigg|_{z = \pm n} = 1.$$

The point $z = 0$ is not a singularity since

$$\frac{\pi z \cos \pi z - \sin \pi z}{z \sin \pi z} = \frac{\pi z \left[1 - \frac{\pi^2 z^2}{2!} + O(z^4)\right] - \left[\pi z - \frac{\pi^3 z^3}{3!} + O(z^5)\right]}{\pi z^2 + O(z^4)}$$

$$= O(1).$$

Let C_n be the square with vertices at the points $(n + \frac{1}{2})(\pm 1 \pm i)$, where $n = 0, \pm 1, \ldots$. We note that C_n is a contour that does not pass through a pole of $f(z)$.

The function $1/z$ is clearly bounded on these squares. We study the behaviour of $|\cot \pi z|$ on these squares.

a) Since

$$|\cot \pi z| = \left|\frac{e^{iz} + e^{-iz}}{e^{iz} - e^{-iz}}\right|,$$

we see that $|e^{iz} - e^{-iz}|$ is bounded away from zero (for all y) on the segment $\frac{1}{2} \pm \frac{1}{2}i$ and the numerator is clearly bounded on this segment. By periodicity, $|\cot \pi z|$ is bounded on the parts of C_n for which $|y| \leq \frac{1}{2}$.

b) For $|y| > \frac{1}{2}$, we can calculate,

$$|\cot \pi z| \leq \frac{e^{-\pi y} + e^{\pi y}}{|e^{-\pi y} - e^{\pi y}|} < \frac{1 + e^{-\pi}}{1 - e^{-\pi}};$$

thus $|\cot \pi z|$ is bounded on all the contours C_n taken as a whole.

Substituting in 11.3, we get

$$f(z) = \sum_{n \neq 0}\left[\frac{1}{z - n} + \frac{1}{n}\right].$$

This series is absolutely and uniformly convergent in any bounded region at positive distance from the integers; thus we can reorder the series to obtain

$$\pi \cot \pi z = \frac{1}{z} + \sum_{n=1}^{\infty} \frac{2z}{z^2 - n^2}.$$

11.4 Expansion of an Entire Function as an Infinite Product

Suppose $f(z)$ has simple zeros at the points $a_1, a_2, \ldots, a_n$. In the neighborhood of a_n we have $f(z) = (z - a_n)g(z)$, where $g(z)$ is analytic and nonzero. Thus

$$\frac{f'(z)}{f(z)} = \frac{1}{z - a_n} + \frac{g'(z)}{g(z)},$$

with $g'(z)/g(z)$ being analytic at a_n. Hence $f'(z)/f(z)$ has a simple pole at $z = a_n$ with residue 1.

Suppose $f'(z)/f(z)$ is a function of the type considered in the expansion of a meromorphic function $f(z)$. Then $P(z) = f'(z)/f(z)$ has poles at $a_1, a_2, \ldots, a_n$ and

$$P(z) = P(0) + \sum_{n=1}^{\infty} b_n \left[\frac{1}{z - a_n} + \frac{1}{a_n} \right].$$

Clearly $P(0) = f'(0)/f(0)$ and $b_n = 1$ for all n. Thus

$$\frac{f'(z)}{f(z)} = \frac{f'(0)}{f(0)} + \sum_{n=1}^{\infty} \left[\frac{1}{z - a_n} + \frac{1}{a_n} \right].$$

Integrating from 0 to z along a path not passing through a pole, we have

$$\log f(z) - \log f(0) = z \frac{f'(0)}{f(0)} + \sum_{n=1}^{\infty} \left[\log(z - a_n) - \log(-a_n) + \frac{z}{a_n} \right],$$

where the value of the logarithms depend upon the path. Taking exponentials, we obtain

$$f(z) = f(0) \exp\left[z \frac{f'(0)}{f(0)} \right] \prod_{n=1}^{\infty} \left[1 - \frac{z}{a_n} \right] \exp \frac{z}{a_n}.$$

As an example, consider

$$f(z) = \frac{\sin z}{z} = \prod_{n \neq 0} \left[1 - \frac{z}{n\pi} \right] \exp \frac{z}{n\pi}$$

or

$$\sin z = z \prod_{n=1}^{\infty} \left[1 - \frac{z^2}{n^2 \pi^2} \right].$$

11.4.1

Exercises

1: Obtain the following expansions:

i. $\operatorname{cosec} z = \dfrac{1}{z} + 2z \sum\limits_{n=1}^{\infty} \dfrac{(-1)^{n-1}}{n^2 \pi^2 - z^2}$

ii. $\tan z = 2z \sum\limits_{n=0}^{\infty} \dfrac{1}{(n+\frac{1}{2})^2 \pi^2 - z^2}$

iii. $\dfrac{1}{e^z - 1} = \dfrac{1}{z} - \dfrac{1}{2} + 2z \sum\limits_{n=1}^{\infty} \dfrac{1}{z^2 + 4n^2 \pi^2}.$

2: Show that $\cos z = \prod\limits_{n=1}^{\infty} \left[1 - \dfrac{4z^2}{(2n-1)^2 \pi^2} \right].$

Infinite Products

3: Show that if $\alpha \neq 0$ or $2k\pi$, $k = \pm 1, \pm 2, \ldots$, then
$$\frac{\cosh z - \cos \alpha}{1 - \cos \alpha} = \prod_{n=-\infty}^{\infty} \left[1 + \frac{z^2}{(2n\pi + \alpha)^2}\right].$$

4: Show that $\dfrac{\cos z}{1 + \sin z} = \dfrac{4}{\pi + 2z} - 8 \sum_{n=1}^{\infty} \dfrac{\pi + 2z}{4^2 n^2 \pi^2 - (\pi + 2z)^2}$, where $1 + \sin z \neq 0$.

5: Prove that $\dfrac{\sin \pi(z + c)}{\sin \pi c} = \dfrac{z + c}{c} \prod_{n=-\infty}^{\infty}{}' [1 - \dfrac{z}{n - c}] e^{z/n}$, c not integral.

6: Prove that $1 - \dfrac{\sin^2 \pi z}{\sin^2 \pi c} = \prod_{n=-\infty}^{\infty} \left[1 - \dfrac{z^2}{(n - c)^2}\right]$, c not integral.

7: Show that $\prod_{n=-\infty}^{\infty} \left[1 - \dfrac{4z^2}{(n\pi + z)^2}\right] = - \dfrac{\sin 3z}{\sin z}$.

8: Prove that the Blaschke product
$$\prod_{\nu=1}^{\infty} \left[e^{-i\alpha_\nu} \frac{z - a_\nu}{\bar{a}_\nu z - 1}\right] \quad \text{where} \quad a_\nu = |a_\nu| e^{i\alpha_\nu}, \ 0 < |a_\nu| < 1, \ \nu = 1, 2, \ldots,$$
converges if and only if the series $\sum_{\nu=1}^{\infty} (1 - |a_\nu|)$ converges.

If the series converges, show that the product defines a function that is analytic in the disk $|z| < 1$ and has zeros at $z = a_\nu$, where $\nu = 1, 2, \ldots$. However, if the series diverges, show that the product is identically zero.

11.5 Meromorphic Functions

We now study several results that prepare us for a representation theorem called the Mittag–Leffler theorem. This theorem essentially represents a meromorphic function by decomposition into partial fractions. In fact, a meromorphic function may be characterized by the nature of its poles only to within an added entire function. Thus we require, in particular, to study the case where infinity is an accumulation point of poles. (Clearly, no finite point can be an accumulation point of poles.)

11.5.1

Theorem *Let $M_0(z)$ be a particular meromorphic function. If $G(z)$ is an arbitrary entire function, then*
$$M(z) = M_0(z) + G(z)$$

is the most general meromorphic function that coincides with $M_0(z)$ in its poles and corresponding principal parts.

PROOF. If $M_0(z)$ and $M(z)$ are two meromorphic functions that coincide in their poles and corresponding principal parts, then their difference $M_0(z) - M(z)$ is evidently an entire function, say, $G(z)$ and $M(z) = M_0(z) + G(z)$. □

For example, $\cot z$ and $2i/(e^{2iz} - 1)$ are two meromorphic functions that coincide in poles and corresponding principal parts; thus they differ only by an additive entire function. The poles of $\cot z$ are the zeros of $\sin z = 0$, that is, $z = n\pi$, $n = 0, \pm 1, \ldots$. The poles of $2i/(e^{2iz} - 1)$ are the roots of $e^{2iz} = 1 = e^{2n\pi i}$, that is, $z = n\pi$, $n = 0, \pm 1, \ldots$. The difference is i, an entire function.

11.5.2

Lemma *Suppose $f(z)$ is analytic in the entire finite complex plane except for poles $z_1, z_2, \ldots, z_k$. Let $P(z, z_r)$ be the principal part of $f(z)$ at $z = z_r$, $r = 1, \ldots, k$. Then there exists a function $\phi(z)$, analytic in the entire finite complex plane, such that*

$$f(z) = \sum_{r=1}^{k} P(z, z_r) + \phi(z).$$

Moreover, $\phi(z)$ has the same principal part at $z = \infty$ as $f(z)$.

PROOF. Consider

$$g(z) = \sum_{r=1}^{k} P(z, z_r), \quad \text{where} \quad P(z, z_r) = \sum_{(j)} \frac{b_{jr}}{(z - z_r)^j}.$$

Clearly, $g(z)$ is analytic in the entire finite complex plane except at points $z_1, \ldots, z_k$, where it has poles. Since $P(z, z_s)$ is analytic at $z = z_r$ when $s \neq r$, the principal part of $g(z)$ at $z = z_r$ is $P(z, z_r)$. Hence $g(z)$ has the same poles and principal parts at $z_1, \ldots, z_k$ as $f(z)$ and is analytic everywhere else.

Consider $\phi(z) = f(z) - g(z)$, where $z \neq z_r$, $r = 1, 2, \ldots$. This function is analytic in the entire complex plane except for possible poles at $z_1, z_2, \ldots, z_k$. Since $g(z)$ and $f(z)$ have the same poles and principal parts, the difference $\phi(z)$ is such that the principal part at each pole is zero. Thus for

$$f(z) = \frac{b_{-n}}{(z-a)^n} + \cdots + \frac{b_{-1}}{(z-a)} + b_0 + b_1(z-a) + \cdots,$$

$$g(z) = \frac{b_{-n}}{(z-a)^n} + \cdots + \frac{b_{-1}}{(z-a)},$$

we have

$$\phi(a) = b_0$$

and $\phi(z)$ has removable singularities at $z_1, \ldots, z_k$. Thus $\phi(z)$ is analytic. Since $g(z) \to 0$ when $z \to \infty$, $g(z)$ has a removable singularity at $z = \infty$. Thus $\phi(z)$ has the same principal part at $z = \infty$ as $f(z)$. Writing $f(z) = g(z) + \phi(z)$, we establish the lemma. □

11.5.3

Mittag–Leffler Theorem *Let $z_0, z_1, \ldots, z_n, \ldots$ be any sequence of distinct points tending to infinity. Suppose that each z_n is associated with a polynomial $P_n(1/(z - z_n))$ in the variable $1/(z - z_n)$. It is possible to find a meromorphic function $f(z)$ with poles at the points z_n but no other points and with corresponding principal parts, $P_n(1/(z - z_n))$. Then $f(z)$ may be represented in the form*

$$f(z) = \omega(z) + \sum_{v=0}^{\infty} \left[P_v\left(\frac{1}{z - z_v} \right) - q_v(z) \right],$$

where $q_v(z)$ are polynomials and $\omega(z)$ is an entire function of z.

PROOF. Unlike the finite case, we must ensure that the given representation converges. Suppose the sequence $\{z_n\}$ is ordered. Thus $|z_0| \leq |z_1| \leq \cdots$ since the only point of accumulation is infinity. Possibly $z_0 = 0$, but all other points differ from zero. We assume that $z_0 \neq 0$, and the function $P_v(1/(z - z_v))$, being analytic everywhere except at z_v, must itself be analytic at the origin. Thus P_v has a Taylor expansion at the origin given by

$$P_v\left(\frac{1}{z - z_v} \right) = a_0^{(v)} + a_1^{(v)}z + a_2^{(v)}z^2 + \cdots.$$

The radius of convergence is clearly $|z_v|$. The series converges uniformly in the circle C_v: $|z| \leq \tfrac{1}{2}|z_v|$. Thus $P_v(1/(z - z_v))$ can be approximated in C_v by a finite sum as closely as we please. In particular, for

$$q_v(z) = a_0^{(v)} + a_1^{(v)}z + \cdots + a_{k_v}^{(v)} z^{k_v}$$

we have

$$\left| P_v\left(\frac{1}{z - z_v} \right) - q_v(z) \right| < \frac{1}{2^v}$$

throughout C_v. The series

$$\sum_{v=0}^{\infty} \left[P_v\left(\frac{1}{z - z_v} \right) - q_v(z) \right]$$

converges to the desired meromorphic function in every circle about the origin, since any such circle can be contained within one of the C_v.

In C_v

$$\sum_{n=0}^{v-1} \left[P_n\left(\frac{1}{z - z_n} \right) - q_n(z) \right]$$

is well behaved. It is an analytic function with no singularities but the prescribed poles. Thus

$$\sum_{n=v}^{\infty}\left[P_n\left(\frac{1}{z-z_n}\right)-q_n(z)\right]$$

is analytic in C_v and dominated by $\sum_{n=v}^{\infty}(1/2^n)$. Then the series $\sum_{n=v}^{\infty}$ converges uniformly in C_v. Thus, since a uniformly convergent series of analytic functions converges to an analytic function, the second part of the series introduces no new singularities into C_v. If $z_0 = 0$, we add on $P_0(1/z)$. Thus the theorem is proved. □

11.5.4

In general, the polynomials $q_v(z)$ will not be uniformly bounded (to ensure convergence). However, in special circumstances all the $q_v(z)$ may be chosen of the same finite degree. It is sufficient to take the degree k_v of the polynomial $q_v(z)$ [that is, the sum of the first k_v terms of the power series for $P_v(1/(z-z_v))$] so large, that for a chosen arbitrary $R > 0$ the terms

$$\left|P_v\left(\frac{1}{z-z_v}\right)-q_v(z)\right|$$

of the series for all $|z| \leq R$ and large v remain less than the terms of a convergent series of positive terms.

EXAMPLE. The convergence-producing terms $q_v(z)$ are not always necessary. If, for example, the points $0, 1, 4, \ldots, v^2, \ldots$ are to be poles of order one with respective principal parts $1/(z - v^2)$, then

$$f(z) = \frac{1}{z} + \sum_{v=1}^{\infty} \frac{1}{z-v^2} = \sum_{v=0}^{\infty} \frac{1}{z-v^2}$$

is a solution. Note that for $R > 0$ and $m > \sqrt{2R}$, the series from $v = m+1$ with $|z| \leq R$ converges uniformly because

$$\left|\frac{1}{(z-v^2)}\right| \leq \frac{1}{(v^2-R)} < \frac{1}{(v^2-\frac{1}{2}v^2)} = \frac{2}{v^2}.$$

Similarly, if the function has only simple poles

$$P_v\left(\frac{1}{z-z_v}\right) = \frac{a_v}{z-z_v},$$

then the $q_v(z)$ may all be chosen of degree n if the series

$$\sum_{v=0}^{\infty} \frac{|a_v|}{|z_v|^{n+2}}$$

converges. Since

$$\left|\frac{a_v}{z-z_v}-q_v(z)\right|=\left|-\frac{a_v}{z_v}\left[1+\frac{z}{z_v}+\frac{z^2}{z_v^2}+\cdots\right]-q_v(z)\right|,$$

Infinite Products

by choosing $q_v(z)$ as a polynomial of degree n, namely,

$$-\frac{a_v}{z_v}\left[1 + \frac{z}{z_v} + \cdots + \frac{z^n}{z_v^n}\right],$$

we get

$$\left|\frac{a_v}{z - z_v} - q_v(z)\right| \leq \frac{|a_v||z|^{n+1}}{|z_v|^{n+2}} + \frac{|a_v||z|^{n+2}}{|z_v|^{n+3}} + \cdots.$$

Thus if

$$\sum_{v=0}^{\infty} \frac{|a_v|}{|z_v|^{n+2}}$$

converges, it assures the convergence of

$$\sum_{(v)}\left[\frac{a_v}{z - z_v} - q_v(z)\right].$$

This case is the one that arises in most applications. Although the Mittag–Leffler theorem can be used to expand a function with simple poles into partial fractions, the entire function $\omega(z)$ is still to be determined. This special case has a more direct approach. Let C be any simple closed curve not passing through any pole of $f(z)$. If z is a regular point inside C, then

$$f(z) + \sum \text{Res}\{g(z)\} = \frac{1}{2\pi i}\int_C \frac{f(\xi)}{\xi - z}d\xi \quad \text{and} \quad g(z) = \frac{f(\xi)}{\xi - z},$$

that is,

$$f(z) = \frac{1}{2\pi i}\int_C \frac{f(\xi)}{\xi - z}d\xi - \sum_v \text{Res}\left\{\frac{f(z_v)}{z_v - z}\right\},$$

where $\text{Res}\{f(z_v)/(z_v - z)\}$ means the residue of $f(\xi)/(\xi - z)$ at $\xi = z_v$, the sum taken over all singularities z_v of $f(z)$ in C.

Note: $f(z)$ is the residue of $f(\xi)/(\xi - z)$ at the pole $\xi = z$.
Since $f(z)$ is assumed to have simple poles, we get

$$\text{Res}\left\{\frac{f(z_v)}{z_v - z}\right\} = \frac{\text{Res}\{f(z_v)\}}{z_v - z}$$

and

$$f(z) = \frac{1}{2\pi i}\int_C \frac{f(\xi)}{\xi - z}d\xi + \sum_v \frac{\text{Res}\{f(z_v)\}}{z - z_v}.$$

Since the poles are isolated, there exists a sequence of closed curves C_n such that $C_1 \supset C_2 \supset C_3 \supset \cdots \supset C_n \cdots$, each avoiding all the poles of $f(z)$, and such that the distance of C_n from the origin tends to infinity when $n \to \infty$. If for some such sequence

$$\lim_{n \to \infty}\int_{C_n}\frac{f(\xi)}{\xi - z}d\xi = 0$$

and the poles in the annulus between C_{n-1}, C_n are denoted by $z_v^{(n)}$, then the series

$$f(z) = \sum_{n=1}^{\infty} \sum_v \left[\frac{\text{Res}\{f(z_v^{(n)})\}}{z - z_v^{(n)}} \right]$$

converges and gives the decomposition of the function into partial fractions.

As an example of the theorem and the above method we consider $\pi \cot \pi z$.

11.5.5 $\pi \cot \pi z$

The poles are the zeros of $\sin \pi z$, namely, poles of order 1 at $z = 0, \pm 1, \pm 2, \ldots$. The residue at $z = n$ is

$$\lim_{z \to n} (z - n) \cdot \pi \frac{\cos \pi z}{\sin \pi z} = 1.$$

Thus, the principal parts are

$$P_v\left(\frac{1}{z - z_v}\right) = \frac{1}{z - z_v} = -\frac{1}{z_v} - \frac{z}{z_v^2} - \cdots,$$

where $z_0 = 0$, $z_{2v-1} = v$, $z_{2v} = -v$ for $v = 1, 2, 3, \ldots$.

We may take the degree k_v of the polynomials $q_v(z)$ to be zero, hence $q_v(z) = -1/z_v$ and

$$\left| P_v\left(\frac{1}{z - z_v}\right) - q_v(z) \right| = \left| \frac{1}{z - z_v} + \frac{1}{z_v} \right| = \frac{|z|}{|z_v||z - z_v|} \leq \frac{R}{|z_v||z - z_v|}.$$

Also, $|z - z_v| = |z_v - z| \geq |z_v| - |z| \geq |z_v| - R$. Thus,

$$\frac{R}{|z_v||z - z_v|} \leq \frac{R}{|z_v|(|z_v| - R)}$$

and for large $|z_v|$, namely, $|z_v| = v > 4R$, we have

$$R < \frac{1}{4}|z_v| \quad \text{and} \quad |z_v| - R > \frac{3}{4}|z_v|.$$

Therefore,

$$\frac{R}{|z_v||z_v - R|} < \frac{R}{|z_v|} \cdot \frac{4}{3|z_v|} < \frac{2R}{|z_v|^2}$$

and hence $|P_v - q_v|$ is less than the terms of a convergent series of positive terms. Thus,

$$f(z) = \pi \cot \pi z = \omega(z) + \frac{1}{z} + \sum_{v=1}^{\infty} \left[\frac{1}{z - z_v} + \frac{1}{z_v} \right]$$

or, rewriting,

$$f(z) = \omega(z) + \frac{1}{z} + \sum_{n \neq 0} \left[\frac{1}{z - n} + \frac{1}{n} \right].$$

Infinite Products

Further, $\omega(z)$ is an entire function and the sum is uniformly and absolutely convergent in any bounded region at positive distance from the integers or, more precisely, on every set of the form

$$\{z \mid |z| \leq R, |z - n| \geq \delta > 0, n \in Z\}.$$

For this reason the terms of the series can be reordered to give

$$\pi \cot \pi z = \omega(z) + S(z),$$

where

$$S(z) = \frac{1}{z} + \sum_{n=1}^{\infty} \frac{2z}{z^2 - n^2}.$$

Note that $\sum_{n=1}^{\infty}[1/(z-n)]$ is not convergent. The problem of determining $\omega(z)$ can sometimes be difficult. If we know that

$$\sin \pi z = \pi z \prod_{n=1}^{\infty}\left[1 - \frac{z^2}{n^2}\right],$$

then, taking logarithms and differentiating, we have

$$\pi \cot \pi z = \frac{1}{z} + \sum_{n=1}^{\infty} \frac{2z}{z^2 - n^2},$$

showing that $\omega(z) = 0$. However, this is a rather circular argument and we consider the following. $S(z)$ is periodic with period 1, since

$$S(z) - S(z+1) = \frac{1}{z} - \frac{1}{z+1}$$

$$+ \lim_{M \to \infty} \sum_{n=1}^{M} \left[\frac{1}{z-n} + \frac{1}{z+n} - \frac{1}{z-(n-1)} - \frac{1}{z+(n+1)}\right]$$

$$= \lim_{M \to \infty}\left[\frac{1}{z-M} - \frac{1}{z+M+1}\right] = 0.$$

Also, $S(z)$ is uniformly convergent in the region

$$\{z = x + iy \mid |y| \leq R, |z-n| \geq \delta > 0, n \in Z\}.$$

However, it is not uniformly convergent for unbounded y, since, if any $0 < \epsilon < \frac{1}{2}$ is given and if an $N = N(\epsilon)$ could then be found such that

$$\left|\sum_{n=N+1}^{\infty} \frac{2z}{z^2 - n^2}\right| < \epsilon \quad \text{for all} \quad z,$$

we would have, by putting $z = 2iN$, that

$$\left|\sum_{n=N+1}^{\infty} \frac{2z}{z^2 - n^2}\right| = \left|(-i) \sum_{n=N+1}^{\infty} \frac{4N}{n^2 + 4N^2}\right|$$

$$> \sum_{n=N+1}^{2N} \frac{4N}{n^2 + 4N^2} \geq \sum_{n=N+1}^{2N} \frac{4N}{4N^2 + 4N^2} = \frac{1}{2}.$$

Then $\pi \cot \pi z$ is periodic with period 1, and so $\omega(z)$ must be also periodic with period 1. Further

$$\pi \cot \pi z = \pi \frac{\cos \pi x \cosh \pi y - i \sin \pi x \sinh \pi y}{\sin \pi x \cosh \pi y + i \cos \pi x \sinh \pi y}$$

$$\to \pm \pi \frac{e^{-i\pi x}}{ie^{-i\pi x}} = \mp i\pi \quad \text{as} \quad y \to \pm \infty$$

and thus $\pi \cot \pi z$ is bounded as $|y| \to +\infty$. Similarly $|S(z)|$ is bounded as $|y| \to +\infty$, since, using

$$|z^2 - n^2|^2 - (n^2 + y^2 - 1)^2 \geqslant 0$$

or observing that $|f| \geqslant -R\{f\}$, we get

$$\left| \frac{2z}{z^2 - n^2} \right| \leqslant \frac{2(|y| + 1)}{n^2 + y^2 - 1} \quad \text{for} \quad z \in \{z \mid 0 \leqslant x \leqslant 1\}$$

and

$$\sum_{n=1}^{\infty} \frac{2(y+1)}{n^2 + y^2 - 1} \leqslant \sum_{n<y} \frac{2(y+1)}{y^2 - 1} + \sum_{n>y} \frac{2(y+1)}{n^2 - n}$$

$$\leqslant 2 + \frac{2(y+1)}{y-1} \leqslant 8 \quad \text{for} \quad y \geqslant 2.$$

Note also that $S(iy) = -i\pi \coth \pi y \to \mp i\pi$ as $y \to \pm \infty$. Thus $\omega(z)$ is periodic with period 1 and bounded as $|y| \to \infty$; hence it is bounded in $\{z = x + iy \mid 0 \leqslant x \leqslant 1\}$ and so is bounded in the whole plane. However, it is an entire function, thus it is a constant.

Since $\omega(-z) = -\omega(z)$, we have that $\omega(z) = 0$ for all z [or examine $\omega(iy)$]. Hence

$$\pi \cot \pi z = \frac{1}{z} + \sum_{n=1}^{\infty} \frac{2z}{z^2 - n^2}.$$

Alternatively, let

$$I_n = \frac{1}{2\pi i} \int_{R_n} \frac{\pi \cot \pi z}{z - \xi} dz.$$

Let R_n be the square $|R(z)| = n + \frac{1}{2}, |I(z)| = n + \frac{1}{2}, n = 0, \pm 1, \ldots,$ the vertical sides passing between the poles at $n, n+1$. Then

$$I_n = \frac{1}{4\pi i} \left[\int_{R_n} \frac{\pi \cot \pi z}{z - \xi} dz + \int_{R_n} \frac{\pi \cot \pi (-z)}{-z - \xi} d(-z) \right]$$

$$= \frac{1}{4\pi i} \int_{R_n} \pi \cot \pi z \left[\frac{1}{z - \xi} - \frac{1}{z + \xi} \right] dz$$

$$= \frac{1}{4\pi i} \int_{R_n} \frac{2\xi}{z^2 - \xi^2} \pi \cot \pi z \, dz.$$

Thus,

$$\left| \frac{1}{4\pi i} \int_{R_n} \frac{2\xi \pi \cot \pi z}{z^2 - \xi^2} dz \right|$$

$$\leq \frac{1}{4\pi} \cdot (\text{Bound for } |\pi \cot \pi z| \text{ on } R_n) \cdot \frac{2|\xi|}{(n+\tfrac{1}{2})^2 - |\xi|^2} \, 8\left(n+\frac{1}{2}\right)$$

$\to 0$ as $n \to \infty$ and $\pi \cot \pi \xi = S(\xi)$.

11.5.6
Exercises

1: Prove that $\coth z = \dfrac{1}{z} + \sum\limits_{n=1}^{\infty} \dfrac{2z}{z^2 + n^2\pi^2}$ and deduce that

$$\frac{2}{e^z - e^{-z}} = \frac{1}{z} - \frac{2z}{z^2 + \pi^2} + \frac{2z}{z^2 + 4\pi^2} - \frac{2z}{z^2 + 9\pi^2} + \cdots.$$

2: Prove that $\sec z = \sum\limits_{n=1}^{\infty} (-1)^{n+1} \dfrac{(2n-1)\pi}{(n-\tfrac{1}{2})^2 \pi^2 - z^2}$.

3: Prove that $\operatorname{cosec}^2 z = \sum\limits_{n=-\infty}^{\infty} \dfrac{1}{(z - n\pi)^2}$.

We now study one of the more important "nonelementary" functions that has many applications in higher analysis. A study will be made of the simple properties only, since a vast amount of literature is available on the subject.*

11.6

We now study the so-called gamma function $\Gamma(z)$. It is convenient to develop properties of this function for a real argument since the formula may be extended by analytic continuation to complex values (provided they do not introduce any other singularities).

11.6.1 Gamma Function and Its Properties
Define

$$\Gamma(x) = \int_{0^+}^{\infty} e^{-t} t^{x-1} dt, \quad \text{where } x \text{ is real.}$$

---

*See, for example, E. T. Whittaker and G. N. Watson, *A Course of Modern Analysis*. London and New York: Cambridge University Press, 1962.

11.6.2

The integral converges at the upper limit since for all x we have
$$t^{x-1}e^{-t} = t^{-2}t^{x+1}e^{-t} = O(t^{-2}) \quad \text{as} \quad t \to \infty.$$
Then $\int_\delta^1 e^{-t}t^{x-1}\,dt$ does not converge for $x < 0$ if $0 < \delta < 1$, since $e^t < e$ for $t \in (0, 1)$. Thus $e^{-t} > e^{-1}$ and
$$\int_\delta^1 t^{x-1}e^{-t}\,dt > \int_\delta^1 t^{x-1}e^{-1}\,dt = \frac{1}{e}\frac{t^x}{x}\Big]_\delta^1 = \frac{1}{xe}(1 - \delta^x),$$
which diverges as $\delta \to 0$ for $x < 0$. Also, since $e^{-t} < 1$ when $t > 0$, we have for $x > 0$ that
$$\int_\delta^1 t^{x-1}e^{-t}\,dt < \int_\delta^1 t^{x-1}\,dt,$$
which remains bounded.

11.6.3

The integral converges uniformly for $0 < a \leqslant x \leqslant b$, since
$$\int_0^\infty t^{x-1}e^{-t}\,dt = \int_0^1 + \int_1^\infty = O\left[\int_0^1 t^{a-1}\,dt\right] + O\left[\int_1^\infty t^{b-1}e^{-t}\,dt\right] = O(1)$$
independently of x. Hence the integral represents a continuous function for $x > 0$.

11.6.4

If z is complex, $\int_0^\infty t^{z-1}e^{-t}\,dt$ is again uniformly convergent over any finite region in which $R(z) \geqslant a > 0$, since
$$|t^{z-1}| = t^{x-1} \quad \text{if} \quad z = x + iy,$$
and we use 11.6.3. Hence $\Gamma(z)$ is analytic for $R(z) > 0$.

11.6.5

For $x > 1$, integration by parts gives
$$\Gamma(x) = (x - 1)\Gamma(x - 1), \qquad \Gamma(1) = \int_0^\infty e^{-t}\,dt = 1,$$
and $\Gamma(n) = (n - 1)!$ for positive integral n.

11.6.6

We have $\Gamma(0^+) = +\infty$, since
$$\Gamma(x) = \int_{0+}^1 t^{x-1}e^{-t}\,dt > \frac{1}{e}\int_{0+}^1 t^{x-1}\,dt = \frac{1}{ex} \to \infty \quad \text{as} \quad x \to 0^+.$$
Also $\lim_{x\to 0^+} x\Gamma(x) = 1$ since $x\Gamma(x) = \Gamma(x + 1)$, and, since $\Gamma(x)$ is continuous, we get
$$\lim_{x\to 0} \Gamma(x + 1) = \Gamma(1) = 1.$$

11.6.7
For $x > 0$, $y > 0$, we have
$$\frac{\Gamma(x)\Gamma(y)}{\Gamma(x+y)} = \int_0^\infty \frac{t^{y-1}}{(1+t)^{x+y}} dt.$$

Since for $x > 0$, $y > 0$,
$$\Gamma(x)\Gamma(y) = \int_0^\infty t^{x-1} e^{-t} dt \int_0^\infty s^{y-1} e^{-s} ds,$$

put $s = tv$ and obtain
$$\Gamma(x)\Gamma(y) = \int_0^\infty t^{x-1} e^{-t} dt \int_0^\infty t^y v^{y-1} e^{-tv} dv$$
$$= \int_0^\infty v^{y-1} dv \int_0^\infty t^{x+y-1} e^{-t(1+v)} dt.$$

Letting $u = t(1+v)$, we get
$$\Gamma(x)\Gamma(y) = \int_0^\infty v^{y-1} dv \int_0^\infty u^{x+y-1} e^{-u} (1+v)^{-x-y} du$$
$$= \Gamma(x+y) \int_0^\infty \frac{v^{y-1}}{(1+v)^{x+y}} dv.$$

The inversion of integrals is justified, since the individual integrals converge uniformly for $x \geq \epsilon > 0$ and $y \geq \epsilon > 0$. Sometimes $\Gamma(x)\Gamma(y)/\Gamma(x+y)$ is called the *Beta function* $B(x, y)$. It can be shown by a suitable transformation that
$$B(x, y) = \int_0^1 t^{x-1} (1-t)^{y-1} dt.$$

(Find the transformation.)

11.6.8
Putting $x = y = \frac{1}{2}$, $v = \tan^2\theta$, we obtain
$$\left[\Gamma\left(\frac{1}{2}\right)\right]^2 = 2\int_0^{\pi/2} d\theta = \pi.$$

Since $\Gamma(\frac{1}{2}) > 0$, we have $\Gamma(\frac{1}{2}) = \sqrt{\pi}$. Also, putting $y = 1 - x$, we get
$$\Gamma(x)\Gamma(1-x) = \int_0^\infty \frac{u^{-x}}{1+u} du$$

and
$$\int_0^\infty \frac{x^{a-1}}{1+x} dx = \frac{\pi}{\sin a\pi} \quad \text{for} \quad 0 < a < 1$$

by contour integration. Thus,
$$\Gamma(x)\Gamma(1-x) = \frac{\pi}{\sin \pi x} \quad \text{for} \quad 0 < x < 1.$$

11.6.9 Asymptotic Behavior of $\Gamma(x)$: Stirling's Formula.

Consider $\Gamma(x)$, where x is an integer, say, n; then $\Gamma(n) = (n-1)!$ We have

$$\log(n-1)! = \sum_{v=1}^{n-1} \log v,$$

also,

$$\int_{v-\frac{1}{2}}^{v+\frac{1}{2}} \log t\, dt = \int_0^{\frac{1}{2}} [\log(v+t) + \log(v-t)]\, dt$$

$$= \int_0^{\frac{1}{2}} [\log v^2 + \log(1 - t^2/v^2)]\, dt$$

$$= \log v + C_v, \quad \text{where} \quad C_v = O\left(\frac{1}{v^2}\right).$$

Thus,

$$\log \Gamma(n) = \log(n-1)!$$

$$= \int_{\frac{1}{2}}^{n-\frac{1}{2}} \log t\, dt - \sum_{v=1}^{n-1} C_v$$

$$= \left(n - \frac{1}{2}\right)\log\left(n - \frac{1}{2}\right) - \left(n - \frac{1}{2}\right) - \frac{1}{2}\log\frac{1}{2} + \frac{1}{2} - \sum_{v=1}^{\infty} C_v + o(1)$$

$$= \left(n - \frac{1}{2}\right)\log n - n + C + o(1),$$

where C is a constant.

Before establishing the nonintegral case, we prove the following lemma.

11.6.10

Lemma For n large, $\Gamma(n)/\Gamma(n+a) \approx n^{-a}$.

PROOF.

$$\lim_{n\to\infty} \frac{n^a \Gamma(a)\Gamma(n)}{\Gamma(a+n)} = \lim_{n\to\infty} n^a \int_0^1 t^{a-1}(1-t)^{n-1}\, dt.$$

Transform $t = v/n$. Then the right-hand side is

$$\lim_{n\to\infty} \int_0^n v^{a-1}\left(1 - \frac{v}{n}\right)^{n-1} dv = \int_0^\infty v^{a-1} e^{-v} = \Gamma(a).$$

Thus, for large n,

$$\lim_{n\to\infty} \frac{n^a \Gamma(n)}{\Gamma(a+n)} = 1 \quad \text{and} \quad \frac{\Gamma(n)}{\Gamma(n+a)} \approx n^{-a}. \quad \square$$

If x is not an integer, let $x = n + a$, where n is integral and $0 < a < 1$.

Since $\Gamma(n+a) \approx \Gamma(n)n^a$, from the previous lemma we have that
$$\log \Gamma(x) = \log \Gamma(n+a) = \log \Gamma(n) + a \log n + o(1)$$
$$= \left(n - \frac{1}{2}\right)\log n - n + C' + a \log n + o(1)$$
$$= \left(x - a - \frac{1}{2}\right)\log(x-a) - x + a + C' + a\log(x-a) + o(1)$$
$$= \left(x - \frac{1}{2}\right)\log x - x + C + o(1).$$

To evaluate C, consider $\Gamma(2x)\Gamma(\frac{1}{2}) = 2^{2x-1}\Gamma(x)\Gamma(x+\frac{1}{2})$, a recurrence relation obtained by considering, for example, $B(x,x)$. Taking the logarithms, we obtain

$$(2x - \tfrac{1}{2})\log 2x - 2x + C + o(1) + \log\sqrt{\pi}$$
$$= (2x - 1)\log 2 + (x - \tfrac{1}{2})\log x - x + C + o(1)$$
$$+ x\log(x + \tfrac{1}{2}) - x - \tfrac{1}{2} + C + o(1).$$

Thus,
$$2x\log 2 + 2x\log x - \tfrac{1}{2}\log 2x - 2x + C + o(1) + \log\sqrt{\pi}$$
$$= 2x\log 2 - \log 2 + x\log x - \tfrac{1}{2}\log x$$
$$- 2x + 2C - \tfrac{1}{2} + x\log x \cdot \left[1 + \frac{1}{2x}\right] + o(1)$$

and
$$\log\sqrt{\pi} + \tfrac{1}{2}\log 2 = C - \tfrac{1}{2} + x \cdot \frac{1}{2x} + o(1) = C + o(1),$$

giving $C = \log\sqrt{2\pi} + o(1)$. Hence,
$$\log \Gamma(x) = (x - \tfrac{1}{2})\log x - x + \log\sqrt{2\pi} + o(1),$$
and, since $e^{o(1)} = 1 + o(1)$, for x not a negative integer we get
$$\Gamma(x) \approx x^{x-\frac{1}{2}} e^{-x} \sqrt{2\pi} \left[1 + o(1)\right].$$

Stirling's formula for complex z can be proven, the result holding uniformly for $-\pi + \delta \leq \arg z \leq \pi + \delta$, $\delta > 0$. The negative real axis is excluded, since $\Gamma(z)$ has an infinite number of poles on it.

11.7 Analytic Continuation of $\Gamma(z)$

$\Gamma(z) = \int_0^\infty e^{-t} t^{z-1} dt$ is an analytic function for $R(z) > 0$. We require now to find the analytical extension into the rest of the complex z-plane. Consider
$$I(z) = \int_C e^{-\xi}(-\xi)^{z-1} d\xi,$$
where C is the real axis from infinity to $\delta > 0$, $|\xi| = \delta$ in the positive sense, and the real axis from δ to infinity.

Define $(-\xi)^{z-1} = e^{(z-1)\log(-\xi)}$ and chose $\log(-\xi)$ to be real for $\xi = -\delta$. The integral converges uniformly in any finite region of the z-plane (since the integral depends upon C and the circle does not pass through the origin, that is, δ is fixed). Thus $I(z)$ is analytic for all finite values of z.

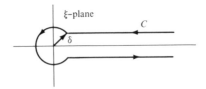

To evaluate $I(z)$, set $\xi = \rho e^{i\varphi}$. Then

$$\log(-\xi) = \log\rho + i(\varphi - \pi)$$

on the contour [so as to make $\log(-\xi)$ real for $\xi = -\delta$]. The integrals on the portions of C consisting of (∞, δ) and (δ, ∞) give

$$\int_\infty^\delta e^{-\rho+(z-1)(\log\rho - i\pi)}\,d\rho + \int_\delta^\infty e^{-\rho+(z-1)(\log\rho + i\pi)}\,d\rho$$

$$= \int_\delta^\infty e^{-\rho+(z-1)\log\rho}\bigl[e^{(z-1)i\pi} - e^{-(z-1)i\pi}\bigr]d\rho$$

$$= -2i\sin\pi z \int_\delta^\infty e^{-\rho}\rho^{z-1}\,d\rho.$$

Also, on the circle $|\xi| = \delta$ we have

$$|(-\xi)^{z-1}| = |e^{(z-1)\log(-\xi)}| = |e^{(z-1)[\log\delta + i(\varphi - \pi)]}|$$
$$= e^{(x-1)\log\delta - y(\varphi - \pi)} = O(\delta^{x-1}),$$

and the integral around the circle $|\xi| = \delta$ gives $O(\delta^x) = o(1)$ as $\delta \to 0$ if $x > 0 (\leq \delta^{x-1}\int_0^{2\pi} d\varphi)$. Letting $\delta \to 0$, we obtain for $R(z) > 0$:

$$I(z) = -2i\sin\pi z \int_0^\infty e^{-\rho}\rho^{z-1}\,d\rho = -2i\sin\pi z\,\Gamma(z).$$

Since $I(z)$ is analytic for all finite z, the function $\tfrac{1}{2}iI(z)\operatorname{cosec}\pi z$ is analytic, except possibly for poles of $\operatorname{cosec}\pi z$. Further,

$$\tfrac{1}{2}iI(z)\operatorname{cosec}\pi z = \Gamma(z), \quad R(z) > 0.$$

Hence $\tfrac{1}{2}iI(z)\operatorname{cosec}\pi z$ is the analytic continuation of $\Gamma(z)$ in the entire z-plane.

Since the poles of $\operatorname{cosec}\pi z$ are $z = 0, \pm 1, \pm 2, \ldots$, and $\Gamma(z)$ is analytic

Infinite Products

at $z = 1, 2, \ldots$, the only possible poles of $\frac{1}{2} iI(z) \operatorname{cosec} \pi z$ are $z = 0, -1, -2, \ldots$. These are actually poles of $\Gamma(z)$, for if z is one of these numbers, say, $-n$, then $(-\xi)^{z-1}$ is single-valued in C and $I(z)$ can be calculated directly by Cauchy's theorem:

$$(-1)^{n+1} \int_C \frac{e^{-\xi}}{\xi^{n+1}} d\xi = \frac{2\pi i}{n!} (-1)^{n+n+1} = -\frac{2\pi i}{n!}.$$

Thus $I(-n) = -2\pi i/n!$ and the poles of $\operatorname{cosec} \pi z$ at $z = 0, -n$ are actually poles of $\Gamma(z)$, where

$$\Gamma(-n) = I(-n) \cdot \frac{1}{-2i \sin n\pi}.$$

The residue at $z = -n$ is

$$\lim_{z \to -n} \left[\frac{-2\pi i}{n!} \frac{z+n}{-2i \sin \pi z} \right] = \frac{(-1)^n}{n!};$$

actually,

$$\lim_{z \to -n} I(z) \frac{i}{2} \operatorname{cosec} \pi z \cdot (z + n) = I(-n) \frac{i}{2} \lim_{z \to -n} \frac{z+n}{\sin \pi z}.$$

The formula $\Gamma(x)\Gamma(1 - x) = \pi \operatorname{cosec} \pi x$ (and others) can now be justified for complex values. Thus $\Gamma(z)\Gamma(1 - z) = \pi \operatorname{cosec} \pi z$ for all nonintegral z (since the left-hand side is $\pi \operatorname{cosec} \pi z$ for $0 < z < 1$ and $\Gamma(z)$ has an analytic continuation to the whole plane). Hence $1/\Gamma(z)$ is an entire function, since the poles of $\Gamma(1 - z)$ are cancelled by zeros of $\sin \pi z$ or

$$\frac{1}{\Gamma(z)} = \frac{1}{\pi} \sin \pi z \frac{1}{2} i \cdot \frac{I(1-z)}{\sin \pi (1-z)} = \frac{i}{2\pi} I(1 - z)$$

and $I(1 - z)$ is analytic everywhere in the finite plane.

11.8

There are several alternative methods for defining the gamma function, although it can be reasonably difficult to establish that they are all indeed the same function. For example, according to Gauss,

$$\Gamma(z) = \lim_{n \to \infty} \frac{n^z n!}{z(z+1) \cdots (z+n)},$$

and it is easily seen that $\Gamma(z + 1) = z\Gamma(z)$. Similar formulas can be easily established.

Another definition gives

$$\frac{1}{\Gamma(z)} = e^{\gamma z} z \prod_{n=1}^{\infty} \left[1 + \frac{z}{n} \right] e^{-z/n},$$

where

$$\gamma = \lim_{n\to\infty}\left[1 + \frac{1}{2} + \cdots + \frac{1}{n} - \log n\right] \sim 0.577$$

and is called *Euler's constant*. This definition shows immediately that, since $1/\Gamma(z)$ has simple zeros at $z = 0, -1, -2, \ldots$, then $\Gamma(z) \in A$, except at simple poles $z = 0, -1, \ldots$. Furthermore, $1/\Gamma(z)$ is clearly an entire function.

11.8.1
Exercises

1: Show that $\int_0^\infty t^{z-1} \cos t\, dt = \Gamma(z)\cos\frac{\pi z}{2}$, where $0 < R(z) < 1$.

2: Prove that $\{\Gamma(z)\}^2 = \dfrac{B(z,\frac{1}{2})\Gamma(2z)}{2^{2z-1}}$.

3: Show that $\int_0^\infty \dfrac{\sin t}{t^z}\, dt = \dfrac{\pi}{2\Gamma(z)\sin(\pi z/2)}$, where $0 < R(z) < 2$.

4: Prove that $\gamma = \int_0^\infty \left[\dfrac{1}{e^t - 1} - \dfrac{1}{te^t}\right] dt$.

5: Prove that the Gauss definition of the gamma function is equivalent to the reciprocal of $e^{\gamma z} \cdot z \prod_{n=1}^{\infty} (1 + \frac{z}{n}) e^{-z/n}$.

6: Verify that $\int_0^\infty x^{p-1} \cos ax\, dx = \dfrac{\Gamma(p)\cos(\pi p/2)}{a^p}$, where $a > 0, 0 < p < 1$.

7: $\int_0^\infty \cos x^p\, dx = \dfrac{1}{p}\Gamma(\frac{1}{p})\cos\dfrac{\pi}{2p}$, where $p > 1$.

8: $\int_0^\infty \dfrac{\sin x^p}{x^p}\, dx = \dfrac{1}{p-1}\Gamma(\frac{1}{p})\cos\dfrac{\pi}{2p}$, where $p > \frac{1}{2}$. Show that this result is consistent for $p = 1$.

9: Show that $B(m,n) = 2\int_0^{\pi/2} \sin^{2m-1}\theta \cos^{2n-1}\theta\, d\theta = B(n,m)$. Hence, evaluate $\int_0^{\pi/2} \sqrt{\sin\theta}\, d\theta$.

10: Find a) $\lim_{n\to\infty} (n!)^{1/(n\log n)}$ b) $\lim_{n\to\infty}\left[\dfrac{\log n!}{n} - \log n\right]$.

11: Show that $\lim_{n\to\infty} \dfrac{1}{n}\sqrt[n]{n!} = \dfrac{1}{e}$.

12: Prove that, if $b > a > -1$, then
$$\int_0^{\pi/2} \cos^a\theta \cos b\theta \, d\theta = \dfrac{\pi\Gamma(a+1)}{2^{a+1}\Gamma\left(\dfrac{a}{2}+\dfrac{b}{2}+1\right)\Gamma\left(\dfrac{a}{2}-\dfrac{b}{2}+1\right)}.$$

12
Infinite Product Representation: Order and Type

A study will now be made of the infinite product representation of an entire function. Several distinctions between functions with an infinite number of zeros and functions with a finite number of zeros will emerge. The concept of "order" will be dealt with in more detail, leading to a further object for studying these functions, namely "type." We start with a well-known and very illuminating theorem of Weierstrass.

12.1 The Weierstrass Factorization Theorem

Since an entire function is an analytic function with no singularities except at infinity, consider a polynomial $f(z)$ with zeros at $z_1, z_2, \ldots, z_n$. The polynomial can then be factorized as

$$f(z) = f(0)\left[1 - \frac{z}{z_1}\right]\left[1 - \frac{z}{z_2}\right] \cdots \left[1 - \frac{z}{z_n}\right].$$

Thus, given any finite set of points $\{z_1, z_2, \ldots, z_n\}$, there always exists an entire function with zeros $z_1, z_2, \ldots, z_n$.

Consider now an infinite sequence of points $z_1, z_2, \ldots, z_n, \ldots$ such that

$$0 < |z_1| \leqslant |z_2| \leqslant \cdots \leqslant |z_n| \leqslant \cdots,$$

whose sole limit point is infinity, that is, $\lim_{n \to \infty} |z_n| = \infty$.

We wish to construct an entire function that vanishes at these points and nowhere else; unfortunately, $\prod_{n=1}^{\infty}[1 - (z/z_n)]$ may diverge. This is, in effect, accomplished by Weierstrass's factorization theorem, which demonstrates the construction and existence of such a function. We shall prove a form of the theorem stated.

12.1.1

Theorem *Let $\{z_n\}$ be an arbitrary sequence of complex numbers, all different from zero, whose sole limit is ∞ and let M be a given nonnegative integer. Then there exist an entire function $G(z)$ having roots at $z_1, z_2, \ldots$ (and at these points only), and a root of multiplicity m at $z = 0$. Furthermore, $G(z)$ has the representation*

$$G(z) = e^{g(z)} z^m \prod_{n=1}^{\infty} \left(1 - \frac{z}{z_n}\right) e^{Q_n(z/z_n)}$$

where $g(z)$ is an arbitrary entire function, $Q_n(z) = \sum_{k=1}^{v(n)} z^k/k$ and $v(n)$ are positive integers to be defined later.

PROOF. Let R be an arbitrarily large positive number and let $|z| \leq R$. Let q be a positive integer such that $|z_q| > 2R$ (we know that such q exists since $|z_n| \to \infty$). For $n \geq q$ and $|z| \leq R$ consider $(1 - z/z_n) \exp Q_n(z/z_n) = 1 + U_n(z)$. Since $|z/z_n| < \frac{1}{2}$ we have,

$$1 - \frac{z}{z_n} = \exp\left(-\sum_{k=1}^{\infty} \frac{1}{k}(z/z_n)^k\right)$$

and thus,

$$U_n(z) = \exp\left(-\sum_{k=v(n)+1}^{\infty} \frac{1}{k}(z/z_n)^k\right) - 1.$$

Now we observe that

$$|e^u - 1| \leq e^{|u|} - 1$$

for any complex n, and

$$\sum_{k=v(n)+1}^{\infty} \frac{1}{k}\left|\frac{z}{z_n}\right|^k \leq 2\left|\frac{z}{z_n}\right|^{v(n)+1},$$

since $|z/z_n| < \frac{1}{2}$. It follows that

$$|U_n(z)| \leq \exp\left(2\left|\frac{z}{z_n}\right|^{v(n)+1}\right) - 1$$

$$\leq 2e\left|\frac{z}{z_n}\right|^{v(n)+1}$$

since $e^u - 1 \leq ue^u$ for nonnegative u.

Two cases arise:

I. either there exist a positive integer p such that $\sum_{n=1}^{\infty} 1/|z_n|^p < \infty$, or
II. there does not exist such an integer.

CASE I. Take $v(n) = p - 1$ and

$$|U_n(z)| < \frac{R^p}{|z_n|^p} \cdot 2e$$

since $|z| \leq R$, and $\sum_{(n)} |U_n(z)| < \infty$ for $|z| \leq R$. Thus, for $|z| \leq R$, $\sum_{n=q+1}^{\infty} (1 + U_n(z))$ converges absolutely and uniformly.

CASE II. Take $v(n) = n - 1$ so that

$$|U_n(z)| < 2e \left| \frac{z}{z_n} \right|^n \quad \text{for} \quad n > q,$$

$$\left| \frac{z}{z_n} \right| < 1, \quad |z| \leq R, \quad |z_n| \to \infty.$$

Thus by the root test $\sum_{(n)} |U_n(z)| < \infty$ with the same result as before. Thus $\prod_{n=1}^{\infty} (1 - (z/z_n)) e^{Q_n(z/z_n)}$ is analytic in $|z| \leq R$, and since R is arbitrary, the product represents an entire function. If further, $z = 0$ is a zero of order m of $G(z)$, then $G(z)/z^m G_1(z)$ has no zeros and equals $e^{g(z)}$, say, where $G_1(z)$ is the above infinite product, and

$$G(z) = e^{g(z)} z^m \prod_{n=1}^{\infty} \left(1 - \frac{z}{z_n} \right) e^{Q_n(z/z_n)}.$$

Here $g(z)$ is an arbitrary entire function. □

If $G(z)$ is subject to further restrictions, it should be possible to say more about $g(z)$. The factor $e^{Q_n(z/z_n)}$ is clearly inserted to product convergence of the infinite product. The factors

$$\left(1 - \frac{z}{z_n} \right) \exp\left(\frac{z}{z_n} + \cdots + \frac{1}{v} \left(\frac{z}{z_n} \right)^v \right)$$

are called "primary factors" and the product formed with these factors is called the "canonical product," provided that v is the smallest integer for which $\sum_{(n)} 1/|z_n|^{v+1}$ converges, if such v exists.

12.2 Order of an Entire Function

A function $f(z)$ is said to be of *finite order*, if there exists a constant λ such that $|f(z)| < \exp r^\lambda$ for $|z| = r > r_0$. If f is nonconstant and of finite order, then $\lambda > 0$. If the inequality is true for a certain λ, then it is true for $\lambda' > \lambda$, thus there exists an infinity of λ values of $\lambda > 0$ satisfying this inequality.

The lower bound of these λ values is called the *order of* f. Denote this lower bound by ρ. Then given $\epsilon > 0$, there exists r_0 such that

$$|f(z)| < \exp r^{\rho+\epsilon} \quad \text{for} \quad |z| = r > r_0.$$

This implies

$$M(r) = \max_{|z|=r} |f(z)| < \exp r^{\rho+\epsilon} \quad \text{for} \quad r > r_0,$$

while $M(r) > \exp r^{\rho-\epsilon}$ for an infinite number of $r \to +\infty$. Thus

$$\frac{\log \log M(r)}{\log r} < \rho + \epsilon \quad \text{and} \quad \frac{\log \log M(r)}{\log r} > \rho - \epsilon$$

for an infinite number of $r \to +\infty$; hence

$$\rho = \varlimsup_{r \to \infty} \frac{\log \log M(r)}{\log r}.$$

Alternatively, $f(z)$ is of finite order if there exists $A > 0$ such that

$$|f(z)| = O(\exp r^A) \quad \text{as} \quad |z| = r \to \infty.$$

Thus $f(z)$ is of order ρ if $|f(z)| = O(\exp r^{\rho+\epsilon})$ for every $\epsilon > 0$, but not for any negative ϵ. The constant implied in O depends in general on ϵ; otherwise ϵ could be replaced by zero in the formula. A few theorems will now be proved by using the above expression for the order.

12.2.1

Theorem *If $f_1(z)$ and $f_2(z)$ are entire functions of order ρ_1 and ρ_2, respectively, and if $\rho_1 < \rho_2$, then the order of $F(z) = f_1(z) + f_2(z)$ is equal to ρ_2.*

PROOF. We assume that ρ_2 is finite. Since for $r > r_0(\epsilon)$

$$M(r; f_1 + f_2) \leqslant M(r; f_1) + M(r; f_2)$$
$$\leqslant \exp r^{\rho_1+\epsilon} + \exp r^{\rho_2+\epsilon}$$
$$< 2 \exp r^{\rho_2+\epsilon},$$

we have that $\rho \leqslant \rho_2 + \epsilon$ and hence $\rho \leqslant \rho_2$. On the other hand, there exists a sequence of numbers $r_n \to \infty$ such that $M(r_n; f_2) > \exp r_n^{\rho_2-\epsilon}$. Thus

$$M(r_n; f_1 + f_2) \geqslant \exp r_n^{\rho_2-\epsilon} - \exp r_n^{\rho_1+\epsilon}$$
$$= \exp r_n^{\rho_2-\epsilon}\left[1 - \exp(r_n^{\rho_1+\epsilon} - r_n^{\rho_2-\epsilon})\right]$$
$$> \frac{1}{2} \exp r_n^{\rho_2-\epsilon},$$

provided ϵ is so small that $\rho_1 + \epsilon < \rho_2 - \epsilon$ and n is sufficiently large. Thus $\rho \geqslant \rho_2$, and the order of the sum $f_1 + f_2$ equals ρ_2. □

Note: The proof also applies without significant changes when $\rho_2 = \infty$. It should be noted that the theorem is sometimes false when $\rho_1 = \rho_2$. Taking $f_1(z) = e^z$ and $f_2(z) = -e^z$, we get $\rho_1 = \rho_2 = 1$ and $\rho = 0$. We can say that, if $\rho_1 \leqslant \rho_2$, then $\rho \leqslant \rho_2$.

12.2.2

Theorem *If $f_1(z)$ and $f_2(z)$ are entire functions of order ρ_1 and ρ_2, respectively, where $\rho_1 \leqslant \rho_2$, then the order ρ of $F(z) = f_1(z)f_2(z)$ is such that $\rho \leqslant \rho_2$.*

PROOF. Given $\epsilon > 0$ and r sufficiently large, we have
$$M(r; f_1 f_2) \leqslant M(r; f_1) M(r; f_2)$$
$$\leqslant \exp r^{\rho_1 + \epsilon} \exp r^{\rho_2 + \epsilon}$$
$$\leqslant \exp 2 r^{\rho_2 + \epsilon}.$$
Thus $\rho \leqslant \rho_2 + \epsilon$ and hence $\rho \leqslant \rho_2$. □

12.2.3

Theorem *If $f(z)$ is an entire function of order ρ and $P(z)$ is a nonzero polynomial, then the product $f(z)P(z)$ is of order ρ. If the quotient $f(z)/P(z)$ is an entire function, then it also is of order ρ.*

PROOF. From the previous theorem we know that the order of $f(z)P(z)$ does not exceed ρ. Since $|P(z)| > 1$ for z sufficiently large, for these values of z we get $|f(z)P(z)| \geqslant |f(z)|$. Thus the order of $f(z)P(z)$ is not less than ρ, which proves the first part of the theorem. We note also that if $f(z)/P(z)$ is entire, then its order is the same as $P(z)f(z)/P(z) = f(z)$, and the second part of the theorem is proved. □

12.3 Type

Entire functions may be further subdivided as follows. Given an entire function of order ρ (finite), suppose there exists a positive k such that $M(r) < \exp k r^\rho$ for $r > R_0$. Then $f(z)$ is said to be of *finite type*. The greatest lower bound $\sigma = \inf k \geqslant 0$ of the k values for which $M(r) < \exp k r^\rho$ holds (starting from some sufficiently large r), is called the *type* of $f(z)$; for example, e^z has order $\rho = 1$ and type $\sigma = 1$.

12.3.1

Definition $f(z)$ is called *normal*, *mean*, or *finite type* if $0 < \sigma < \infty$; $f(z)$ is called *minimum type* if $\sigma = 0$; $f(z)$ is called *maximum* or *infinite type* if $\sigma = \infty$ (that is, $M(r)$ exceeds $\exp k r^\rho$ for arbitrary large r).

As a consequence we have the following result.

12.3.2

Theorem *The type σ of an entire function of order ρ, where $0 < \rho < \infty$, is given by the formula*
$$\sigma = \varlimsup_{r \to \infty} \frac{\log M(R)}{r^\rho}.$$

PROOF. Since $\sigma = \inf k$, given $\epsilon > 0$, there exits $R(\epsilon) > 0$ such that
$$M(r) < \exp[(\sigma + \epsilon) r^\rho] \quad \text{for} \quad r > R(\epsilon).$$
Also, there is a sequence $\{r_n\}$ such that
$$0 < r_1 < \cdots < r_n < \cdots \quad \text{and} \quad M(r_n) > \exp[(\sigma - \epsilon) r_n^\rho].$$

Thus,
$$\frac{\log M(r)}{r^\rho} < \sigma + \epsilon \quad \text{for} \quad r > R(\epsilon),$$
while
$$\frac{\log M(r_n)}{r_n^\rho} > \sigma - \epsilon \quad \text{for arbitrary large } r_n.$$
This is precisely what is meant by $\sigma = \overline{\lim}_{r\to\infty}[\log M(r)]/r^\rho$. □

EXAMPLE.
$$\frac{e^r - 1}{2} \leq M(r) = \max_{|z|=r}|\sin z| \leq \frac{e^r + 1}{2}$$
and $\sin z$ is of order $\rho = 1$ and type $\sigma = 1$.

12.4

Note that the maximum modulus $M(r)$ describes the growth of an entire function in the neighborhood of the point at infinity but gives no information about the behavior of $f(z)$ in various unbounded subdomains. Consider, for example, e^z (order 1, type 1):
$$|e^z| = e^x = e^{r\cos\theta}.$$
In the closed angle range
$$-\frac{\pi}{2} + \epsilon \leq \theta \leq \frac{\pi}{2} - \epsilon, \quad \text{where} \quad \epsilon > 0,$$
we have
$$e^{r\sin\epsilon} \leq |e^z| \leq e^r$$
and thus $e^z \to \infty$ as $r \to \infty$. However, in every closed angle range
$$\frac{\pi}{2} + \epsilon \leq \theta \leq \frac{3\pi}{2} - \epsilon, \quad \text{where} \quad \epsilon > 0,$$
we have $|e^z| \leq e^{-r\sin\epsilon}$ and $e^z \to 0$ as $r \to \infty$. Thus there exist two open angle ranges, each of π radians, viz., the right half-plane $R(z) > 0$ and the left half-plane $R(z) < 0$, such that $\lim_{r\to\infty} e^z = \infty$ in every angle in $R(z) > 0$, while $\lim_{r\to\infty} e^z = 0$ in every angle in $R(z) < 0$.

12.5

There are several types of "enumerative" functions.* The simplest of these is the function $n(r)$, namely, the number of zeros of $f(z)$ in $|z| \leq r$. We now give several theorems dealing with $n(r)$.

---

*They may be studied, for example, in *Introduction to the Theory of Entire Functions* by A.S.B. Holland, New York: Academic Press, 1973.

12.5.1

Theorem *If $f(z)$ is an entire function of order $\rho < \infty$ and if it has an infinity of zeros with $f(0) \neq 0$, then, given $\epsilon > 0$, there exists R_0 such that for $R \geq R_0$*

$$n\left(\frac{R}{3}\right) \leq \frac{1}{\log 2} \cdot \log \frac{\exp R^{\rho+\epsilon}}{|f(0)|},$$

where $n(R)$ denotes the number of zeros of $f(z)$ in $|z| \leq R$.

PROOF. If $f(z)$ is analytic in $|z| \leq R$ and $a_1, a_2, \ldots, a_n$ are zeros of $f(z)$ in $|z| \leq R/3$, then for

$$g(z) = \frac{f(z)}{\left[1 - \dfrac{z}{a_1}\right] \cdots \left[1 - \dfrac{z}{a_n}\right]}$$

we have $|g(z)| \leq M/2^n$, where $|f(z)| \leq M$ for $|z| = R$ and $|a_p| \leq R/3$, $p = 1, \ldots, n$. Thus, for $|z| = R$,

$$\left|\frac{z}{a_p}\right| \geq 3 \quad \text{and} \quad \left|1 - \frac{z}{a_p}\right| \geq 2.$$

By the maximum-modulus theorem, $|g(0)| \leq M/2^n$. Thus, $|f(0)| \leq M/2^n$ and

$$n \equiv n\left(\frac{R}{3}\right) \leq \frac{1}{\log 2} \cdot \log \frac{M}{|f(0)|}.$$

If, further, $f(z)$ is of order ρ, then for $r > r_0$, we have $M(r) < \exp r^{\rho+\epsilon}$ and the result follows. □

12.5.2

Corollary

I. $n(R) = O(R^{\rho+\epsilon})$. Since

$$n\left(\frac{R}{3}\right) \leq \frac{1}{\log 2}\left[R^{\rho+\epsilon} - \log|f(0)|\right] \quad \text{and} \quad -\log|f(0)| < R_1^{\rho+\epsilon};$$

then for $R \geq R_1$ we have

$$n\left(\frac{R}{3}\right) < \frac{2}{\log 2} R^{\rho+\epsilon} \quad \text{and} \quad n(R) = O(R^{\rho+\epsilon}).$$

II. Since $n(R)$ denotes the number of zeros for which $|a_n| \leq R$, then $n(R)$ is a nondecreasing function of R that is constant in intervals. It is zero for $R < |a_1|$ if $f(0)$ is not zero. By virtue of Jensen's formula, we have

$$\int_0^R \frac{n(x)}{x} dx = \frac{1}{2\pi} \int_0^{2\pi} \log|f(Re^{i\theta})| d\theta - \log|f(0)|.$$

Since $f(z)$ is of order ρ, that is, $|f(Re^{i\theta})| < k_1 \exp R^{\rho+\epsilon}$, $\epsilon > 0$, then

$$\log|f(Re^{i\theta})| < kR^{\rho+\epsilon}.$$

Thus,
$$\int_0^{2R} \frac{n(x)}{x}\,dx < k_2 R^{\rho+\epsilon}.$$

Since $n(R)$ is nondecreasing,
$$\int_0^{2R} \frac{n(x)}{x}\,dx \geqslant n(R)\int_0^{2R} \frac{dx}{x} = n(R)\log 2$$

and
$$n(R) \leqslant \frac{1}{\log 2}\int_0^{2R} \frac{n(x)}{x}\,dx < k_3 R^{\rho+\epsilon}.$$

Thus, roughly, the higher the order of a function, the more zeros it may have in a given region.

12.6

Theorem If $f(z)$ is of order $\rho < \infty$ and $r_1, r_2, \ldots$ are the moduli of its infinite number of zeros, then $\sum_{(n)} 1/r_n^\alpha < \infty$ for $\alpha > \rho$.

PROOF. Let β be a number between α and ρ, that is, $\rho < \beta < \alpha$. Then $n(r) < Ar^\beta$. Putting $r = r_n$, we have
$$n < Ar_n^\beta \quad \text{and} \quad r_n^{-\alpha} < A_1 n^{-\alpha/\beta}.$$
Since $\alpha/\beta > 1$, we get $\sum_{(n)} 1/r_n^\alpha < \infty$. □

Note: Clearly the result is trivial for a finite number of zeros.

12.7

Definition The lower bound of the positive numbers α for which $\sum_{(n)} 1/r_n^\alpha$ is convergent, is called the *exponent of convergence of the zeros* and is denoted by ρ_1. Formally, the empty set has $\rho_1 = 0$ and if the series diverges for all positive α, then $\rho_1 = \infty$. Thus
$$\sum_{(n)} \frac{1}{r_n^{\rho_1+\epsilon}} < \infty \quad \text{and} \quad \sum_{(n)} \frac{1}{r_n^{\rho_1-\epsilon}} = \infty, \quad \epsilon > 0.$$

We have proved that $\rho_1 \leqslant \rho$ since it may be possible to find numbers less than ρ for which the series converges.

12.7.1

Lemma The number defined by the equations
$$\rho_1 = \varlimsup_{n\to\infty} \frac{\log n}{\log r_n} = \varlimsup_{r\to\infty} \frac{\log n(r)}{\log r}$$

has the following property: if
$$\rho_1(\text{finite}) = \varlimsup_{n\to\infty} \frac{\log n}{\log r_n} \quad \text{or} \quad \varlimsup_{r\to\infty} \frac{\log n(r)}{\log r},$$

then ρ_1 is the exponent of convergence of the zeros of $f(z)$, that is,

$$\sum_{(n)} \frac{1}{r_n^{\rho_1+\epsilon}} < \infty \quad \text{and} \quad \sum_{(n)} \frac{1}{r_n^{\rho_1-\epsilon}} = \infty, \quad \epsilon > 0.$$

PROOF. The limit implies that

$$\frac{\log n}{\log r_n} < \rho_1 + \epsilon \quad \text{for} \quad n > N \tag{12.1}$$

and

$$\frac{\log n}{\log r_n} > \rho_1 - \epsilon \quad \text{for} \quad n > N. \tag{12.2}$$

From inequality (12.2) we have $n = r_n^{\rho_1-\epsilon}$ and thus $1/n < 1/r_n^{\rho_1-\epsilon}$, but since $\sum_{(n)} 1/n = \infty$, we have $\sum_{(n)} 1/r_n^{\rho_1-\epsilon} = \infty$.

Using inequality (12.1), let $\epsilon' > 0$ and $\epsilon = \epsilon'/2$; then there exists N such that $n < r_n^{\rho_1+\epsilon}$ for all $n > N$. Define

$$\delta \equiv \frac{\epsilon' - \epsilon}{\rho_1 + \epsilon} = \frac{\epsilon'/2}{\rho_1 + (\epsilon'/2)}$$

and thus $\delta > 0$. Hence for all $n > N$ we obtain

$$\epsilon' = \epsilon + \epsilon\delta + \rho_1\delta \to \rho_1 + \epsilon' = (\rho_1 + \epsilon)(1 + \delta)$$

$$\to r_n^{(\rho_1+\epsilon)(1+\delta)} > n^{1+\delta} \to r_n^{\rho_1+\epsilon'} > n^{1+\delta}$$

$$\to \frac{1}{r_n^{\rho_1+\epsilon'}} < \frac{1}{n^{1+\delta}} \to \sum_{(n)} \frac{1}{r_n^{\rho_1+\epsilon'}} < \infty. \qquad \square$$

The series $\sum_{(n)} 1/r_n^{\rho_1}$ may either converge or diverge. For example, take $r_n = n$ or $r_n = n(\log n)^2$. If the zeros of $f(z)$ are finite or nil, then $\rho_1 = 0$. Thus $\rho_1 > 0$ implies the existence of infinitely many zeros.

Note: We can have $\rho_1 < \rho$, for example, if $f(z) = e^z$ and $\rho = 1$, but since there are no zeros of e^z, then $\rho_1 = 0$. Let $f(z)$ be an entire function of order $\rho < \infty$, $f(0) \neq 0$, $f(z_n) = 0$, $n = 1, 2, \ldots$. Then there exists an integer $p + 1$ such that $\sum_{n=1}^{\infty} 1/|z_n|^{p+1} < \infty$. By the previous theorem, any integer exceeding ρ will serve as $p + 1$.

12.8

Definition The smallest integer p for which $\sum_{n=1}^{\infty} 1/|z_n|^{p+1} < \infty$, is called the *genus* (rank) of the canonical product. The genus of the general entire function

$$f(z) = z^m e^{g(z)} \prod_{n=1}^{\infty} \left[1 - \frac{z}{z_n}\right] e^{Q_n(z/z_n)}$$

will be defined later. Sometimes the two will coincide. Thus, by

Weierstrass's theorem,

$$f(z) = e^{g(z)} z^m \prod_{n=1}^{\infty} \left[1 - \frac{z}{z_n}\right] \exp\left[\frac{z}{z_n} + \frac{z^2}{2z_n^2} + \cdots + \frac{z^v}{v \cdot z_n^v}\right],$$

where $v = p$. (We were previously looking for v such that $\sum_{n=1}^{\infty} |z/z_n|^{v+1}$ would converge.) If the values of z_n are finite, define $p = 0$ and the product as $\prod_{m=1}^{n} [1 - (z/z_n)]$.

EXAMPLE. If $z_n = n$, then $p = 1$, as in $\sum_{(n)} 1/|z_n|^2 < \infty$; $z_n = e^n$, then $p = 0$, and if $z_1 = \frac{1}{2}\log 2$, and $z_n = \log n$, $n \geq 2$, then no finite p exists.

12.8.1

Summarizing: For $\rho_1 < \sigma$, we have

$$\sum_{n=1}^{\infty} \frac{1}{|z_n|^\sigma} < \infty \quad \text{and} \quad \sum_{n=1}^{\infty} \frac{1}{|z_n|^{p+1}} < \infty,$$

but $\sum_{n=1}^{\infty} 1/|z_n|^p = \infty$, since p is the smallest integer for which the preceding equation holds. Thus, if ρ_1 is not an integer, $p = [\rho_1]$. If ρ_1 is an integer, then either

$$\sum_{n=1}^{\infty} \frac{1}{|z_n|^{\rho_1}} < \infty \quad \text{or} \quad \sum_{n=1}^{\infty} \frac{1}{|z_n|^{\rho_1}} = \infty.$$

12.8.2

Note that $\sum_{n=1}^{\infty} 1/|z_n|^{\rho_1 + \epsilon} < \infty$ for $\epsilon > 0$, but we have no information for $\sum_{n=1}^{\infty} 1/|z_n|^{\rho_1}$. It is useful to subdivide entire functions into two further classes, depending on whether the function $f(z)$ has zeros such that $\sum_{n=1}^{\infty} 1/r_n^{\rho_1}$ converges or diverges, but we shall not pursue this subject any further.

We have two cases to consider:

$$p + 1 = \rho_1 \quad \text{and} \quad p = \rho_1.$$

Hence $p \leq \rho_1$, but $\rho_1 < \rho$, thus

$$p \leq \rho_1 \leq \rho \quad \text{and} \quad p \leq \rho_1 \leq p + 1$$

since $\sum_{n=1}^{\infty} 1/|z_n|^{p+1} < \infty$.

12.9 Hadamard's Factorization Theorem

If $f(z)$ is an entire function of order ρ with zeros $z_1, z_2, \ldots$, and $f(0) \neq 0$, then $f(z) = e^{Q(z)} P(z)$, where $P(z)$ is the canonical product formed with the zeros of $f(z)$ and $Q(z)$ is a polynomial of degree not greater than ρ. (The canonical product of course includes the exponential convergence-producing factor which may be unity.)

PROOF. Since $f(z)$ is an entire function, then $f(z) = f(0) P(z) e^{Q(z)}$, where $P(z)$ is a product of primary factors and $Q(z)$ is an entire function. We require to prove that $Q(z)$ is a polynomial. Let $v = [\rho]$. Thus $p \leqslant v$. Taking the logarithms and differentiating $v + 1$ times, we obtain

$$\frac{d^v}{dz^v}\left[\frac{f'(z)}{f(z)}\right] = Q^{(v+1)}(z) - v!\sum_{n=1}^{\infty}\frac{1}{(z_n - z)^{v+1}}.$$

Note that

$$\frac{d^{v+1}}{dz^{v+1}}\sum_{(n)}\left[\frac{z}{z_n} + \cdots + \frac{1}{p}\left(\frac{z}{z_n}\right)^p\right] = 0.$$

To show that $Q(z)$ is a polynomial of degree at most v, we require to show that $Q^{(v+1)}(z) = 0$. Let

$$g_R(z) = \frac{f(z)}{f(0)}\prod_{|z_n|<R}\left[1 - \frac{z}{z_n}\right]^{-1}.$$

For $|z| = 2R$ and $|z_n| \leqslant R$, we get $|1 - (z/z_n)| \geqslant 1$ and thus

$$|g_R(z)| \leqslant \frac{|f(z)|}{|f(0)|} = O(\exp(2R)^{\rho+\epsilon}) \quad \text{for} \quad |z| = 2R.$$

Since $g_R(z)$ is entire, $\{f(0) \neq 0, f(z)$ is entire and $\prod_{|z_n|<R}[1 - (z/z_n)]^{-1}$ cancels with factors in $f(z)\}$, then $|g_R(z)| = O[\exp(2R)^{\rho+\epsilon}]$ also for $|z| < 2R$ (by the maximum-modulus theorem).

Let $h_R(z) = \log g_R(z)$, the logarithms being determined for $h_R(0) = 0$. Then $h_R(z)$ is analytic for $|z| \leqslant R$ since $g_R(z) \neq 0$ in $|z| \leqslant R$ and $R\{h_R(z)\} < KR^{\rho+\epsilon}$. (We have absorbed $2^{\rho+\epsilon}$ in K.) The real part may be negative but cannot be $-\infty$ in $|z| \leqslant R$.

By a previous result (Theorem 7.5.1), we have

$$|h_R^{(v+1)}(z)| \leqslant \frac{2^{v+3}(v+1)!R}{(R-r)^{v+2}} \cdot KR^{\rho+\epsilon} \quad \text{for} \quad |z| = r < R.$$

and thus

$$h_R^{(v+1)}(z) = O(R^{\rho+\epsilon-v-1}) \quad \text{for} \quad |z| = \frac{1}{2}R = r.$$

Hence

$$Q^{(v+1)}(z) = h_R^{(v+1)}(z) + v!\sum_{|z_n|>R}\frac{1}{(z_n - z)^{v+1}}.$$

Since

$$h_R(z) = \log g_R(z) = \log f(z) - \log f(0) - \sum_{|z_n|<R}\log\left[1 - \frac{z}{z_n}\right],$$

then

$$h_R^{(v+1)}(z) = \frac{d^v}{dz^v}\left[\frac{f'(z)}{f(z)}\right] + v!\sum_{|z_n|<R}\frac{1}{(z_n - z)^{v+1}}.$$

Thus

$$Q^{(v+1)}(z) = O[R^{\rho+\epsilon-v-1}] + O\left[\sum_{|z_n|>R}|z_n|^{-v-1}\right]$$

for $|z| = R/2$ and so also for $|z| < R/2$. Since $v = [\rho]$, we get $v + 1 > \rho$. Terms $O[R^{\rho+\epsilon-v-1}] \to 0$ as $R \to \infty$, provided ϵ is small. Also, since $\sum_{n=1}^{\infty}|z_n|^{-v-1}$ converges, terms $O[\sum_{|z_n|>R}|z_n|^{-v-1}] \to 0$ as $R \to \infty$ and $\sum_{|z_n|>R}|z_n|^{-v-1}$ becomes in effect the remainder term for R sufficiently large. Since $Q^{(v+1)}(z)$ is independent of R, it must be zero, and the theorem follows. □

What we have shown is that $f(z) = e^{Q(z)}P(z)$, where $Q(z)$ is a polynomial of degree $v \leq \rho$ and

$$P(z) = \prod_{n=1}^{\infty}\left[1 - \frac{z}{z_n}\right]\exp\left[\frac{z}{z_n} + \cdots + \frac{1}{p}\left(\frac{z}{z_n}\right)^p\right],$$

where p is the smallest integer for which $\sum_{n=1}^{\infty} 1/|z_n|^{p+1} < \infty$.

As an example of Weierstrass's theorem and Hadamard's theorem we express $\sin \pi z$ as an infinite product. The zeros are $z = \pm n$ and all are simple. Arrange the zeros in a sequence $0, +1, -1, +2, -2, \ldots$.

I. We consider

$$z \prod_{n=1}^{\infty}\left[1 - \frac{z}{n}\right]e^{z/n}\left[1 + \frac{z}{n}\right]e^{-z/n}; \qquad (12.3)$$

the polynomial

$$Q_n\left(\frac{z}{z_n}\right) = \frac{z}{z_n} + \cdots + \frac{1}{v}\left(\frac{z}{z_n}\right)^v$$

has $v = p$ such that $\sum_{n=1}^{\infty} 1/|z_n|^{p+1}$ converges and p is integral, $p = 1$, and $v = 1$ and $Q_n(z/z_n) = z/n$. Then (12.3) becomes $z\prod_{n=1}^{\infty}[1 - (z^2/n^2)]$ and

$$\sin \pi z = e^{g(z)} \cdot z \prod_{n=1}^{\infty}\left[1 - \frac{z^2}{n^2}\right].$$

Taking the logarithmic derivative, we get

$$\frac{\pi \cos \pi z}{\sin \pi z} = g'(z) + \frac{1}{z} + \sum_{n=1}^{\infty}\frac{2z}{z^2 - n^2}.$$

If we use the fact that

$$\pi \cot \pi z = \frac{1}{z} + \sum_{n=1}^{\infty}\frac{2z}{z^2 - n^2},$$

then $g'(z) = 0$ and $g(z) =$ constant. Since $(\sin \pi z)/\pi z \to 1$ as $z \to 0$, then

$$e^{g(z)} = \pi \qquad \text{and} \qquad \sin \pi z = \pi z \prod_{n=1}^{\infty}[1 - z^2/n^2].$$

II. By Hadamard's theorem, since the order of $\sin \pi z$ is $\rho = 1$, then $Q(z)$ is a polynomial of degree less than or equal to $\rho = 1$ and hence

$Q(z) = A + Bz$. Since

$$e^{Q(z)} = \frac{\sin \pi z}{z \prod_{n=1}^{\infty}\left[1-(z^2/n^2)\right]},$$

$e^{Q(z)}$ is an even function and $e^{A+Bz} = e^{A-Bz}$, $e^{2Bz} = 1$, or $B = 0$. Taking the limit as $z \to 0$, we have $e^A = \pi$. Thus

$$e^{Q(z)} = \pi \quad \text{and} \quad \sin \pi z = \pi z \prod_{n=1}^{\infty}\left[1-(z^2/n^2)\right].$$

With Hadamard's theorem, we are now in a position to prove the following result.

12.9.1

Theorem *If $f(z)$ is an entire function of order ρ and $g(z)$ is an entire function of order $\rho' \leqslant \rho$ and if the zeros of $g(z)$ are all zeros of $f(z)$, then $H(z) = f(z)/g(z)$ is of order ρ, at most.*

PROOF. Writing $P_1(z)$ and $P_2(z)$ to be the canonical product of $f(z)$ and $g(z)$, respectively, we have,

$$f(z) = P_1(z)e^{Q_1(z)} \quad \text{and} \quad g(z) = P_2(z)e^{Q_2(z)},$$

Q_1, Q_2 being appropriate polynomials. Thus

$$H(z) = P(z)e^{Q_1(z)-Q_2(z)},$$

where $P(z) = P_1(z)/P_2(z)$ is the canonical product formed from the zeros of $P_1(z)$ that are not zeros of $P_2(z)$. Since the exponent of convergence of a sequence is not increased by removing some of the terms, the exponent of convergence, and hence the order of $P(z)$, does not exceed ρ. Further, $Q_1(z) - Q_2(z)$ is a polynomial of degree not exceeding ρ, thus the order of $H(z) = f(z)/g(z)$ is ρ, at most. □

12.10

Theorem *The order of a canonical product equals the exponent of convergence of its zeros.*

PROOF. Since for any entire function $\rho_1 \leqslant \rho$, we must prove that $\rho \leqslant \rho_1$ for a canonical product. Let the zeros be $z_1, z_2, \ldots$, and k be a constant greater than 1. Let $P(z)$ be the canonical product, and we have

$$\log|P(z)| = \sum_{|z_n| \leqslant kr} \log\left|\left[1-\frac{z}{z_n}\right]\exp\left[\frac{z}{z_n}+\cdots+\frac{1}{p}\left(\frac{z}{z_n}\right)^p\right]\right|$$

$$+ \sum_{|z_n| > kr} \log\left|\left[1-\frac{z}{z_n}\right]\exp\left[\frac{z}{z_n}+\cdots+\frac{1}{p}\left(\frac{z}{z_n}\right)^p\right]\right|$$

$$= \Sigma_1 + \Sigma_2.$$

For Σ_2, since $|z| = r$ and $|z_n| > kr$, we have $|z/z_n| < 1$ and

$$\log\left[1 - \frac{z}{z_n}\right]\exp\left[\frac{z}{z_n} + \cdots + \frac{1}{p}\left(\frac{z}{z_n}\right)^p\right] = -\frac{1}{p+1}\left(\frac{z}{z_n}\right)^{p+1} - \cdots.$$

Thus

$$\left|\log\left[1 - \frac{z}{z_n}\right]\exp\left[\frac{z}{z_n} + \cdots + \frac{1}{p}\left(\frac{z}{z_n}\right)^p\right]\right|$$
$$< \frac{1}{p+1}\left[\left|\frac{z}{z_n}\right|^{p+1} + \left|\frac{z}{z_n}\right|^{p+2} + \cdots\right]$$
$$= \frac{1}{p+1}\frac{|z/z_n|^{p+1}}{1 - |z/z_n|} \le K\left|\frac{z}{z_n}\right|^{p+1}.$$

Also, $\log|f| = R\{\log f\} \le |\log f|$. Thus

$$\log\left|\left[1 - \frac{z}{z_n}\right]\exp\left[\frac{z}{z_n} + \cdots + \frac{1}{p}\left(\frac{z}{z_n}\right)^p\right]\right| \le A\left(\frac{z}{z_n}\right)^{p+1}$$

and

$$\sum_{|z_n|>kr}\log|\cdots| = \Sigma_2 = O\left[\sum_{|z_n|>kr}\left|\frac{z}{z_n}\right|^{p+1}\right] = O\left[|z|^{p+1}\sum_{|z_n|>kr}\frac{1}{|z_n|^{p+1}}\right].$$

If $p + 1 = \rho_1$, then

$$\Sigma_2 = O[|z|^{p+1}] = O[|z|^{\rho_1}] = O[r^{\rho_1}],$$

(we recall that $p \le \rho_1 \le p + 1$). If $p + 1 > \rho_1 + \epsilon$ (recall that $\sum_{n=1}^\infty r_n^{-(\rho_1+\epsilon)}$ converges) and ϵ is small enough, then

$$|z|^{p+1}\sum_{|z_n|>kr}|z_n|^{-p-1} = |z|^{p+1}\sum_{|z_n|>kr}|z_n|^{\rho_1+\epsilon-p-1}|z_n|^{-\rho_1-\epsilon}$$
$$< |z|^{p+1}(kr)^{\rho_1+\epsilon-p-1}\sum_{|z_n|>kr}|z_n|^{-\rho_1-\epsilon}$$
$$= O[|z|^{\rho_1+\epsilon}],$$

since $1/|z_n| < 1/kr$, $r = |z|$, and $\rho_1 + \epsilon - p - 1 < 0$.

In Σ_1, $|z/z_n| \ge 1/k$. Note that $|z/z_n|$ can be large but cannot be small. Since

$$\log\left|\left[1 - \frac{z}{z_n}\right]\exp\left[\frac{z}{z_n} + \cdots + \frac{1}{p}\left(\frac{z}{z_n}\right)^p\right]\right|$$
$$\le \log\left[1 + \left|\frac{z}{z_n}\right|\right] + \left|\frac{z}{z_n}\right| + \cdots + \frac{1}{p}\left|\frac{z}{z_n}\right|^p$$

and

$$\log\left[1 + \left|\frac{z}{z_n}\right|\right] < \left|\frac{z}{z_n}\right| \quad \text{since } 1 + |x| < e^{|x|},$$

we obtain

$$\left| \log\left[1 - \frac{z}{z_n}\right] \exp\left[\frac{z}{z_n} + \cdots + \frac{1}{p}\left(\frac{z}{z_n}\right)^p\right] \right| < K \left|\frac{z}{z_n}\right|^p,$$

where K depends on k only. Thus,

$$\Sigma_1 = O\left[\sum_{|z_n| \leqslant kr} \left|\frac{z}{z_n}\right|^p\right] = O\left[|z|^p \sum_{|z_n| \leqslant kr} |z_n|^{\rho_1+\epsilon-p} |z_n|^{-\rho_1-\epsilon}\right]$$

$$= O\left[|z|^p (kr)^{\rho_1+\epsilon-p} \sum_{|z_n| \leqslant kr} |z_n|^{-\rho_1-\epsilon}\right]$$

$$= O[|z|^{\rho_1+\epsilon}],$$

since $|z_n|$ are bounded, $\sum_{|z_n| \leqslant kr} |z_n|^{-\rho_1-\epsilon}$ is a finite series, $p \leqslant \rho_1, p < \rho_1 + \epsilon$, and $|z_n| \leqslant k|z| = kr$. Thus

$$\log|P(z)| = O[|z|^{\rho_1+\epsilon}] \quad \text{and} \quad |P(z)| = O[\exp r^{\rho_1+\epsilon}],$$

from which we conclude that the order of $P(z)$, namely, ρ is such that $\rho \leqslant \rho_1$, and since $\rho_1 \leqslant \rho$, we get $\rho_1 = \rho$. □

A particularly useful result is the following lemma.

12.10.1

Lemma *If ρ is not an integer, then $\rho_1 = \rho$.*

PROOF. In any case, $\rho_1 \leqslant \rho$. Suppose $\rho_1 < \rho$. Then $P(z)$ is of order ρ_1, that is, the order of $P(z)$ is less than ρ. If $Q(z)$ is of degree q, then $e^{Q(z)}$ is of order $q \leqslant \rho$; but $q < \rho$ since q is integral and ρ is not. Hence $f(z)$ is the product of two functions, each of order less than ρ. Thus $f(z)$ is of order less than ρ, which contradicts the hypothesis that $f(z)$ is an entire function of order ρ. □

A consequence is that a function of nonintegral order must have an infinity of zeros. (Since, if the number of zeros is finite, $\rho_1 = 0 = \rho$.) Also, if the order is not integral, the function is dominated by $P(z)$, whereas if the order is integral, $P(z)$ may reduce to a polynomial or a constant, and the order depends entirely on the factor $e^{Q(z)}$.

12.10.2

In any case, since $P(z)$ is of order ρ_1 and $e^{Q(z)}$ is of order q, then $\rho = \max[q, \rho_1]$.

12.11

Definition The *genus* of the entire function $f(z)$ is the greater of the two integers p and q and is therefore an integer. Since $p \leqslant \rho$ and $q \leqslant \rho$, the genus does not exceed the order.

EXAMPLE 1. For the function
$$\sin z = z \prod_{n=1}^{\infty}\left[1 - \frac{z^2}{n^2\pi^2}\right],$$
actually $\prod_{n=1}^{\infty} z[1 \pm (z/n\pi)]e^{\pm z/n}$, the order of $e^{Q(z)}$ is $q = 0$,
$$e^{Q_n(z/z_n)} = e^{\pm z/n}, \quad \text{and} \quad p = 1.$$
The genus is $\max(p,q) = 1$. Also $\rho_1 = 1$ since the series
$$\sum_{n=1}^{\infty} \frac{1}{(n\pi)^{1+\epsilon}} < \infty, \quad \epsilon > 0,$$
and the order $\rho = \max(q, \rho_1) = 1$. Hence the genus is 1 and the order is 1.

EXAMPLE 2. For the function
$$f(z) = \prod_{n=2}^{\infty}\left[1 - \frac{z}{n(\log n)^2}\right]$$
the order of $e^{Q(z)}$ is $q = 0$. The genus if $\max(p,q) = 0$ since $e^{Q_n(z/z_n)} = 1$. The order is $\max(0,1) = 1$ since $\rho_1 = 1$ for $\sum_{n=2}^{\infty} 1/n(\log n)^2 < \infty$. We need to establish that
$$\sum_{n=2}^{\infty} \frac{1}{[n(\log n)^2]^r} < \infty \quad \text{for} \quad r \geqslant 1 \text{ only}.$$
(We can use, for example, Gauss's test or a suitable integral test for infinite series.) Hence the genus is 0. And the order is 1.

12.12

If we have the power series representation of an entire function, the calculation of order and type is fairly simple, as shall be illustrated by the next two theorems. In order to study more sophisticated functions we will need Stirling's approximation for the gamma function.

12.12.1

Theorem A necessary and sufficient condition that $f(z) = \sum_{n=0}^{\infty} a_n z^n$ be an entire function of order ρ, is that
$$\lim_{n \to \infty} \frac{\log(1/|a_n|)}{n \log n} = \frac{1}{\rho}.$$

PROOF. We use the fact that $\sum_{n=0}^{\infty} |a_n z^n|$ does not differ much from its greatest term and that $|f(z)|$ lies between the two. Let
$$\lim_{n \to \infty} \frac{\log(1/|a_n|)}{n \log n} = \mu,$$

where μ is zero, positive, or ∞. Thus for every $\epsilon > 0$

$$\log \frac{1}{|a_n|} (\mu - \epsilon) n \log n \quad \text{for} \quad n > n_0,$$

that is, $|a_n| < n^{-n(\mu-\epsilon)}$. If $\mu > 0$, then $\sum_{n=0}^{\infty} a_n z^n$ converges for all z so that $f(z)$ is the entire function. If μ is finite, then

$$|f(z)| < Ar^{n_0} + \sum_{n=n_0+1}^{\infty} r^n n^{-n(\mu-\epsilon)}, \quad r > 1.$$

Let Σ_1 denote the part of the last series for which $n \leq (2r)^{1/\mu-\epsilon}$ and let Σ_2 be the remainder. Then

$$\Sigma_1 < \exp[(2r)^{1/\mu-\epsilon} \log r] \sum_{(n)} n^{-n(\mu-\epsilon)} < K \exp[(2r)^{1/\mu-\epsilon} \log r].$$

In Σ_2, we have $n > (2r)^{1/\mu-\epsilon}$, thus

$$rn^{-(\mu-\epsilon)} < r[(2r)^{1/\mu-\epsilon}]^{-(\mu-\epsilon)} = \frac{1}{2} \quad \text{and} \quad \Sigma_2 < \Sigma_n \left(\frac{1}{2}\right)^n < 1.$$

Thus $|f(z)| < B \exp[(2r)^{1/\mu-\epsilon} \log r]$ and $\rho \leq 1/(\mu - \epsilon)$. Making $\epsilon > 0$, we get $\rho \leq 1/\mu$. If $\mu = \infty$, the same argument with an arbitrarily large μ shows that $\rho = 0$. On the other hand, given ϵ, there exists a sequence of values of n for which

$$\log \frac{1}{|a_n|} < (\mu + \epsilon) n \log n,$$

that is,

$$|a_n| > n^{-n(\mu+\epsilon)} \quad \text{or} \quad |a_n| r^n > [rn^{-(\mu+\epsilon)}]^n.$$

Take $r = (2n)^{\mu+\epsilon}$; solving for n, we obtain

$$|a_n| r^n > 2^{(\mu+\epsilon)n} = \exp\left[\frac{1}{2}(\mu+\epsilon) r^{1/\mu+\epsilon} \log 2\right].$$

Since Cauchy's inequality gives $M(r) \geq |a_n| r^n$, then for a sequence of values of r tending to ∞, we have $M(r) > \exp[Ar^{1/\mu+\epsilon}]$; thus $\rho \geq 1/(\mu + \epsilon)$ and for $\epsilon \to 0$, $\rho \geq 1/\mu$, that is, if $f(z)$ is an entire function, its order $\rho = 1/\mu$ or

$$\lim_{n \to \infty} \frac{\log(1/|a_n|)}{n \log n} = \frac{1}{\rho}. \qquad \square$$

Further if $\mu = 0$, then $f(z)$ is of infinite order. Let $f(z)$ be a function of finite order ρ. Then $a_n \to 0$ and μ is nonnegative and the argument has shown that $\mu = 1/\rho$.

Note: If $f(z)$ is entire and if

$$\lim_{n \to \infty} \frac{\log(1/|a_n|)}{n \log n} = 0,$$

then $f(z)$ is of infinite order, since the limit can be zero as $n \to \infty$ without $f(z)$ being entire.

A similar theorem for the type follows from the following lemma:

12.13

Lemma Let $f(z)$ have a Taylor series expansion $\sum_{n=0}^{\infty} a_n z^n$. Suppose there exist mumbers $\mu > 0$, $\lambda > 0$, and an integer $N = N(\mu, \lambda) > 0$, such that $|a_n| < (e\mu\lambda/n)^{n/\mu}$ for all $n > N$. Then $f(z)$ is an entire function and, given any $\epsilon > 0$, there is a number $R = R(\epsilon) > 0$ such that

$$M(r) < \exp[(\lambda + \epsilon)r^\mu] \quad \text{for all} \quad r > R.$$

PROOF. Since $|a_n| < (e\mu\lambda/n)^{n/\mu}$,

$$\sqrt[n]{|a_n|} < \left(\frac{e\mu\lambda}{n}\right)^{1/\mu} \quad \text{for all} \quad n > N.$$

Thus $\sqrt[n]{|a_n|} \to 0$ as $n \to \infty$, and $f(z)$ is entire. Further,

$$\sqrt[n]{|a_n|r^n} < \left(\frac{e\mu\lambda}{n}\right)^{1/\mu} r < \frac{1}{2}$$

if $n > n_0 = n_0(r) = 2^\mu e\mu\lambda r^\mu$. Choosing $R^1 = R^1(\mu, \lambda) > 1$ so large that $n_0(r) > N$, we obtain, if $r > R^1$ and $n > n_0$, that

$$\sqrt[n]{|a_n|r^n} < \frac{1}{2} \quad \text{or} \quad |a_n|r^n < \left(\frac{1}{2}\right)^n,$$

We now find the upper bound for $M(r)$:

$$M(r) = \max_{|z|=r}\left|\sum_{n=0}^{\infty} a_n z^n\right| \leq \sum_{n=0}^{\infty} |a_n|r^n$$

$$= \sum_{n=0}^{n_0} |a_n|r^n + \sum_{n=n_0+1}^{\infty} |a_n|r^n$$

$$< \sum_{n=0}^{n_0} |a_n|r^n + \sum_{n=n_0+1}^{\infty} \left(\frac{1}{2}\right)^n$$

$$< \sum_{n=0}^{n_0} |a_n|r^n + 1 \quad \text{if} \quad r > R^1.$$

However,

$$\sum_{n=0}^{n_0} |a_n|r^n = \sum_{n=0}^{N} |a_n|r^n + \sum_{n=N+1}^{n_0} |a_n|r^n$$

$$< r^N \sum_{n=0}^{N} |a_n| + (n_0 - N) \max_{N+1 \leq n \leq n_0} |a_n|r^n,$$

and

$$\max_{N+1\leqslant n\leqslant n_0}|a_n|r^n \leqslant \max_{N+1\leqslant n}|a_n|r^n < \max_{N+1\leqslant n}\left(\frac{e\mu\lambda}{n}\right)^{n/\mu}$$

$$\leqslant \max_{1\leqslant n}\left(\frac{e\mu\lambda}{n}\right)^{n/\mu} r^n = \exp(\lambda r^\mu).$$

The maximum is achieved for $n = \mu\lambda r^\mu$, thus

$$\max_{N+1\leqslant n\leqslant n_0}|a_n|r^n < \exp(\lambda r^\mu).$$

Hence, if $r > R^1$,

$$M(r) < r^N \sum_{n=0}^{N}|a_n| + (n_0 - N)\exp(\lambda r^\mu) + 1$$

$$= r^N \sum_{n=0}^{N}|a_n| + (2^\mu e\mu\lambda r^\mu - N)\exp(\lambda r^\mu) + 1$$

$$= \exp(\lambda r^\mu)\left[2^\mu e\mu\lambda r^\mu - N + r^N\exp(-\lambda r^\mu)\sum_{n=0}^{N}|a_n| + \exp(-\lambda r^\mu)\right].$$

Given any $\epsilon > 0$, there exists a number $R = R(\epsilon) > R^1$ such that the expression in brackets is less than $\exp \epsilon r^\mu$, provided $r > R$. Hence

$$M(r) < \exp(\lambda + \epsilon)r^\mu \quad \text{for all} \quad r > R. \qquad \square$$

12.13.1

Theorem *If $f(z)$ is an entire function of finite order ρ ($0 < \rho < \infty$) and type σ, then*

$$\sigma = \frac{1}{e\rho}\varlimsup_{n\to\infty} n|a_n|^{\rho/n}. \tag{12.4}$$

PROOF. Suppose σ is finite. Then, given any $k > \sigma$, there exists a number $R = R(k) > 0$ such that $M(r) < \exp kr^\rho$ for $r > R$. According to Cauchy's inequality,

$$|a_n| \leqslant \frac{M(r)}{r^n} < \frac{\exp(kr^\rho)}{r^n} \quad \text{for all} \quad r > R.$$

The minimum value of $\exp(kr^\rho)/r^n$ occurs for $r = (n/k\rho)^{1/\rho}$; thus

$$|a_n| < \left(\frac{e\rho k}{n}\right)^{n/\rho} \quad \text{if} \quad n > N \quad \text{and} \quad r = \left(\frac{n}{k\rho}\right)^{1/\rho} > R(k).$$

Rewriting, we obtain

$$k > \frac{1}{e\rho}n|a_n|^{\rho/n} \quad \text{or} \quad k \geqslant \frac{1}{e\rho}\varlimsup_{n\to\infty}n|a_n|^{\rho/n}.$$

Since k is an arbitrary number exceeding σ, we get

$$\sigma \geqslant \frac{1}{e\rho} \varlimsup_{n\to\infty} n|a_n|^{\rho/n},$$

where the right-hand side is clearly finite. Now, let k^1 be any number exceeding the right-hand side of (12.4). Then there exists a number $N = N(k^1) > 0$ such that

$$|a_n| < \left(\frac{e\rho k^1}{n}\right)^{n/\rho} \quad \text{for all} \quad n > N.$$

Applying the lemma with $\lambda = k^1$ and $\mu = \rho$, we obtain that, given any ϵ, there exists a number $R = R(\epsilon) > 0$ [not to be confused with $R(\lambda)$], such that

$$M(r) < \exp\left[(k^1 + \epsilon)r^\rho\right] \quad \text{for all} \quad r > R.$$

Thus $\sigma \leqslant k^1$ and because of the choice of k^1, we have

$$\sigma \leqslant \frac{1}{e\rho} \varlimsup_{n\to\infty} n|a_n|^{\rho/n}.$$

Hence the result. Also, if the right-hand side of (12.4) is finite, so is σ, and if σ is infinite, so is the right-hand of (12.4). □

EXAMPLE 1. The function

$$f(z) = \sum_{n=1}^{\infty} \left(\frac{e\rho\sigma}{n}\right)^{n/\rho} z^n$$

is of order ρ and type σ.

EXAMPLE 2. Since

$$\lim_{n\to\infty} \frac{\log n}{\log\left[1/\sqrt[n]{|a_n|}\right]} = 0$$

characterizes an entire function of order zero, any function with coefficients $|a_n| = 1/n^{n/\epsilon_n}$, where $\{\epsilon_n\}$ is a sequence of positive numbers converging to zero, is of order zero. For example,

$$f(z) = \sum_{n=1}^{\infty} \frac{z^n}{n^{n^{1+\delta}}}, \quad \delta > 0,$$

has ρ 0.
[Hint: Examine $\log[1/\sqrt[n]{|a_n|}]$].

EXAMPLE 3. The condition

$$\varlimsup_{n\to\infty} \frac{\log n}{\log\left[1/\sqrt[n]{|a_n|}\right]} = \infty, \quad \text{or} \quad \varlimsup_{n\to\infty} \frac{\log[1/|a_n|]}{n\log n} = 0,$$

together with $\lim_{n\to\infty} \sqrt[n]{|a_n|} \to 0$, characterizes an entire function of infinite order. For example, consider $|a_n| = 1/n^{n\epsilon_n}$, where $\{\epsilon_n\}$ is a sequence of positive numbers converging to zero slowly enough, so that $\lim_{n\to\infty}(\epsilon_n \log n) = \infty$, since we require

$$\lim_{n\to\infty}\left[1/\sqrt[n]{|a_n|}\right] \to \infty.$$

The sequence $\epsilon_n = 1/(\log n)^{1-\delta}$, $n = 1, 2, \ldots$, meets these requirements if $0 < \delta < 1$, since $\epsilon_n \to 0$ but $\lim_{n\to\infty}(\epsilon_n \log n) \to \infty$. Thus the series

$$f(z) = \sum_{n=0}^{\infty} \frac{z^n}{\exp(n\delta \log n)}, \quad 0 < \delta < 1,$$

represents an entire function of infinite order.

12.13.2

Exercises

1: Determine the order and type of $\sin z$.

2: Show that all of the following series represent entire functions of order $\rho = 1/\alpha$ and type α/e, α, and 1, respectively:

 i. $\sum_{1}^{\infty} \dfrac{z^n}{n^{n\alpha}}$

 ii. $\sum_{1}^{\infty} \dfrac{z^n}{(n!)^{\alpha}}$

 iii. $\sum_{1}^{\infty} \dfrac{z^n}{\Gamma(\alpha n + 1)}.$

3: Find an infinite product expansion for the entire function

$$f(z) = \cosh z - \cos z;$$

hence find its order and genus.

4: The differential equation $w'' - zw = 0$ has a solution $f(z)$ with $f(0) = 1$ and $f'(0) = 0$. Show that $f(z)$ is an entire function of order $3/2$ and type $2/3$.

5: Let $f(z)$ denote a transcendental entire function of order ρ, that is finite but nonintegral; let $P(z)$ be a nonzero polynomial. Show that the equation $f(z) = AP(z)$ has infinitely many roots for every complex number A.

6: Let f and P be as in Problem 5. If $M(r; P) = \max[|P(z)|; |z| \leq r]$ and

$M(r; f) = \max[|f(z)|; |z| \leq r]$, then show that
$$\lim_{r \to \infty} \frac{M(r; P)}{M(r; f)} = 0.$$

7: Deduce the exponent of convergence and the order of the following entire functions:

i. $f(z) = \prod_{n=1}^{\infty} \left[1 + \frac{z}{e^n}\right]$

ii. $f(z) = e^{z^2/2} \prod_{n=1}^{\infty} \left[1 - \frac{z}{n^2}\right]$.

8: Use Stirling's approximation for the gamma function (or other approach) to show that the orders of the entire functions

i. $\sum_{n=0}^{\infty} \frac{(-1)^n z^{2n}}{n! \Gamma(n + v + 1)} = z^{-v} J_v(2z), \quad v > -1$

ii. $\int_0^1 \exp(zt^2) \, dt$

are both given by $\rho = 1$.

9: Find the genus and order of the entire function
$$f(z) = \sum_{n=2}^{\infty} \left[1 - \frac{z}{n(\log n)^2}\right]$$

10: Prove that $1/\Gamma(z)$ is an entire function and determine its order and type.

11: If $f(z)$ is a canonical product of genus ρ, prove that
$$\frac{f'(z)}{f(z)} = \sum_{n=1}^{\infty} \frac{z^\rho}{z_n^\rho (z - z_n)}.$$

12: Let $f(z) = \prod_{n=0}^{\infty}[1 - q^n z]$, $0 < q < 1$. Determine $n(r)$ and show that
$$\log M(r, f) < \frac{1}{2}\left[\log \frac{1}{q}\right]^{-1}\left[\log^2(er) + 1\right],$$
$$M(r, f) = \max|f(re^{i\theta})|, \quad 0 \leq \theta \leq 2\pi.$$

Author Index

A
Abel, N.H., 93, 94, 95
Ahlfors, L.V., 120, 241

B
Bazilyevich, I.E., 249
Bell, E.T., 183
Bessel, F.W., 60, 146
Bieberbach, L., 249
Blaschke, W., 259
Bonnet, P., 168
Bowman, F., 146

C
Copson, E.T., 63

E
Esterman, T., 61

G
Gauss, C.F., 60, 273, 274, 291
Goursat, E., 61

H
Hadamard, J., 131
Hardy, G.H., 133
Hayman, W.K., 247
Hille, E., 249
Holland, A.S.B., 109, 157, 281
Hurwitz, A., 210

K
Knopp, K., 63, 121

L
Landau, E., 63
Levin, B.J., 200
Littlewood, J.E., 249

N
Nehari, Z., 241, 247
Nevanlinna, R., 50

P
Paatero, V., 50
Picard, E., 151
Pierpont, J., 42
Pollard, S., 43, 61

S
Saks, S., 115
Springer, G., 135

T
Tauber, A., 95
Titchmarsh, E.C., 95, 123, 131

V
Valiron, G., 63

W
Watson, G.N., 267
Whittaker, E.T., 267
Widder, D.V., 168

Y
Young, W.H., 100

Z
Zygmund, A., 115

Subject Index

A
Abel's theorem, 93–95
Absolute value, 4
Accumulation point, 22
Almost uniformly, 241; *see also* Convergence, locally uniform
Amplitude, 10
Analytic continuation, 51, 121, 123, 124, 152
  of $\Gamma(Z)$, 271
  functions, 45, 60, 81, 86, 101, 108, 114, 124, 127–129, 137
Arc, 28, 185
Arcwise connected, 28
Argument, 10, 204, 207, 265, 267
Argument principle, theorem, 202

B
Bernoulli numbers, 148
  lemniscate of, 216
Bessel function of order n, 144
Beta-function, 176, 269
Bieberbach conjecture, 249
Bilinear transformation, 229, 230, 232–237, 240; *see also* Linear fractional transformation, 230 *or* Möbius transformation, 229, 230
Binary operations, 1
Binomial theorem, 134
Blaschke product, 259
Bolzano–Weierstrass property, 22
Borel and Carathéodory, theorem, 111
Boundary points, 17, 18, 63, 247, 248
Bounded, 17, 108, 139, 257, 266
  locally uniformly, 241–244, 246
    sequence, 21
    variation, 31
    region, 265
    uniformly, 262
Branch, 134–139, 191, 203
  cut, 135, 139, 191
  points, 134, 138, 190

C
Canonical product, 278, 284, 285, 288, 297
Cantor diagonal argument, 243
Cauchy criterion, 26
  formula, 77, 81
  inequality, 81, 89, 102, 144, 148, 150, 294
  integral formula, 68, 69, 78, 81, 86, 101, 142
  principal value, 166, 167
  product, 26, 90–94
  Riemann equations, 43, 45, 47, 50, 60, 130
    complex form, 229
  root test, 83
  sequence, 20, 22
  theorem, 60, 63, 65, 66, 68, 80, 142, 160, 171, 183, 184, 186, 273
    strong, 64
Circle of convergence, 82, 83, 85, 86, 105, 130, 131, 156
Closed interval, 17, 116
  Jordan curve, 136
  polygon, 224
  rectifiable Jordan curve, 59, 64, 74, 87, 160, 204
  region, 105
  set, 18, 70, 101
Closure, 18
Compact sets, 23, 28, 41
Completeness property, 22
Conformal, 226, 236, 237, 240, 241
  mapping, 234, 245, 247
Conjugate, 4
  function, 51
  harmonic functions, 50
Connected set, 18, 28
Continuous, 27, 41, 44, 45, 56, 59
  function, 66, 100, 135, 143
  mapping, 27
  one-to-one mapping, 27
Contour, 29
Convergence, 25, 97, 126, 139, 167, 241, 251, 252, 254, 264, 278
  absolute, 253, 278
  exponent of, 283–285, 288, 297
  locally uniform, 241, 242, 244, 246
  radius of, 82, 83, 85, 90, 105, 153
  uniform, 83, 95, 101, 103, 110, 143, 151, 152, 242, 243, 255–257, 265, 268, 269, 272
  series, 262
Convex function, 114, 115
Cross ratio, 240
Curve, 27, 28
Curvilinear integrals, 55

301

D

d'Alembert, ratio test, 83
Darboux theorem, 229, 232
Dedekind cut, 22
  property, 22
De Moivre's theorem, 11
Diameter, 17
Differentiability, 42–45, 48, 60, 61, 74, 101
Diverge to zero, 251, 253
Divergence class, 285
Domain, 27, 29, 45, 51, 58, 65, 67, 87, 95, 107, 108, 121, 124, 134, 135, 137, 139, 160, 192, 210, 224
  of definition, 40

E

Enestrom's theorem, 14
Entire, 45, 47, 52, 105, 125, 126, 151, 156, 204, 256
  function, 257, 266, 274, 276–280, 284, 285, 288, 291, 295, 296
Enumerative functions, 281
Essential singularity, 154, 158, 159
Euclidean spaces, 23
  n-space, 24
Euler's constant, 274
  formula, 10
  numbers, 148
Exponential function, 10, 132, 254
Extended complex (closed) plane, 224
  plane, 229
Extremal function, 249

F

Factor theorem, 89
Fixed points, 232
Fresnel integrals, 183
Function, 45, 47
  analytic, *see* analytic functions
    of complex variable, 40
  harmonic, 50, 52, 115, 129
  many-valued, 124, 132, 134, 135, 138
  multivalued, 40
  single-valued, 40, 42, 73, 123, 136, 139
  trigonometric, 52, 53
Fundamental theorem of algebra, 89

G

Gamma function, 267, 273, 274, 291, 297
Gaps, 130
Genus, 284
  of the entire function, 290, 296, 297
Greatest lower bound, 17
Green's theorem, 60

H

Hadamard's factorization theorem, 285
  three-circle theorem, 114
Harnack's inequality, 113
Heine–Borel theorem, 24, 122, 203
Hölder's inequality, 6
Homeomorphism, 27
Hyperplanes, 24

I

Image, 27, 222, 223
Improper integrals, 166
Infinite products, 251, 252, 254, 257, 276, 278, 287
Inner-product, 6
Interior point, 29
Inverse-function theorem, 228
  image, 27
  mapping, 230
  points, 232, 236
Inversion, 215, 232, 269
Isolated essential singularity, 155, 156
  singularity, 152, 154, 155, 157, 158, 160, 166
Isomorphism, 2

J

Jacobian, 181
Jensen's formula, 117, 282
  theorem, 116, 119, 120
Jordan arc, 55–57
  curve, 29, 41
  theorem, 29
  lemma, 170–172, 187
  polygon, 67

K

Koebe's constant, 247, 249
  theorem, 250

L

Laplace's equation, 50, 52
Laurent expansion, 146, 150, 152, 154–156, 161, 162, 165
  series, 142, 147, 148, 151
  theorem, 142, 144, 149
Least upper bound, 17
L'Hôpital's rule, 163
Limit, 20, 22, 40, 108, 140, 243
  inferior, 26
  point, 88, 157–159, 210, 276
  superior, 26
Linear fractional transformation, 230
  functions, 212
Liouville's theorem, 81, 89, 149, 150, 155, 156
Lipschitz condition, 244
Littlewood, J.E., 249
Logarithmic derivative, 287
  function, 131
  series, 134
Logarithms, 191, 203, 219, 254, 258, 265, 271, 286
l^p-norm, 6

M

Maclaurin expansion, 124
Mapping, 27, 135, 212, 215−220, 223, 224, 227−229, 232, 233, 237, 240, 247
Maximum modulus, 107, 114, 281
  principle, 109, 114
  theorem, 107, 108, 282, 286
Meromorphic functions, 158, 159, 174, 200, 256, 259−261
  expansion, 198, 258
Mertens' theorem, 91, 92, 94
Minimum principle, 108
  value, 294
Minkowski theorem, 7
Mittag−Leffler theorem, 251, 259, 261, 263
Möbius, 229
  transformation, 230
Modulus, 4
Morera's theorem, 80, 100
Multiple point, 29, 67
Multiply connected, 30, 114
  domains, 65, 108
Multivalued functions, 191

N

Natural boundaries, 130
Negative variations, 33
Neighborhood, 17
Nested-interval property, 22, 25
  sequence, 23
Normal families, 241
Null set, 18

O

One-to-one, 27
Open, covering of, 23
  interval, 17, 115
  -mapping theorem, 225
  set, 17
Order, 278, 281, 283, 291, 296, 297
  of zero, 88
Ordered pairs, 1
Orthogonally, 233

P

Partial fractions, 159, 163, 164, 259, 264
Path, 28, 137
Pathwise connected, 28
Picard's theorem, 157
Point of accumulation, 18
  at infinity, 155, 164, 216, 224, 245, 281
Poisson−Jensen formula, 119, 120
Poisson's integral formula, 110, 111, 113
Polar form of, 9
Pole, 153−157, 159−162, 185, 197, 201, 202, 259, 260, 262, 263
Pollard conditions, 68
Positive variations, 33

Power series, 82, 127, 130, 131, 142, 153, 156, 163
Primary factors, 278, 285
Principal of reflections, 126
  part, 152, 153, 159, 260−262, 264
  value, 10, 132, 134, 166, 180, 254

R

Rational functions, 159, 167, 174, 178, 197
Rectifiable curves, 34, 63, 226
  Jordan arcs, 58, 67, 69, 71−73, 84, 85, 127, 133, 137
Reflections, 215
Region, 29, 96, 255, 257, 265, 268, 283
Removable singularity, 153, 154, 158, 159, 161, 173, 193, 261
Residue at infinity, 163−165
  theorem, 160, 169, 178, 183, 188, 199, 201
Riemann integral, 36
  mapping theorem, 241
  Stieltjes integral, 35, 37
  surfaces, 135, 190
  zeta-function, 106
Rolle's theorem, 44
Roots, 296
  of unity, 12
Rouché's theorem, 200, 206−210, 244

S

Schlicht, 226, 250
Schwarz, inequalities of, 5, 7
  lemma, 109, 110, 112
  principle of reflections, 127
Second mean-value theorem, 167, 168, see Bonnet
Separation, 18
Sequence, 20, 276, 277, 279, 287, 288, 295
Simple closed curve, 29, 263
  conformal mappings, 226, 227
  curve, 29
Simply connected, 30, 67, 224, 229
  domain, 72, 73, 129, 130, 132, 136, 202, 241, 245, 247
Single-valued, 40, 45, 61, 74, 114, 131, 135, 136, 187, 190, 194, 226, 227, 245, 273
  analytic function, 134, 135, 220
Singularities, 142, 152, 171, 192, 256, 257, 262, 263, 267, 276
Stereographic projection, 224
Stieltjes integral, 56
Stirling's approximation, 291, 297
  formula, 270, 271
Subfield, 2
Subsequence, 20
Summation of series, 197
Symmetric, 233, 236, 239

T

Taylor's expansion, 125, 130, 131, 161, 228, 261, 293
  series, 99, 121, 144, 149, 162
  theorem, 86–90, 134, 222
Total variation, 31
Triangle inequality, 5
Trigonometric functions, 52, 53, 178
Type, 280, 293, 296, 297
  finite, 280
  infinite, 280
  maximum, 280
  mean, 280
  minimum, 280
  normal, 280

U

Uniformly bounded, 244
  continuous, 41, 56, 64, 67
  convergent infinite product, 254
Union, 16
Univalent, 226–229, 241, 246–250

V

Variations, 33
  negative, 33, 34
  positive, 33, 34

W

Weierstrass M-test, 85
  factorization theorem, 276
  theorem, 150, 151, 156, 157, 285, 287

Z

Zero of an analytic function, 129
  modulus, 117
  order m, 88